Analysis of Structural Systems

John F. Fleming

**Professor of Civil Engineering
University of Pittsburgh**

and

**President
Structural Software Systems**

**Prentice Hall
Upper Saddle River, New Jersey 07458**

Library of Congress Cataloging-in-Publication Data

Fleming, John F.
    Analysis of structural systems / by John F. Fleming.
      p. cm.
    Includes bibliographical references and index.
    ISBN 0-13-325986-2
    1. Structural analysis (Engineering)--Data processing.
TA647.F5595 1997
624.1'71--dc20   96-322
                              CIP

Acquisitions editor: **WILLIAM STENQUIST**
Edition-in-Chief: **MARCIA HORTON**
Production editor: **IRWIN ZUCKER**
Managing editor: **BAYANI MENDOZA DE LEON**
Copy editor: **PETER ZURITA**
Cover designer: **WENDY ALLING DESIGN**
Buyer: **JULIA MEEHAN**
Editorial assistant: **MARGARET WEIST**
Composition: **PREPARÉ Inc.**

© 1997 by Prentice-Hall, Inc.
Simon & Schuster / A Viacom Company
Upper Saddle River, New Jersey 07458

All right reserved. No part of this book may be
reproduced, in any form or by any means,
without permission in writing from the publisher.

The author and publisher of this book have used their best efforts in preparing this book. These efforts include the development, research, and testing of the theories and programs to determine their effectiveness. The author and publisher make no warranty of any kind, expressed or implied, with regard to these programs or the documentation contained in this book. The author and publisher shall not be liable in any event for incidental or consequential damages in connection with, or arising out of, the furnishing, performance, or use of these programs.

Printed in the United States of America

10 9 8 7 6 5 4 3 2 1

ISBN 0-13-325986-2

Prentice-Hall International (UK) Limited, London
Prentice-Hall of Australia Pty. Limited, Sydney
Prentice-Hall Canada Inc., Toronto
Prentice-Hall Hispanoamericana, S.A., Mexico
Prentice-Hall of India Private Limited, New Delhi
Prentice-Hall of Japan, Inc., Tokyo
Simon & Schuster Asia Pte. Ltd., Singapore
Editora Prentice-Hall do Brasil, Ltda., Rio de Janeiro

This book is dedicated to my wife Lois, without whose help and understanding it would merely be a dream.

Contents

PREFACE   viii

COMPUTER PROGRAMS   ix

1 STRUCTURAL ANALYSIS AND STRUCTURAL LOADS   1

Introduction and Assumed Reader Background   1
Units for Structural Engineering   2
Mathematical Model of a Structure   3
Sign Convention   7
Structural Loads   8
    Dead Loads, 10
    Li e Loads, 11

2 EQUILIBRIUM AND REACTIONS   14

Equations of Equilibrium   14
    Global Equilibrium of Plane Trusses and Plane Frames, 15
    Global Equilibrium of Space Trusses and Space Frames, 17
Structural Supports   17
    Supports for Plane Structures, 18
    Supports for Space Structures, 22
Stability and Static Determinacy   23
    Unstable Structures, 24
    Statically Determinate Structures, 24
    Statically Indeterminate Structures, 27
Computation of Reactions   27
Equations of Condition   34
    Stability, Determinacy, and Indeterminacy, 39
Computer Programs SDPTRUSS and SDPFRAME   41
    Computer Solution for Example Problem 2.1, 44
    Computer Solution for Example Problem 2.2, 45
    Computer Solution for Example Problem 2.4, 47
Influence Lines fo Beam Reactions   50
    Basic Definition of an Influence Line, 51
    Properties of Influence Lines, 54
Computations of Reactions Using Influence Lines   58
Computer Program BEAMIL   61
Suggested Problems   63

3 ANALYSIS OF STATICALLY DETERMINATE PLANE TRUSSES   69

Mathematical Model for a Plane Truss   69
Stability and Static Determinacy of Plane Trusses   70
Analysis of a Plane Truss by Joint Equilibrium   72
Method of Joints   80
Method of Sections   83
Classification of Plane Trusses   86
    Simple Trusses, 86
    Compound Trusses, 88
    Complex Trusses, 93
Zero-Force Members   94
Computer Program SDPTRUSS   96
Influence Lines for Plane Truss Member Forces   99
Computer Program PTRUSSIL   103
References   106
Suggested Problems   106

4 ANALYSIS OF STATICALLY DETERMINATE SPACE TRUSSES   115

Mathematical Model of a Space Truss   115
Stability and Static Determinacy of Space Trusses   116
Analysis of a Space Truss by Joint Equilibrium   117
Method of Joints   120
Computer Program SDSTRUSS   124
Suggested Problems   127

5 ANALYSIS OF STATICALLY DETERMINATE BEAMS   130

Definition of a Beam   130
Axial Force, Shear Force, and Bending Moment   131
Relationship between Load, Shear Force, and Bending Moment   138
V and M Expressions by Direct Integration   140

Shear-Force and Bending-Moment
   Diagrams   146
Computer Program BEAMVM   156
Influence Lines for Shear Force and Bending
   Moment in Beams   158
Computer Program BEAMIL   164
Maximum Bending Moment in Beams   167
Shear-Force and Bending-Moment
   Envelopes   171
References   175
Suggested Problems   175

6 ANALYSIS OF STATICALLY DETERMINATE PLANE FRAMES   180

Shear Force and Bending Moment in Plane
   Frames   180
Computer Program SDPFRAME   189
   Sign Con ention for Program SDPFRAME, 191
   *Stability and Static Determinacy of Plane
      Frames, 192*
   Method of Analysis, 193
   Member Distributed Forces, 194
   Computer Solution of Example Problem 6.1, 198
   Computer Solution of Example Problem 6.2, 198
Suggested Problems   203

7 ANALYSIS OF STATICALLY DETERMINATE SPACE FRAMES   206

Shear Force and Bending Moment in Space
   Frames   206
Computer Program SDSFRAME   211
   Sign Con ention for SDSFRAME, 212
   *Stability and Static Determinacy of Space
      Frames, 214*
   Method of Analysis, 215
   Computer Analysis of Example Problem 7.1, 216
Suggested Problems   221

8 DEFLECTION OF BEAMS   222

Deflection of Structural Systems   222
Differential Equation of Bending   222
   Analysis of Statically Determinate Beams, 227
   Analysis of Statically Indeterminate Beams, 236
   Analysis of Nonprismatic Beams, 241
The Moment Area Method   244
   Moment Area Theorems, 244
   Bending-Moment Diagrams by Cantile er Parts, 255
   Nonprismatic Beams, 267
The Conjugate Beam Method   269
   Support Conditions for the Conjugate Beam, 273
   *Analysis of Statically Indeterminate Beams by the
      Conjugate Beam Method, 279*
Computer Program BEAMVM   281
Suggested Problems   284

9 DEFLECTION OF TRUSSES AND FRAMES   291

Deflection Computations by Conservation of
   Energy   291
   External Work, 292
   Internal Work, 293
Deflection Computations by Castigliano's
   Theorem   308
   Superposition: A Word of Warning, 313
Deflection Computations by the Principle of
   Virtual Work   314
Castigliano's Theorem Revisited   330
FATPAK II—Student Edition   332
   *Computer Solution for Example Problems 9.1 and
      9.5, 333*
   Computer Solution for Example Problem 9.2, 334
   Computer Solution for Example Problem 9.9, 335
   Deformed Plots of Space Structures, 340
Betti's Law and Maxwell's Theorem   340
   Betti's Law, 340
   Maxwell's Theorem, 343
References   345
Suggested Problems   345

10 THE MÜLLER-BRESLAU PRINCIPLE   350

Development of the Müller-Breslau
   Principle   350
   Influence Line for Reacti e Force for a Beam, 352
   Influence Line for Shear Force for a Beam, 353
   *Influence Line for Bending Moment for a
      Beam, 353*
   *Influence Line for Bending Moment for a
      Frame, 353*
Influence Lines for Statically Determinate
   Beams   355
Influence Lines for Statically Indeterminate
   Beams   358
Influence Lines for Trusses   367
Application of the Müller-Breslau Principle to
   3D Structures   379
Experimental Generation of Influence
   Lines   380

Contents

References 381
Suggested Problems 381

11 THE METHOD OF CONSISTENT DEFORMATIONS 384

Statically Determinate Versus Statically Indeterminate Analysis 384
  Equilibrium Conditions, 385
  Geometric Compatibility Conditions, 385
  Load-Displacement Relationships, 386
Analysis of Statically Indeterminate Structures 386
  One-Degree Statically Indeterminate Structures, 388
  Multi-Degree Statically Indeterminate Structures, 399
Suggested Problems 402

12 THE FLEXIBILITY METHOD 405

Basic Structural Matrices 405
  Global Joint Load Matrix and Global Joint Displacement Matrix, 406
  Global Flexibility Matrix, 408
  Local Member End Load Matrix and Local Member Deformation Matrix, 408
  Local Member Flexibility Matrix, 413
  Total Structure Member End Load Matrix and Member Deformation Matrix, 415
  Total Structure Member Flexibility Matrix, 417
  Global Equilibrium Matrix, 417
Matrix Formulation of the Principle of Virtual Work 418
Formulation of the Flexibility Method 420
  Computation of Redundant Loads, 421
  Computation of Member End Loads, 425
  Computation of Joint Displacements, 425
  Geometric Compatability Check, 426
Summary of Flexibility Method for Statically Indeterminate Structures 427
Analysis of Statically Determinate Structures by the Flexibility Method 429
Computer Program FLEX 429
Development of Influence Lines Using FLEX 446
Modifications to Program FLEX 451
Reference 452
Suggested Problems 452

13 THE SLOPE-DEFLECTION METHOD 454

Basic Principle of the Slope-Deflection Method 454
  Geometric Compatibility Relationships, 455
  Load-Deformation Relationships, 460
  Equilibrium Relationships, 467
Suggested Problems 489

14 THE MOMENT DISTRIBUTION METHOD 493

Basic Analysis Procedure 493
  Sign Con ention, 495
  Fixed-End Moments, 495
  Member Stiffness, 495
  Distribution Factor, 496
  Carry-O er Factor, 497
Application of the Moment Distribution Method to the Analysis of Beams 497
Application of the Moment Distribution Method to the Analysis of Frames 507
Reference 511
Suggested Problems 511

15 THE STIFFNESS METHOD 513

Basic Principles of the Stiffness Method 513
  Global Joint Load Matrix and Global Joint Displacement Matrix, 514
  Global Stiffness Matrix, 515
  Local Member End Load Matrix and Local Member Deformation Matrix, 517
  Local Member Stiffness Matrix, 520
The Assembly Process 523
  Global Member End Load Matrix and Global Member Deformation Matrix, 524
  Global Member Stiffness Matrix, 526
  Generation of the Global Stiffness Matrix, 529
Computation of Member End Loads 531
Example Problem 532
Computer Programs PTRUSS and PFRAME 539
  Sample Plane Truss for Program PTRUSS, 540
  Sample Plane Frame for Program PFRAME, 540
Moment End Releases for Plane Frame Members 542
References 544
Suggested Problems 545

INDEX 547

Preface

This book is intended to be used in a set of introductory courses dealing with the analysis of statically determinate and statically indeterminate structures. These courses normally consist of a two-term sequence taught at the junior level in most civil engineering programs. It is assumed that the students have completed a set of introductory courses that covered the basic concepts of static equilibrium and the mechanics of deformable bodies. These courses are usually taught at the sophomore level in an accredited civil engineering or mechanical engineering curriculum. It is also assumed that the students have had the equivalent of 2 years of college mathematics that covered a basic introduction to calculus and differential equations. This book also should be useful to engineers who have been out of school for several years and wish to refresh their background in structural analysis.

The purpose of this book is to introduce the student to the basic theory of structural analysis and to demonstrate the application of this theory to various types of structures. The primary difference between this and the numerous other structural analysis books is the integration of the computer as an analysis tool throughout. The topics that are presented range from the calculation of the reactions for a statically determinate structure by a simple equilibrium analysis to the computation of the joint displacements and the member end loads for a statically indeterminate structure by the Stiffness Method. A set of suggested problems is included at the end of each chapter that the student can solve to gain a better understanding of the analysis concepts presented in the chapter. Most course instructors will probably supplement these problems with additional problems that they have found to be useful to reinforce the specific concepts that they might emphasize in their classroom discussions.

A set of programs to accompany this book are available on Prentice Hall's web site for use on a desktop computer running under MS-DOS. These programs may be used by the student for various types of structural analysis. The programs can perform tasks ranging from the computation of the member forces and reactive forces for a statically determinate plane truss, by solving a set of joint equilibrium equations, to the computation of joint displacements, member end loads, and reactive loads for a statically indeterminate frame by the Stiffness Method. The output for some of the programs is limited to numeric output, while the output of other programs consists of screen plots such as the shear-force and bending-moment diagrams for a beam, the influence lines for the member forces in a plane truss, and the deflected shape of a plane truss, space truss, plane frame, or space frame due to a set of applied loads.

One of the purposes for supplying these programs is to provide the students a means to verify manual solutions for the suggested problems at the end of the chapters. The programs have been used in the structural analysis courses at the University of Pittsburgh and have proved to be very beneficial. One of the advantages of having these programs available is that they can help to reduce the frustration that students often feel when they have difficulty in solving a homework problem. Although the programs do not demonstrate the actual solution procedure while they are running, they do give the advantage of seeing the correct solution for the problem. In many cases, by working backwards from the correct solutions, students are able to determine where their solutions went wrong. This can be extremely helpful at 1:00 A.M., when the course instructor is not available to answer questions.

Another advantage of these programs is that they give the ability to analyze a structure for a variety of conditions in order to determine how changes in the properties of the structure or the applied loads affect the response of the structure. By performing various parametric studies, the student can gain a better understanding of structural behavior. The author has discovered from experience that it is not practical to expect students to perform any type of meaningful parametric studies manually. By the time they have performed a few manual solutions, they are so exhausted and frustrated that they have no interest in trying to compare and make sense out of the results of the analysis. The computer programs eliminate the drudgery involved in performing the analysis and permit concentration on the results obtained from the analysis. However, the programs are not intended to eliminate the need to perform manual solutions. The only way in which the students can gain a real understanding of the analysis process is to solve problems manually.

Only an executable binary file is supplied for most programs, because the purpose of this book is to teach the concepts of structural analysis and not computer programming. However, the source codes for several programs are supplied to demonstrate how the various analysis procedures that have been presented in the text can be implemented on the computer. The source codes for these particular programs are supplied in ASCII disk files. By studying these programs, the student can gain an understanding of the many decisions that must be made when developing a structural analysis program. By breaking down the analysis process into a set of systematic steps, for implementation on the computer, the student can gain a much better understanding of the analysis procedure.

COMPUTER PROGRAMS

All of the programs that are supplied for this book (available on Prentice Hall's web site http://www.prenhall.com) are written in QBasic, which is supplied with MS-DOS Version 5 and newer versions. The QBasic language was used because it is now available on essentially any computer running under MS-DOS. The executable files were generated by compiling the programs with Version 7.1 of the Microsoft Basic Professional Development System. The programs should run on any personal computer with a minimum of 640 kilobytes of memory running under Version 5 or a newer version of MS-DOS. It is suggested that the programs be run from a hard disk although individual programs also can be run from a floppy disk as long as there is additional space on the disk for data files and temporary files that are generated while the programs are running, as explained in the program instructions. The instructions for using each program are supplied in an ASCII disk file with the same name as the program and a .DOC extension. A hard copy of the instruction file for any program can be obtained on the printer by using the command COPY NAME.DOC PRN: in MS-DOS, where NAME is the name of the program. These files are already formatted to give a 1-inch left margin on the printer for insertion into a three-ring binder. The printer should be set to give 1-inch top and bottom margins. The instructions for setting the printer should be contained in the manual for the particular printer that is being used.

The following is a list of the programs which are supplied for this book. Some of the analysis programs have companion programs that perform supplemental tasks such as interactively preparing the input data file or plotting the geometry of the structure. The names of these companion programs are indented in this list The programs whose names are followed by an asterisk are those for which both an executable file, with a .EXE extension, and an ASCII source code file, with a .BAS extension, are supplied. Only executable files are supplied for the other programs.

SDPTRUSS*—statically determinate plane truss analysis using equilibrium equations
    SDPTDATA—create an input data file for SDPTRUSS
    PLOTSDPT—plot plane truss geometry from the SDPTRUSS data file

SDPFRAME*—statically determinate plane frame analysis using equilibrium equations
    SDPFDATA—create an input data file for SDPFRAME
    PLOTSDPF—plot plane frame geometry from the SDPFRAME data file
    PLOTPFVM—plot shear-force and bending-moment diagrams for plane frame members after analysis by SDPFRAME

BEAMIL—plot influence lines for reactions, shear force, and bending moment for beams
    BMILDATA—create an input data file for BEAMIL

SEQSOLVE—solve a set of linear simultaneous equations by Gauss-Jordan Elimination

PTRUSSIL—plot influence lines for member forces in plane trusses
    PTILDATA—create an input data file for PTRUSSIL

SDSTRUSS*—statically determinate space truss analysis using equilibrium equations
    SDSTDATA—create an input data file for SDSTRUSS
    PLOTSDST—plot space truss geometry from the SDSTRUSS data file

BEAMVM—plot shear-force and bending-moment diagrams and deflected shape for beams
    BMVMDATA—create an input data file for BEAMVM

SDSFRAME*—statically determinate space frame analysis by equilibrium equations
    SDSFDATA—create an input data file for SDSFRAME
    PLOTSDSF—plot space frame geometry from the SDSFRAME data file
    PLOTSFVM—plot shear-force and bending-moment diagrams for space frame members after analysis by SDSFRAME

FLEX*—statically determinate or statically indeterminate analysis by the Flexibility Method

PTRUSS*—plane truss analysis by the Stiffness Method

PFRAME*—plane frame analysis by the Stiffness Method

FATPAK II—This is a student edition of a general-purpose structural analysis package. It can perform the analysis of statically determinate and statically indeterminate planes trusses, space trusses, plane frames, and space frames. In addition to computing joint displacements, member end loads, and support reactions, the package can plot the original shape and the deformed shape of the structure and the shear-force and bending-moment diagrams for any member in a plane frame or space frame. The analysis is performed by the Stiffness Method. The individual analysis programs are T2DII for the analysis of plane trusses; T3DII for the analysis of space trusses; F2DII for the analysis of plane frames; and F3DII for the analysis of space frames. In addition, a set of companion programs named T2D2DATA, T3D2DATA, F2D2DATA, and F3D2DATA are available for interactively creating the input data files for the analysis programs. FATPAK II is the student edition of the commercial structural analysis package FATPAK V, which is in use in a number of structural engineering offices in the United States and several foreign countries.

The screen plots for any of the plotting programs may be generated on a CGA monitor at 640 by 200 pixels, an EGA monitor at 640 by 350 pixels, or a VGA monitor at 640

Preface

by 480 pixels. The program user must specify the type of plot that is desired in response to a question from the program. This capability is included because not all students own the most up-to-date computer systems.

The easiest way to obtain a hard copy of the source code for any program for which a .BAS file is supplied is to load the program into QBasic with the **File/Load** command and then to print it with the **File/Print** command. The printer should be set to give 1-inch left, top, and bottom margins and to print at 12 characters per inch. An alternate approach is to load the ASCII source code file into a word processor program such as WordPerfect and then to print it using the formatting capabilities of that program.

The programs available on Prentice Hall's web site are licensed to the purchasers of this book for educational and demonstration use only. They are not intended for commercial use and any such use is a violation of the user license. All of the programs are protected by the U.S. copyright laws. The copyright notice will be displayed by each program when it is executed.

The author and the publisher assume no responsibility for any damages resulting from the use of these programs. No warranty of any type is given or implied concerning the correctness or accuracy of any results obtained from the programs. It is the responsibility of the program user to independently verify any analysis results. The programs have been carefully checked for reliability and do not contain any known errors. Please contact the author by mail at

John F. Fleming
Structural Software Systems
4440 Gateway Drive
Monroeville, Pennsylvania 15146

or by E-Mail at

fleming@civeng1.civ.pitt.edu

if any errors are discovered. The program files, the instruction files and the sample data files that are supplied for this book are available on Prentice Hall's web site. A total of 67 files are supplied, which require approximately 2.8 megabytes of disk space. The names of the individual files are

SDPTRUSS.BAS	SDPTRUSS.EXE
SDPTRUSS.DOC	SDPTRUSS.DPT
SDPTDATA.EXE	PLOTSDPT.EXE
SDPFRAME.BAS	SDPFRAME.EXE
SDPFRAME.DOC	SDPFRAME.DPF
SDPFDATA.EXE	PLOTSDPF.EXE
PLOTPFVM.EXE	BEAMIL.EXE
BEAMIL.DOC	BEAMIL.BIL
BMILDATA.EXE	SEQSOLVE.EXE
SEQSOLVE.DOC	SEQSOLVE.SEQ
PTRUSSIL.EXE	PTRUSSIL.DOC
PTRUSSIL.TIL	PTILDATA.EXE
SDSTRUSS.BAS	SDSTRUSS.EXE

SDSTRUSS.DOC	SDSTRUSS.DST
SDSTRUSS.EXE	PLOTSDST.EXE
BEAMVM.EXE	BEAMVM.DOC
BEAMVM.BVM	BMVMDATA.EXE
SDSFRAME.BAS	SDSFRAME.EXE
SDSFRAME.DOC	SDSFRAME.DSF
SDSFDATA.EXE	PLOTSDSF.EXE
PLOTSFVM.EXE	FLEX.BAS
FLEX.EXE	FLEX.DOC
FLEX.FLX	PTRUSS.BAS
PTRUSS.EXE	PTRUSS.DOC
PTRUSS.PTR	PFRAME.BAS
PFRAME.EXE	PFRAME.DOC
PFRAME.PFR	T2DII.EXE
T2D2DATA.EXE	T2DII.INP
T3DII.EXE	T3D2DATA.EXE
T3DII.INP	F2DII.EXE
F2D2DATA.EXE	F2DII.INP
F3DII.EXE	F3DII2.EXE
F3D2DATA.EXE	F3DII.INP
FATPAKII.DOC	

To access the software directly from the web, connect to http://www.prenhall.com, click on "Custom Catalog," and search for either the author or the title of this book, click on the proper response, scroll down to the end of the page, and click on "download library." From there you should be able to download the software. To access the software directly from the ftp site, ftp to ftp ftp.prenhall.com, change directory to pub/esm/civil_engineering.s-044/fleming/anal_struct_systems, and download the software.

It is assumed that essentially all students either have Internet access on their computer or have access to such a computer in a computing laboratory at their university. If necessary, a copy of the programs on floppy disks can be obtained from

Structural Software Systems
4440 Gateway Drive
Monroeville, Pennsylvania 15146

The price is $15 prepaid in U.S. funds by check or money order. All checks must be drawn on a U.S. bank. An additional $10 are required for shipping for addresses outside of the United States or Canada. All shipments to addresses in Pennsylvania must also include 7% sales tax. All orders must be prepaid in full. Purchase orders and credit cards are not accepted. Make all checks payable to Structural Software Systems and please specify $3\frac{1}{2}$- or $5\frac{1}{4}$-inch high-density disks. If no specification is made, then $3\frac{1}{2}$-inch disks will be supplied. Be sure to include your return address with the order.

1

Structural Analysis and Structural Loads

INTRODUCTION AND ASSUMED READER BACKGROUND

The field of structural engineering predates all other engineering disciplines. Before early humans had even developed any form of mathematics they were building structures. One of the earliest structures was probably a bridge across a stream which was constructed by chopping down a convenient tree. Although no calculations were performed to compute the stresses or deflections, the builders probably did use basic engineering principles in the construction. After a few trials, it was discovered that a large diameter tree provided a much more rigid structure which was less likely to break under the weight of a person than a small sapling.

Today, we use a somewhat different approach in the design of structural systems since we now possess the mathematical tools to perform a complete analysis of the structure before it is built. It is no longer necessary to rely upon a trial-and-error approach to arrive at the correct geometry and member sizes to carry the desired load. In the following chapters, we will develop the mathematical equations and analysis procedures for determining the stresses and deformations in various types of structural systems. This is the first step in the extensive training which is required by any person who desires to be classified as a structural engineer.

In order to fully understand the material presented in this book, the reader must have an adequate background in mathematics and engineering mechanics. It is assumed that the reader has had at least 2 years of calculus and is familiar with the concepts of differentiation, integration, and the techniques for solving simple differential equations. The reader also should be familiar with the requirements for static equilibrium of rigid bodies in two and three dimensions and have a knowledge of the basic theory concerning the deformations of elastic bodies under static loads and the properties of engineering materials in the elastic range. This material is usually presented in courses which are taught at the freshman and sophomore levels in an accredited civil engineering or mechanical engineering curriculum. In addition, the reader should be familiar with the basic rules of matrix algebra and understand the concepts of matrix addition and subtraction, matrix multiplication, matrix transposition and matrix inversion.

Since the computer is now an established tool for engineering analysis, it is also assumed that the reader has some background in the application of computers to engineering problem solving. In particular, an understanding of the basic concepts of how to operate a desktop computer running under MS-DOS will be very useful since the programs which are distributed with this book operate on computers of this type. The reader should know how to format a floppy disk, copy files between disks, execute a program with a .EXE extension, generate a hard copy of a screen plot using the MS-DOS command GRAPHICS, and use an ASCII Editor, such as the program EDIT, which is supplied with MS-DOS Version 5 and newer versions, to create an ASCII data file for input into a program.

UNITS FOR STRUCTURAL ENGINEERING

There are two sets of basic units which are in use in the field of structural engineering: the *American System* (U.S.) (which was formerly called the English System) and the *International System* (SI). The U.S. system was essentially the only system of units which was used in the United States until the 1960s when the International Bureau of Weights and Measures specified that the SI system would become the world standard. The SI system has been officially adopted and is now being used exclusively in most English-speaking countries. However, the switchover to this system of units in the United States has been very slow and has met a great deal of resistance, particularly from older engineers and the building trades, although progress is being made. Federal government agencies such as the Federal Highway Administration, the National Institute of Standards and Technology (formerly the National Bureau of Standards), and the National Science Foundation have been very aggressive in the past few years in attempting to establish the SI system as the standard system of units for all structural design and engineering construction in the United States.

In the U.S. system, the commonly used units for force are either the pound (lb) or the kilopound (kip), where 1 kilopound is equal to 1000 pounds. It is usually more convenient to express the loads acting on a structure in terms of kips since this reduces the size of the numbers which must be carried through in the calculations. The commonly used units for length are either the inch (in) or the foot (ft), where 1 foot is equal to 12 inches. The overall dimensions of a structure are usually expressed in terms of feet, while the dimensions and properties of the cross-sections of the members, such as the area or the moment of inertia, are usually expressed in terms of inches. This can lead to a possible source of error in the analysis of a structure since it is often necessary to perform a conversion between these two length units at some stage in the calculations. Stresses are commonly expressed in either pounds per square inch or kips per square inch.

In the SI system, the commonly used units for force are either the newton (N) or the kilonewton (kN), where 1 kilonewton is equal to 1000 newtons. A newton is defined as the force which is required to cause a mass of 1 kilogram (i.e., 2.2046 lb-s^2/in) to undergo an acceleration equal to 9.8066 m/s^2, which is the standard value for the acceleration of gravity. The commonly used units for length are either the millimeter (mm) or the meter (m), where 1 meter is equal to 1000 millimeters. Stresses are commonly expressed as either pascals (Pa) or megapascals (MPa), where 1 megapascal is equal to 1,000,000 pascals and 1 pascal is equal to one newton per square meter. Some useful relationships between the U.S. system and the SI system are shown in Table 1.1.

TABLE 1.1 RELATIONSHIP BETWEEN US UNITS AND SI UNITS

1 in = 25.400 mm = 0.0254 m	1 m = 39.370 in = 3.281 ft
1 lb = 4.448 N = 0.004448 kN	1 N = 0.225 lb = 0.000225 kip
1 lb/in^2 = 6.894 kN/m^2	1 kN/m^2 = 0.145 lb/in^2

All of the equations which will be derived in this book will be in symbolic form and may be evaluated using any desired system of units. Some of the example problems which will be included in the following chapters to demonstrate the various analysis procedures will be solved using U.S. units while others will be solved using SI units. However, the emphasis in these problems will be on the analysis procedure and not the particular units which are being used. The units will merely affect the numbers which are being used in the analysis but not the order or the type of calculations which are performed.

The computer programs which are supplied with this book make no unit conversions at any step in the analysis. Therefore, the input to these programs may be in any desired system of units. The output from the programs will be in the same units as the input. For example, if the input is expressed in kip and inch units, then the output will be in kip and inch units, while, if the input is expressed in kN and meter units, then the output will be in kN and meter units. It is up to the program user to interpret the units properly for the results of any analysis since no units are shown in the program output listings.

MATHEMATICAL MODEL OF A STRUCTURE

When an engineer analyzes any type of structure, the analysis is not performed for the actual structure, but rather for a *mathematical model* which represents the properties of that structure and the loads acting on it. Therefore, the first step in the analysis must be the development of the mathematical model. Whether the results of the analysis are a true representation of the response of the structure to the applied loads is highly dependent upon the skill of the analyst in developing this model. In many situations, widely varying results can be obtained, depending on the assumptions which are made for the properties of the mathematical model, even though the analysis procedure results in an exact analysis of that model. The analyst must have the experience to choose the correct properties of the mathematical model so that its behavior does represent the actual response of the structure.

For the types of structures which will be considered in this book, the mathematical models will consist of a set of *joints*, located at various points in space, which are connected by long slender elastic *members*. The joints may be either *unrestrained joints*, whose displacements are only restricted by the resistance of the members attached to them, or *support joints*, which are totally restrained against displacement in particular directions. As an example, Figure 1.1(a) shows a two-story structure with two vertical steel wide-flange columns and two horizontal steel wide-flange girders which is subjected to a uniform distributed horizontal wind load ω on the left side. The columns are continuous over the full height of the structure and are welded to base plates at their lower ends. The base plates are bolted to large concrete footings which are buried in the ground. The ends of the girders are welded to the flanges of the columns. Figure 1.1(b) shows a mathematical model for this structure, in which the six small solid rectangles represent the joints and the six lines between the joints represent the members. The joints are located at the support points and at the points where the horizontal girders intersect the vertical columns. Therefore, even though each column is actually continuous over the full height of the structure, they are each represented by two members in the

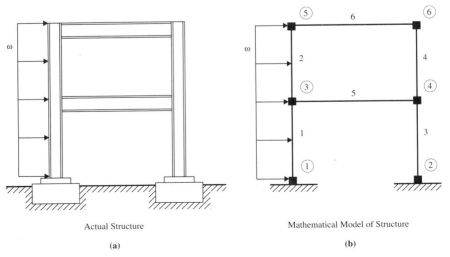

Actual Structure

(a)

Mathematical Model of Structure

(b)

Figure 1.1 Mathematical model.

mathematical model. It would be reasonable to assume that all of the members are rigidly attached to their end joints in this model due to the welded construction in the frame. The two joints at the lower ends of the columns are support joints, while the upper four joints are unrestrained joints. The circled numbers adjacent to each joint in the mathematical model are the joint numbers, while the numbers near the center of each member are the member numbers. The joints and members are numbered for identification purposes during the analysis of the mathematical model.

The loads acting on the structure might consist of concentrated forces and moments acting at any points on the structure and distributed forces, which have the units of force per unit length, acting on the members. The specific types of loads which can be applied to the mathematical model of any structure will depend upon the type of structure the model represents. The concentrated forces and moments acting in the unrestrained directions at various points and the distributed forces acting on the members will be designated as the *acti e loads*, while the forces and moments applied to the support joints, due to the restraints at those joints, will be designated as the *reacti e loads*, or simply the *reactions*. If all of the members are attached to the joints such that no bending or twisting moments can be transmitted between the joints and the members and if all active and reactive loads consist only of concentrated forces on the joints, then the structure will be defined to be a *truss*; otherwise, it will be defined to be a *frame*. In addition, if all of the joints for the structure lie in a plane and all loads act in that plane, it will be designated to be a plane structure. If all joints do not lie in a plane, or if any loads cause displacements of the joints or members out of the plane, it will be designated to be a space structure. If we now consider all possible combinations of the previous definitions, we see that there are four basic types of structures which must be considered: *plane trusses*, *space trusses*, *plane frames*, and *space frames*. Although some textbooks consider a *beam* to be another basic structure type, it is included in the preceding since a beam is actually a plane frame in which all of the members lie in a straight line. The same analysis procedures which would be used for a plane frame can also be used for a beam.

Figure 1.2 shows typical illustrations of the type which will be included in this book for plane trusses, plane frames, space trusses and space frames. The members in the structures are indicated by thin lines which represent the locations of the center lines of the members. The points where the members intersect in a truss structure are represented by small solid circles in the illustrations to differentiate a truss from a frame. The dimensions in the illustrations correspond to the distances between the intersection points of the center lines of the members in the structure. These distances usually can be considered to be the length of the members during the analysis. The width of the members and the size of the connections can be ignored for most analyses, although in some very sophisticated mathematical models, they might be included. We will not consider any mathematical models of this type in this book. We will discuss the rules for locating the joints in the mathematical models for each type of structure in the following chap-

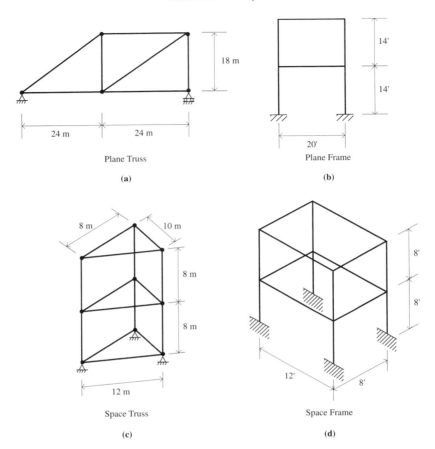

Figure 1.2 Structure types.

ters when we discuss the specialized procedures for the analysis of these structures. These rules are also discussed in the descriptions for the various computer programs which are included with this book.

It is important for the reader to realize that the members in a structure are not considered to be directly connected to other members in the mathematical model. All members in the mathematical model are only attached to their end joints. Any active loads which are applied to the joints are transmitted to the ends of the members which are connected to the joints. Each member then transmits these loads along its length to the joint at its other end. This process is continued until the active loads are finally resisted at the support joints by the reactive loads. The loads which are applied to the ends of the members by the joints will be designated as the *member end loads*. These are the loads which produce stresses in the member material and cause the members to deform, thus causing the joints to displace. The specific type of end loads which will be transmitted to the members from the joints will depend on the type of structure which is being analyzed. This will be discussed in more detail in later chapters.

For the mathematical models which will be considered in this book it will be assumed that any deformations which occur in the structure are small compared to the overall size of the structure. Therefore, during the analysis, the original geometry of the structure will be used when computing any joint locations, member lengths or member orientations. In addition, it will be assumed that the material in the structure remains linear elastic under the applied loads. Finally, it will be assumed that all loads are applied slowly to the structure such that inertia effects may be ignored.

SIGN CONVENTION

Before we can consider the various procedures for performing the analysis of a structure we must establish a sign convention so that we can properly define the loads acting on the structure and also properly interpret the results of the analysis. Although it might be obvious in many situations that a particular reactive load or joint displacement is in a specific direction, there are many cases where it is not obvious. The use of a consistent sign convention is absolutely essential in order to eliminate errors in the analysis. For the analysis procedures presented in this book, the active and reactive loads acting on the structure and the displacements of the joints which are caused by these loads will be expressed as components in a right hand orthogonal *global coordinate system* with the three axes being designated as X, Y, and Z, as shown in Figure 1.3. The system is known as a right hand system because of the relative orientation of the axes with respect to the fingers of the right hand. It is an orthogonal system since the XY, XZ, and YZ planes are at right angles to each other. It is called the global coordinate system since it applies to all active and reactive joint loads and all joint displacements in the structure. Although this coordinate system may be at any orientation in space, it is usually convenient for most structures and loading conditions to define the XZ plane to be horizontal with the positive Y axis extending upward. However, in some situations, it might be convenient to use some other orientation for the coordinate system to correspond to the specific geometry of the structure or the directions of the loads. The direction and magnitude of any active or reactive force or moment or of any joint translation or rotation can be expressed as three components with respect to these three axes.

A force component or a translation component is considered to be positive if it extends in the positive direction of an axis. Otherwise, it is considered to be negative. Note that this means that the gravity loads on a structure, which are always downward forces, will be negative when the positive Y axis is up. This might seem to be unnatural at this time, but it is consistent with essentially every commercial structural analysis computer program now being used in engineering design offices.

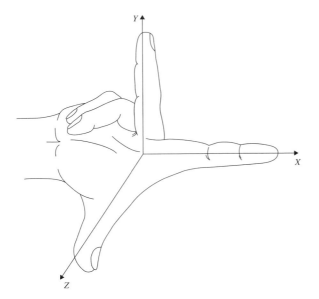

Figure 1.3 Right-hand orthogonal coordinate system.

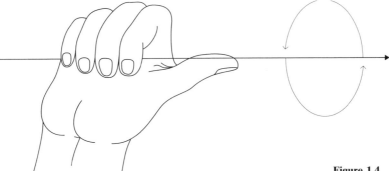

Figure 1.4 Right-hand rotation rule.

All of the computer programs included with this book use this sign convention for loads and displacements, unless specified otherwise.

The positive direction of a moment component or of a rotation component about any of the coordinate axes can be determined by the *right-hand rotation rule*, as defined in Figure 1.4. According to this rule, the positive direction of a moment or rotation about any axis is defined by extending the thumb of the right hand in the direction of the axis and curling the fingers naturally. The curved direction from the knuckles toward the finger tips corresponds to a positive moment or rotation direction about that axis.

STRUCTURAL LOADS

The loads acting on a structure can be considered to fall into two classifications: *dead loads* and *li e loads*. The dead loads are due to the weight of the structure and usually can be considered to be constant in magnitude, position and direction during the life of the structure. These loads act down due to the effects of gravity. The live loads which can act on a structure depend upon the type of structure and the purpose for which it was constructed. These loads are considered to vary in magnitude, position and direction during the life of a structure.

In many situations, the structural engineer has no choice as to the type and magnitude of the loads which are to be considered during the analysis and/or design of a particular structure since they are dictated by various codes and specifications which have been developed by both national and international organizations. Some of the most common codes and specifications which are used in the United States are:

"American National Standard Building Code," American National Standards Institute

"Basic Building Code," Building Officials and Code Administrators

"Building Code Requirements for Reinforced Concrete," American Concrete Institute

"Load and Resistance Factor Design Specifications for Structural Steel for Buildings," American Institute of Steel Construction, Inc.

"Manual for Railway Engineering," American Railway Engineering Association

"National Design Specifications for Stress-Grade Lumber and Its Fasteners," National Forest Products Association

"Requirements for Minimum Design Loads in Buildings and Other Structures," American National Standards Institute

"Specifications for Aluminum Structures," Aluminum Association

"Specification for Structural Steel for Buildings—Allowable Stress Design and Plastic Design," American Institute of Steel Construction, Inc.

"Standard Building Code," South Building Code Congress

"Standard Specifications for Highway Bridges," American Association of State Highway and Transportation Officials

"Uniform Building Code," International Conference of Building Officials

In addition, the Technical Council on Codes and Standards of the American Society of Civil Engineers (ASCE) is in the continuous process of developing and updating a number of different structural design standards. It is the aim of ASCE to update these standards every 5 years. Some of the standards which have been developed are:

"Building Code Requirements for Masonry Structures," ASCE 5-88
"Guideline for Design and Analysis of Nuclear Safety Related Earth Structures," ASCE 1-82 N-725
"Minimum Design Loads for Buildings and Other Structures," ASCE 7-88
"Seismic Analysis of Safety Related Nuclear Structures," ASCE 4-86
"Specification for the Design and Construction of Composite Slabs," ASCE 3-84
"Specifications for Masonry Structures," ASCE 6-88

These standards may be purchased from ASCE. Other standards probably will be developed in the future.

There are also many other specialized codes and specifications for very specific types of structures such as electric transmission towers, signboard support structures, crane support girders and many others. The structural engineer must be sure that the correct specifications are followed for the type of structure which is being designed or analyzed.

Dead Loads

The dead load for any structure is dictated by the weight of the material and the size of the structure. Usually, this loading is considered to be composed of a distributed downward force on the individual members in the mathematical model of the structure to represent the weight of the members and supported load-carrying elements such as floor slabs. It might also consist of additional concentrated forces at specific joints to represent loads which are not considered in the weight of the members. Typical loads of this type might be air conditioning duct work or water and sewer pipes, although these might also be considered to be live loads in some cases since they could change positions and change size during the life of a structure. Machinery, such as air conditioning equipment or manufacturing equipment, is usually considered to be a live load since it is not permanent in position or type. The weights of several typical structural materials are shown in Table 1.2.

Live Loads

The live loads which will act on a structure will vary depending on the type of structure and the purpose for which it was constructed. The following sections de-

TABLE 1.2 WEIGHTS OF STRUCTURAL MATERIALS

Material	U.S. (lb/ft^3)	SI (kN/m^3)
Aluminum	165	25.92
Concrete	150	23.56
Steel	490	76.97
Wood	40	6.28

scribe some of the typical types of live loads which might be considered during the design of various types of structures. However, this is not intended to be a complete discussion of live loads on structures. This topic is usually covered in books dealing with the design of structures in which the governing specifications are discussed in detail.

Vertical Loads on Buildings. The vertical live loads on buildings are usually expressed in terms of a uniform downward force per unit area acting on the floors of the building. The magnitudes of these floor loads vary with the type of building. Several examples of typical uniform floor loads on buildings are shown in Table 1.3. The engineer should carefully check the governing specifications for the specific type of structure before performing an analysis or design. Some specifications require additional concentrated loads to represent specific equipment which might be in a building.

Horizontal Wind Loads on Buildings. The horizontal wind load on a building varies with the velocity of the wind, the temperature, the shape of the building, the height above the ground and many other factors. This load is usually considered to be represented by a static pressure that acts on both the windward and leeward sides of the building, although in some cases, it might be necessary to consider the dynamic effects of the wind on the structure. Equations are

TABLE 1.3 LIVE LOADS ON BUILDINGS (lb/ft^2)

Assembly areas and theaters	
Fixed seats	60
Moveable seats	100
Bowling alleys and poolrooms	75
Dance halls and ballrooms	100
Dining rooms and restaurants	100
Hotels	
Private rooms	40
Public rooms	100
Office buildings	
Lobbies	100
Offices	50
Libraries	
Reading rooms	60
Stacks	150 minimum
Schools	
Classrooms	40
Corridors	80
Stores	
First floor	100
Upper floors	75

given in various specifications for computing the static design wind loads for any location in the United States.

Earthquake Loads on Buildings. An important consideration for the design of any building is the effect of an earthquake on the building. As the ground under the building moves, the building will start to vibrate due to the inertia forces which are introduced due to the mass of the structure. The magnitude of these forces will depend upon a number of factors such as the variation of the ground motion with time, the direction of the ground motion, the mass of the building, the nature period of vibration of the building and many others. Considerable research has been performed in the past and is still ongoing in the United States and many other countries to develop safe design procedures for buildings for earthquake motions. A simplified design approach which has been used in the past was to consider the building to be subjected to a horizontal static force which was some percentage of the weight of the building. Another approach, which is now gaining popularity, is to perform a complete dynamic analysis of the structure for a specified design earthquake. There are a number of engineering firms which specialize in providing design and analysis services for structures which might be subjected to earthquakes.

Vertical Loads on Highway Bridges. The vertical traffic loads for highway bridges are specified by the specifications of the American Association of State Highway and Transportation Officials (AASHTO) to consist of a set of standard trucks with the axle spacings and axle loads shown in Figure 1.5. The two standard trucks are designated as H20-44 and HS20-44, where the number 20 represents the weight of the front two axles of the truck in tons and the number 44 represents the year (i.e., 1944) when this loading was adopted by AASHTO. Either truck should be located at the position along the bridge which will result in the maximum value for the particular quantity, such as a reactive load or a member end load, which is being computed. The rear axle spacing of the HS20-44 truck should be varied between the specified limits to produce the maximum condition. AASHTO also specifies an alternate lane loading which can be used instead of the axle loads. This loading consists of a uniform force and a concentrated force, which has a variable magnitude depending upon whether moment or shear is being computed in the bridge, as shown in Figure 1.6. The uniform force can be considered to act over any desired length of the bridge and the concentrated force can be located at any position to produce the maximum effect. The normal procedure is to compute the desired quantity using both loadings and then to choose the maximum of the two values. The AASHTO specifications also give additional information concerning loadings for multiple lanes and multiple span bridges.

An additional factor which must be considered in a highway bridge is the increase in the stresses in the bridge due to the moving loads. AASHTO considers this dynamic effect by specifying an impact factor for increasing the loads on the structure. The expression for this impact factor is

Structural Loads

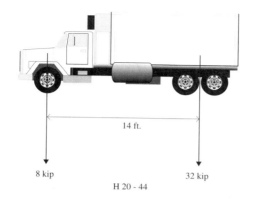

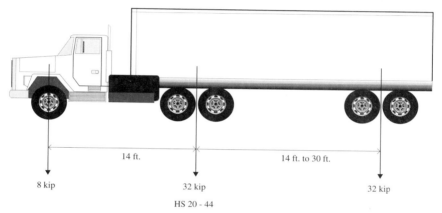

Figure 1.5 Highway bridge truck loadings.

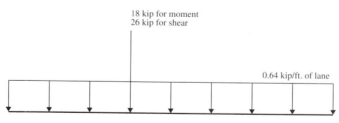

Figure 1.6 Highway bridge equivalent lane loading.

$$I = \frac{50}{L + 125} \leq 0.3 \qquad (1.1)$$

where L is the span length in feet. This factor can be used by either multiplying the loads on the bridge by the quantity $1 + I$ before the analysis is performed or by multiplying the results by this same quantity after the analysis has been completed.

2

Equilibrium and Reactions

EQUATIONS OF EQUILIBRIUM

A rigid body is in equilibrium if it is either at rest or in a state of constant motion (i.e., the acceleration is zero). According to Newton's Law, this requires that the resultant force on the body must be zero to prevent linear acceleration and the resultant moment must be zero to prevent angular acceleration. These two conditions are usually designated by the two equations

$$\Sigma F = 0 \tag{2.1}$$

and

$$\Sigma M = 0 \tag{2.2}$$

where ΣF represents the sum of the forces acting in any direction on the body, and ΣM represents the sum of the moments acting about any axis. Since one of the first steps in the analysis of any structure is usually to compute the reactive loads, we must develop the equations which are needed to compute these quantities. Several different sets of independent equations of equilibrium, which can be used to compute the reactive loads corresponding to any given set of active loads, can be obtained by applying the equilibrium conditions in different ways.

The complete set of active and reactive loads for any stable structure form a load system which is in equilibrium. The reactive loads occur automatically at the support joints as the active loads are applied to the structure. If the active loads change at any time, there will be a corresponding change in the reactive loads to ensure that the total load system is in equilibrium. You can actually compare the

Equations of Equilibrium

structure to a very fast computer. As the active loads are applied, the structure appears to perform an instantaneous set of calculations to determine the magnitude and direction of the required reactive loads. These reactive loads then occur at the supports to ensure that the structure remains in equilibrium.

Global Equilibrium of Plane Trusses and Plane Frames

For analysis purposes, the mathematical model of a plane truss or a plane frame will be assumed to lie in the XY plane of the global coordinate system with the positive X axis extending to the right and the positive Y axis up. All forces acting on the structure will be expressed as components along the X and Y axes and all moments will be about the Z axis. Since the positive Z axis will extend outward from the plane of the structure, positive moments about Z will be counterclockwise as a result of the right hand rotation rule.

Three different sets of global equations of equilibrium can be obtained for a rigid structure in a plane depending upon how the equilibrium conditions are applied. A rigid structure is defined to be a structure which cannot change its shape without deforming the members in the structure. These three sets of equations can be summarized as:

Set 1. If the sum of the moments about Z at any point 1 in a plane structure is zero, then the resultant moment on the structure must be zero and any resultant force must pass through that point. This resultant force now can be shown to be zero by ensuring that the sum of the force components in any two non parallel directions are zero. If the sum of the forces in the X direction is zero, then any resultant force must be parallel to the Y axis. If the sum of forces in the Y direction is now shown to be zero then the resultant force must be zero. These conditions can be expressed as

$$\Sigma_1 M_Z = 0 \quad \Sigma F_X = 0 \quad \Sigma F_Y = 0 \quad (2.3)$$

where $\Sigma_1 M_Z$ represents the sum of the moments about Z at point 1, and ΣF_X and ΣF_Y represent the sum of all of the force components on the structure in the X and Y directions, respectively. This set of conditions is shown in Figure 2.1a.

Set 2. If the sum of the moments about Z at any two points 1 and 2 in a plane structure is zero, then the resultant moment on the structure must be zero and any resultant force must pass through the two points. This resultant force now can be shown to be zero by ensuring that the sum of the force components parallel to the line through points 1 and 2 is zero. These conditions can be expressed as

$$\Sigma_1 M_Z = 0 \quad \Sigma_2 M_Z = 0 \quad \Sigma_{1-2} F = 0 \quad (2.4)$$

where $\Sigma_{1-2} F$ represents the sum of all of the force components on the structure parallel to the line through points 1 and 2. This set of conditions is shown in Figure 2.1b. (Note: The actual sum of forces may be in any direction which is not perpendicular to the line through points 1 and 2.)

Set 3. If the sum of the moments about Z at any three points 1, 2 and 3 in a plane structure, which do not lie on a straight line, is zero, then both the resultant moment

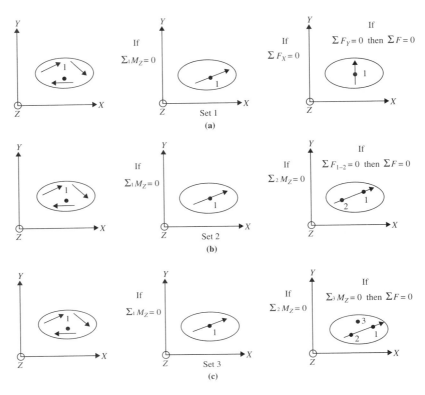

Figure 2.1 Equilibrium conditions for plane rigid body.

and the resultant force on the structure must be zero. As in the previous case, if the sum of the moments about Z at points 1 and 2 is zero, then any resultant force must pass through the two points. If the sum of the moments about any third point, which is not on the line extending through points 1 and 2 is also zero, then the resultant force must be zero. These conditions can be expressed as

$$\Sigma_1 M_Z = 0 \qquad \Sigma_2 M_Z = 0 \qquad \Sigma_3 M_Z = 0 \qquad (2.5)$$

This set of conditions is shown in Figure 2.1c.

For each of these cases, there are three independent equations of equilibrium. Therefore, it is possible to compute three independent reactive load components for a plane truss or plane frame by only using the global equilibrium conditions for the structure. The choice of which set of equations to use will depend upon the specific properties of the structure and the applied loads. If desired, the reactive loads can be computed by using one set of equations and the accuracy of the solution then can be verified by substituting the computed values into one of the other sets. This can be advantageous since many of the calculations in the overall analysis of a structure use the computed values for the reaction loads. If these values are incorrect, then all following calculations which are based on these values also will be incorrect. The analyst should take advantage of any checks which are available before proceeding to the next step in the analysis.

Global Equilibrium of Space Trusses and Space Frames

The forces acting on a rigid body in space may have components in any direction and the moments may have components about any axis. The equations of equilibrium for the body may be expressed in a number of different forms depending upon how the equilibrium conditions are applied. One general form for these equations, which will ensure that there is no resultant force or resultant moment acting on the body, can be obtained by summing the force components along each of the three global coordinate axes and summing the moment components about these same axes. This results in six equilibrium conditions which can be expressed as

$$\Sigma F_X = 0 \qquad \Sigma F_Y = 0 \qquad \Sigma F_Z = 0$$
$$\Sigma M_X = 0 \qquad \Sigma M_Y = 0 \qquad \Sigma M_Z = 0 \tag{2.6}$$

Other forms for the equilibrium equations may also be developed since there are many different ways in which the equilibrium conditions can be satisfied for an object in space.. For example, it is possible to start with any of the previously defined sets of three equations for a plane structure, which ensure that there is no resultant force in the XY plane and no resultant moment about an axis perpendicular to that plane, and then develop three additional equations, which will ensure that there are no additional resultant force components in the other global planes or moment components about the other global axes. The six independent equilibrium equations allow the computation of six independent reactive load components for a space truss or space frame using the global equilibrium conditions.

STRUCTURAL SUPPORTS

The supports for a structure may have various forms, depending on the type of restraints applied to the structure by the support. The purpose of these restraints is to prevent rigid body translations or rigid body rotations of the structure. The number of unknown quantities which must be computed to completely define the reactive loads for the structure will depend on both the number of support joints and the nature of the individual supports. Before we can attempt to compute these reactive loads, we first must understand the properties of the various types of supports which can exist for a structure. We will consider the supports for plane structures and space structures separately, since the supports for these two types of structures are quite different in their actions on the structure.

Supports for Plane Structures

Any reactive forces which act at the support joints in a plane structure must have only components in the plane of the structure (i.e., along the global X and Y axes). All reactive force components along Z must be zero. In addition, any reactive moments must act only about the global Z axis with all moment components about X and Y being zero. If any of these conditions is violated, either due to the geometry of the structure or due to the nature of the active loads, the structure must be analyzed as a space structure.

There are four types of supports which can exist for a plane structure, as shown in Figure 2.2. The difference in these supports is the manner in which they restrain the support joint against translation and rotation.

Fixed Support. A fixed support prevents translation of the support joint in both the X and Y directions and rotation of the joint about the Z axis. The reactive load applied to the support joint by these three restraints corresponds to a force R at an orientation angle α measured from the global X axis and at an eccentricity e from the joint, as shown in Figure 2.2a. Although it is possible to write the equilibrium equations for the structure in terms of the three independent quantities R, α and e, it usually will be more convenient to transfer the eccentric reactive force to act directly on the support joint as an equivalent force and moment. By now resolving this force into components along X and Y, the support load can be expressed as an independent reactive moment M_Z and two independent reactive force components R_X and R_Y.

$$M_Z = Re \tag{2.7}$$

$$R_X = R \cos \alpha \tag{2.8}$$

$$R_Y = R \sin \alpha \tag{2.9}$$

It will be found that the quantities R_X, R_Y and M_Z are usually more convenient to work with during the analysis for the reactions for a plane structure.

Pinned Support. A pinned support prevents translation of the support joint in the X and Y directions while permitting rotation of the joint about the Z axis. The reactive load applied to the support joint corresponds to a force R acting directly on the support joint at an orientation angle β measured from the global X axis, as shown in Figure 2.2b. By resolving this force into two components along X and Y the support load can be expressed as the two independent quantities R_X and R_Y.

$$R_X = R \cos \beta \tag{2.10}$$

$$R_Y = R \sin \beta \tag{2.11}$$

Slide Support. A slide support prevents both translation of the support joint perpendicular to the direction of the slide and rotation of the joint about the

Structural Supports

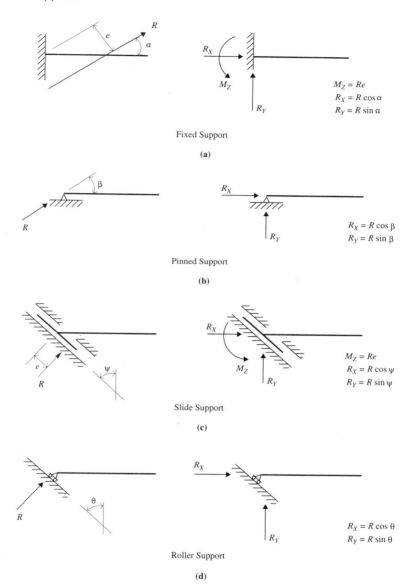

Figure 2.2 Plane structure supports.

Z axis. The joint is free to translate parallel to the direction of the slide. The reactive load consists of a force R, whose line of action is perpendicular to the direction of the slide, at an eccentricity e from the support joint, as shown in Figure 2.2c. The direction of the slide is defined by the orientation angle ψ, measured from the global Y axis. By transferring the line of action of the force R to the support joint, the reactive load can be expressed in terms of R and a moment M_Z.

$$M_Z = Re \tag{2.12}$$

It might now appear that if we would go one step further and resolve the force R into components along X and Y, the reactive force would be expressed in terms of three independent quantities R_X, R_Y and M_Z. However, this is not the case since the force components R_X and R_Y are not independent and can be expressed in terms of the unknown force R and the known orientation angle ψ as

$$R_X = R \cos \psi \tag{2.13}$$

and

$$R_Y = R \sin \psi \tag{2.14}$$

Therefore, the reactive force at a slide support can be expressed in terms of only two independent quantities. However, during any analysis to compute the reactions in a structure, it will often be convenient to work with the two reactive force components R_X and R_Y rather than with the inclined reactive force R.

Roller Support. A roller support prevents translation of the support joint perpendicular to the direction of the roller while permitting both translation of the joint parallel to the direction of the roller and rotation of the joint about the Z axis. The reactive load on the support joint can be expressed in terms of a single quantity which consists of a force R, acting directly on the support joint, whose line of action is perpendicular to the direction of the roller, as shown in Figure 2.2d. As described previously for the slide support, this reactive force can also be resolved into X and Y components. However, these two components are not independent since they can be expressed in terms of the unknown force R and the known roller orientation angle θ measured from the global Y axis, as shown previously for the slide support.

$$R_X = R \cos \theta \tag{2.15}$$

$$R_Y = R \sin \theta \tag{2.16}$$

For the particular case of a horizontal roller support, for which the orientation angle θ is 90°, the vertical force component R_Y is numerically equal to R and the horizontal force component R_X is zero. For a vertical roller, which corresponds to an orientation angle of zero, R_X will be equal to R and R_Y will be zero. During the analysis of any structure it usually can be assumed that a roller support can supply a reactive force to the support joint in either an inward or an outward direction. This will be the case for all roller supports in the following sections of this book unless specified otherwise.

Roller and slide supports are usually inserted into structures in order to permit translation in a specific direction without introducing any forces into the structure in that direction at the support points. A typical situation in which supports of this type would be used is to permit thermal expansion and contraction in a structure, such as a highway bridge, due to seasonal temperature changes. With-

Structural Supports

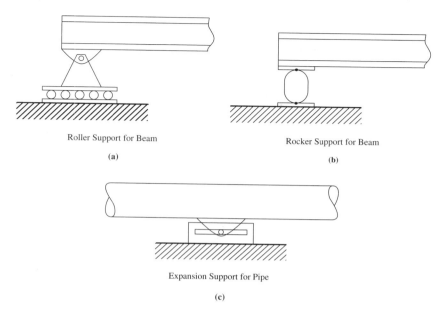

Roller Support for Beam
(a)

Rocker Support for Beam
(b)

Expansion Support for Pipe
(c)

Figure 2.3

out the freedom of movement in the supports the stresses which would be produced in the structure could be very large. Figure 2.3a shows a typical expansion support at the end of a highway bridge in which steel rollers are placed between the upper steel plate attached to the end of the girder and the lower steel plate attached to the bridge abutment to permit translation along the longitudinal axis of the girder. The pin connecting the upper and lower brackets permits rotation of the end of the girder. A variation of this support is also sometimes used in which the steel plates have a permanent graphite coating which provides very little frictional resistance to sliding. Figure 2.3b shows another type of expansion support in which the end of the girder is supported by a steel block with rounded ends which acts like a rocker. This will permit limited longitudinal translation of the end of the girder as the top of the rocker moves back and forth. The end of the girder is also free to rotate. There are many other forms which are used for structural supports. The best way for any student to gain an understanding of the various types of supports and how they function is to visit structures near them and see first hand how the supports are constructed. You can often gain much more from observing the world around you than looking at illustrations in books. Figure 2.3c shows a very simple expansion support which is often used for high temperature piping systems in power plants. The support consists of a steel plate welded to the pipe which is attached to a fixed plate by a bolt through a slotted hole. The bolt can rotate and move back and forth in the hole to provide the effect of a roller support.

It has been assumed in these descriptions that the specified restraints at the support joints are rigid and that they totally prevent translation or rotation. Of course, this might not be a correct assumption in an actual structure, since the supports might exhibit some finite degree of flexibility. This can be extremely important in developing an accurate mathematical model for a structure. The analyst must consider this possibility if a realistic analysis is to be performed. In the following sections in this book it will be assumed that the support restraints are rigid in all example problems and also in all suggested problems at the ends of the chapters unless specified otherwise.

Supports for Space Structures

The supports for a space structure can supply translation and rotation restraints to the support joints in any desired directions in three dimensional space. The number of restraints can vary between one and six, depending upon the nature of the support. The resulting reactive loads can be essentially any combination of the force components R_X, R_Y and R_Z and the moment components M_X, M_Y and M_Z. Figure 2.4 shows three common supports which might be encountered for a space structure.

Fixed Support. A fixed support for a space structure prevents both translation in all directions and rotation about all axes of the support joint. The reactive loads applied to the support can be represented by three force components R_X, R_Y and R_Z and three moment components M_X, M_Y and M_Z acting on the support joint, as shown in Figure 2.4a. For convenience in the illustration, the positive moments M_X, M_Y and M_Z are represented by vectors using the standard symbol of a double headed arrow. The positive rotation direction for each of these moments can be determined by the right hand rotation rule shown previously in Figure 1.4 in Chapter 1 by orienting the thumb of the right hand in the direction of the vector which represents the moment.

Pinned Support. A pinned support for a space structure prevents translation in all directions for the support joint while permitting the joint to rotate about any axis. A physical interpretation of the connection between the support and the support joint is a frictionless ball and socket similar to the human shoulder or hip joint. The reactive loads applied to the support joint can be represented by the three force components R_X, R_Y and R_Z, as shown in Figure 2.4b.

Roller Support. A roller support for a space structure provides a translation restraint to the support joint, perpendicular to the surface of the rollers, while permitting translation parallel to the rollers and rotation about any axis. A physical model of this type of support would consist of two flat plates which can roll with respect to each other on small ball bearings. The reactive load will consist of a single reactive force component acting on the support joint perpendicular to the direction of the rollers, as shown by the horizontal roller in Figure 2.4c. If

Stability and Static Determinacy

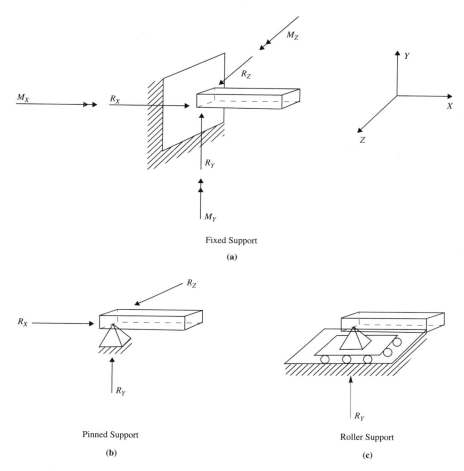

Figure 2.4 Space structure supports.

the roller support is at an orientation such that the reactive force does not act directly along one of the global coordinate axes, it is usually convenient to express the force in terms of its X, Y and Z components.

There are many other possible support configurations for space structures, such as a roller support which can translate in only one direction or even a combination of a roller and a slide support in different directions at the same support joint. The analyst must be very careful when developing the mathematical model for a space structure to ensure that the support conditions are modeled properly.

STABILITY AND STATIC DETERMINACY

Before we proceed with the computation of the reaction loads for any structure, we must first decide whether it is possible to perform the analysis while only using the equations of equilibrium. To accomplish this, we will define NEQ as the number of independent global equilibrium equations for the structure (i.e., three for a plane truss or plane frame and six for a space truss or space frame) and NR as the number of independent reactive load components. We will now consider three conditions which can exist for a structure.

Unstable Structures

If NR < NEQ, it will not be possible to obtain a solution to the set of equilibrium equations for the reactive loads since the number of independent equations is greater than the number of unknowns. For example, consider the following set of equations which might be obtained by summing moments about three different points, which do not lie on a straight line, for a plane truss of the type shown in Figure 2.5a.

$$2R_1 + 3R_2 = 5$$
$$4R_1 + 5R_2 = 7$$
$$7R_1 + 2R_2 = 4$$

There are no sets of values for R_1 and R_2 which will satisfy all three equations. Therefore, there is no set of reactive loads which will balance the active loads on the structure and place the structure in a state of equilibrium. Since equilibrium cannot be satisfied, rigid body translations and/or rotations will occur in the structure. The structure is unstable.

Statically Determinate Structures

If NR = NEQ, it should be possible to determine the reactive loads for the structure using the static equilibrium equations since the number of equations will be equal to the number of unknowns. The structure is statically determinate. However, it is still possible for the structure to be unstable due to its specific geometry. This condition can be demonstrated by the following set of equations which might be obtained by summing moments about three points, which do not lie on a straight line, for a plane truss of the type shown in Figure 2.5b.

$$2R_1 + 3R_2 + 3R_3 = 4$$
$$3R_1 - 3R_2 + 5R_3 = 3$$
$$4R_1 - 9R_2 + 7R_3 = 5$$

If an attempt is made to solve these equations for R_1, R_2 and R_3, it will be found that the three equations are not independent and a finite solution does not exist. Since a finite solution for the reactive loads cannot be found, the reactive loads cannot balance the active loads and the structure is unstable. The reason that this

Stability and Static Determinacy

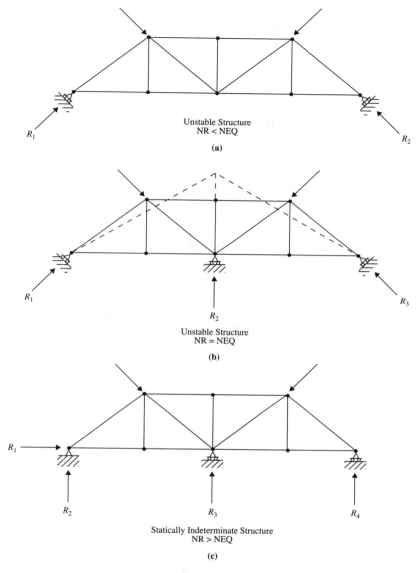

Figure 2.5

situation exists for this particular structure is because of the orientation of its three roller supports, which cause the lines of action of the three reactive forces to intersect at a common point. Therefore, if we would sum moments about this intersection point, all of the reaction forces would be eliminated from the equation, and the only way in which moment equilibrium could exist is if the active loads were restricted to be a set of loads which were in moment equilibrium about the point. This is not a practical restriction since the magnitude and the direction of the loads on a structure must be free to vary during the life of the structure.

Although the reason that the structure shown in Figure 2.5b is geometrically unstable is obvious, this will not always be the case. For many structures, an unstable condition is not easily detected by a visual inspection. However, there is a method for checking the stability of any structure for which the number of unknown reactive loads is equal to the number of equilibrium equations. The method is based upon the classical procedure for solving linear simultaneous equations by determinants known as Cramer's Rule, which is discussed in most undergraduate linear algebra courses. It is assumed that students are familiar with this analysis process from a previous course. To demonstrate this method for checking an unstable condition, we will consider a general set of n linear simultaneous equilibrium equations of the form

$$C_{1,1}R_1 + C_{1,2}R_2 + \cdots + C_{1,n}R_n = B_1$$
$$C_{2,1}R_1 + C_{2,2}R_2 + \cdots + C_{2,n}R_n = B_2 \qquad (2.17)$$
$$\vdots$$
$$C_{n,1}R_1 + C_{n,2}R_2 + \cdots + C_{n,n}R_n = B_n$$

where $C_{i,j}$ represents the coefficient of R_j in equation i and B_i is the right side constant for that equation. These equations can be expressed in standard matrix form as

$$[C]\{R\} = \{B\} \qquad (2.18)$$

where $[C]$ is a square matrix which contains the equation coefficients, $\{R\}$ is a column matrix which contains the unknown reactive quantities and $\{B\}$ is a column matrix which contains the right side constants of the equations. According to Cramer's Rule, the value of any unknown quantity R_j can be computed by

$$R_j = \frac{|C|_j}{|C|} \qquad (2.19)$$

where $|C|$ is the determinant of the coefficient matrix $[C]$ and $|C|_j$ is the determinant of a matrix which is obtained by replacing column j in $[C]$ by the right side column matrix $\{B\}$. No problems should be encountered in using this procedure to solve for each unknown reactive quantity unless the determinant of $[C]$ happens to be zero, in which case R_j will be infinite for one or more values of j. However, if there is no finite set of reactive loads which will satisfy the equilibrium requirements, the structure cannot be in equilibrium. If it is not in equilibrium, then it must be unstable. This leads to a test for checking the stability for any structure. If the determinant of the coefficient matrix of the equilibrium equations is non zero then the structure is stable. If the determinant is zero, then the structure is unstable. A simple computation will show that the determinant of the coefficient matrix is zero for the set of three equilibrium equations for the previous truss in which

Computation of Reactions

the lines of action of the three reactive forces intersect at a common point. Many hand calculators now have the capability to compute the determinant of a matrix.

Statically Indeterminate Structures

If NR > NEQ, it is not possible to determine a unique solution to the equilibrium equations for the reactive loads since the number of independent simultaneous equations is less than the number of unknowns. A situation like this would occur when analyzing a structure of the type shown in Figure 2.5c. If we were to generate three equilibrium equations by summing moments about three points, which do not lie on a straight line, we would have three independent equations with four unknowns. If we would then attempt to generate a fourth equation by either summing moments about another point or by summing forces in any direction, we would find that the new equation would not be independent. The left side coefficients of the fourth equation would be some linear combination of the coefficients of the other three equations. The best that could be accomplished when solving these equations would be to solve for three of the unknowns in terms of the fourth. Since that fourth unknown could then have any arbitrary value, there would actually be an infinite number of solutions to the equations. In order to determine a unique solution, an additional independent equation is required. The structure is defined to be statically indeterminate since we cannot determine the reactions using only the static equilibrium equations. We will discuss several different procedures for analyzing statically indeterminate structures in later chapters. For now, we will restrict our attention to structures which are statically determinate and stable.

COMPUTATION OF REACTIONS

Now that we understand the properties of the various types of supports which can exist for a structure and the various forms for the equations of equilibrium, we are prepared to compute the reactive loads for any statically determinate and stable structure. In this chapter we will only consider the analysis of plane structures. The analysis of space trusses will be discussed in Chapter 4 and the analysis of space frames will be discussed in Chapter 7.

Example Problem 2.1

Compute the reactions for the plane truss shown in Figure 2.6a. The first step in the analysis is to generate the mathematical model for the structure. A joint will be located in the mathematical model of the truss at each support point and at each point where the members intersect. Therefore, for this particular truss, the mathematical model will contain 8 joints and 13 members, as shown in Figure 2.6b. Although it is actually not necessary to number all of the joints and members for the manual analysis for the reactions, they have all been numbered for this example since most analyses of a truss of this type would go beyond the computation of the reactions and would

28 Equilibrium and Reactions   Chap. 2

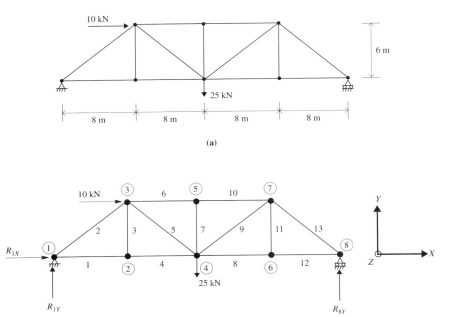

Figure 2.6 Example Problem 2.1.

also include the computation of the member forces. The computation of the member forces for a plane truss will be discussed in the next chapter. The joint and member numbers will be needed for identification purposes for that portion of the analysis. Therefore, we might as well get used to numbering all components in the mathematical model at this stage, even though the present analysis will end with the computation of the reactions. We will also need the joint and member numbers to describe the geometry of the truss in the input data when we analyze this same structure with the computer program SDPTRUSS later in this chapter. The mathematical model that we are using here is consistent with the requirements for the mathematical model for that program. The actual numbering scheme for the joints and members is arbitrary, but it will usually eliminate confusion if both the joints and members are numbered more or less sequentially along the structure rather than in a random order. The directions of the positive X and Y axes of the global coordinate system are shown to the right of the mathematical model in Figure 2.6b.

Since joint 1 is restrained by a pinned support and joint 8 is restrained by a horizontal roller support, there will be three independent reactive quantities for the structure. They will consist of a horizontal and vertical reactive force component at joint 1, which will be designated as R_{1X} and R_{1Y}, and a single vertical reactive force component at joint 8, which will be designated as R_{8Y}. The horizontal reactive force component at joint 8 will be zero due to the horizontal roller. Therefore, the structure will be statically determinate, if it is stable, since the number of unknown reactive quantities is equal to the number of independent equations of equilibrium for a

Computation of Reactions

plane structure. The stability of the structure will be verified during the analysis. Note that all reactive force components in Figure 2.6b have been indicated to act in the positive directions of the global coordinate axes. This is usually a good idea during any analysis, even though it might be obvious that some of the reactions act in the negative direction, since it can eliminate confusion in interpreting the signs in the final results of the analysis.

There are an infinite number of ways in which the equilibrium conditions can be applied to this structure to obtain a set of three independent equilibrium equations for computing R_{1X}, R_{1Y} and R_{8Y}. For example, we could sum moments about Z for three arbitrary points, which do not lie on a straight line, and then solve the resulting equations for the unknown reactive force components. In most cases, this would result in three simultaneous equations in which each unknown would have a non zero coefficient in each equation. The solution of these equations would require a significant amount of computation. However, if we are careful in the manner in which we generate the equilibrium equations, we might be able to simplify the calculations by being able to solve for each unknown reactive quantity in turn without the requirement of solving a combined set of equations. One of the many possible procedures for computing the three reactive forces for this particular structure is:

Summing moments about joint 1 will result in an equation with R_{8Y} as the only unknown.

$$\Sigma_1 M_Z \rightarrow -(10)(6) - (25)(16) + (R_{8Y})(32) = 0 \rightarrow R_{8Y} = 14.375 \text{ kN}$$

Note that the force has been entered first in each of the moment products in this equation since moment is usually considered to be force times distance. The moments have also been entered from left to right along the structure. It is very good practice to attempt to enter the numbers in all calculations in a consistent manner such as this rather than entering them in a haphazard manner. This can help to eliminate errors and it also makes it much easier for the person who will eventually check the calculations. Remember, all critical calculations will be checked by another engineer in most engineering offices.

Next, sum forces in the Y direction to obtain an equation containing R_{1Y} and R_{8Y}. This equation can then be solved for R_{1Y} since R_{8Y} is now known.

$$\Sigma F_Y \rightarrow R_{1Y} - 25 + R_{8Y} = 0 \rightarrow R_{1Y} = 10.625 \text{ kN}$$

Finally, summing forces in the X direction will result in an equation with R_{1X} as the only unknown.

$$\Sigma F_X \rightarrow R_{1X} + 10 = 0 \rightarrow R_{1X} = -10 \text{ kN}$$

No difficulties were encountered in generating or solving the equilibrium equations for unique finite values for the reactive force components. Therefore, the structure is stable. Since the computed values of R_{1Y} and R_{8Y} are positive, these force components act upward. However, since R_{1X} is negative, it acts toward the left.

A convenient method for listing the final results of the analysis is to list each value with a positive sign and an arrow which indicates its direction. This will help to eliminate any errors which might be made by another structural engineer in interpreting the results of the analysis, since all engineers might not use the same sign convention. Using this process, the final results can be summarized as

$$R_{1X} = 10 \text{ kN} \leftarrow$$

$$R_{1Y} = 10.625 \text{ kN} \uparrow$$

$$R_{8Y} = 14.375 \text{ kN} \uparrow$$

Our only problem now is to verify that these are the correct reactive loads for this structure for this set of active loads. One of the advantages of being able to develop more than three global equilibrium equations for any structure, even though they are not independent, is the ability to use these extra equations to check the accuracy of the results of the analysis. For example, we can check the accuracy of this solution by summing moments about Z for any arbitrary point which was not used to generate a moment equilibrium equation during the analysis. Performing this operation for a point midway between joint 2 and joint 3 gives

$$\Sigma M_Z \rightarrow -(R_{1Y})(8) + (R_{1X})(3) - (10)(3) - (25)(8) + (R_{8Y})(24) = 0$$

Since this equilibrium condition is satisfied by the previous numerical values for R_{1X}, R_{1Y} and R_{8Y}, it appears that no errors were made during the analysis and that the computed reactions are correct. It is usually best to choose a point for the moment check which is not on the line of action of any of the reactive force components, otherwise that force will be eliminated from the moment equation. The equilibrium condition could then still be satisfied even though that particular reactive force component was incorrect. More than one check equation can be used to definitely eliminate any possibility of compensating errors in the computed results. It is better to spend the time to verify the accuracy of the results at this stage in the analysis, rather than discovering at a later time that an error existed after many additional hours of calculation had been based upon these computed reactive values. Of course, in many cases, the numbers do not work out as evenly as they did in this problem. Since all calculations must be performed to some finite number of significant figures, we can usually expect some slight round off error to appear when the solution is checked. The size of the unbalance in the check equations will give an indication of the magnitude of the roundoff error in the computed values.

There are a number of other sets of equilibrium equations which could be used to compute the reactions for this structure. Some might lead to fewer calculations than others, but all should lead to the same solution. In some situations, it even might be advantageous to compute the reactions by two completely independent sets of equations to ensure that the correct solution has been obtained. It is the responsibility of the engineers performing the analysis to check their own calculations. They should not count on someone else to discover their errors. Although the ideal situation in an engineering office is to have all calculations independently checked by another engineer as soon as they are performed, this is usually not possible. There are actual cases in which the calculations were not checked until the structure was being constructed. This obviously could lead to serious problems if errors in the analysis were discovered.

Computation of Reactions

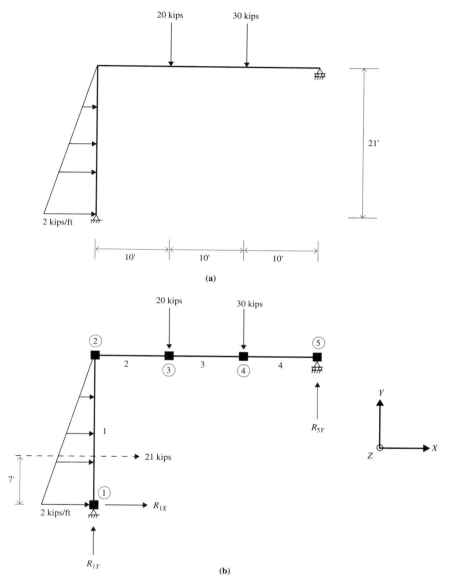

Figure 2.7 Example Problem 2.2.

Example Problem 2.2

Compute the reactions for the plane frame shown in Figure 2.7a. The mathematical model for a frame structure should have a joint at each support point, at each point where members intersect, at each point where a concentrated force or moment is applied and at the two ends of each distributed force. This will result in a mathematical model which is consistent with most structural analysis computer programs, such as those which are supplied with this book. We will discuss the use of the program SDPFRAME to analyze a plane frame later in this chapter. By using these rules, the mathematical model for this frame will have five joints and four members, as shown in Figure 2.7b.

Since joint 1 is restrained by a pinned support and joint 5 is restrained by a horizontal roller support, there will be three independent reactive force components which consist of a horizontal and vertical force at joint 1 and a vertical force at joint 5. The horizontal reactive force component at joint 5 must be zero. Therefore, since there are three unknown reactive quantities for this structure and three independent equations of equilibrium for a plane frame, the structure is statically determinate if it is stable. It appears to be stable.

During the equilibrium analysis, the triangular distributed force on member 1 can be treated as an equivalent concentrated force of 21 kips acting 7 feet up from joint 1, as shown by the dash-line force in Figure 2.7b. The analysis for the reactive force components can be performed as follows:

Summing moments about Z at joint 1 will give an equation with only R_{5Y} as the unknown.

$$\Sigma_1 M_Z \rightarrow -(21)(7) - (20)(10) - (30)(20) + (R_{5Y})(30) = 0 \rightarrow R_{5Y} = 31.567 \text{ kips}$$

$$\rightarrow R_{5Y} = 31.567 \text{ kips} \uparrow$$

The values of R_{1X} and R_{1Y} now can be computed by summing forces in the X and Y directions.

$$\Sigma F_X \rightarrow R_{1X} + 21 = 0 \rightarrow R_{1X} = -21 \text{ kips} \rightarrow R_{1X} = 21 \text{ kips} \leftarrow$$

$$\Sigma F_Y \rightarrow R_{1Y} - 20 - 30 + R_{5Y} = 0 \rightarrow R_{1Y} = 18.433 \text{ kips} \rightarrow R_{1Y} = 18.433 \text{ kips} \uparrow$$

We can now verify the solution by summing moments about a point 10 feet to the left of joint 2 and at the same elevation. This moment check point was arbitrarily chosen so that none of the lines of action of the computed reactive forces passed through it.

$$\Sigma M_Z \rightarrow (R_{1X})(21) + (R_{1Y})(10) + (21)(14) - (20)(20) - (30)(30)$$

$$+ (R_{5Y})(40) = 0.01$$

The solution checks within acceptable roundoff error.

Example Problem 2.3

For a third example, we will consider a slight variation in the plane frame which was analyzed in the previous example, as shown in Figure 2.8a. For this structure, the roller support at the upper right joint is inclined at a three vertical and four horizontal slope.

The mathematical model for the structure is shown in Figure 2.8b, in which the three independent reactive quantities are designated as R_{1X}, R_{1Y} and R_5. The force R_5 will act on joint 5 with a line of action perpendicular to the direction of the roller. Although it is possible to work directly with this inclined force during the analysis, it

Computation of Reactions

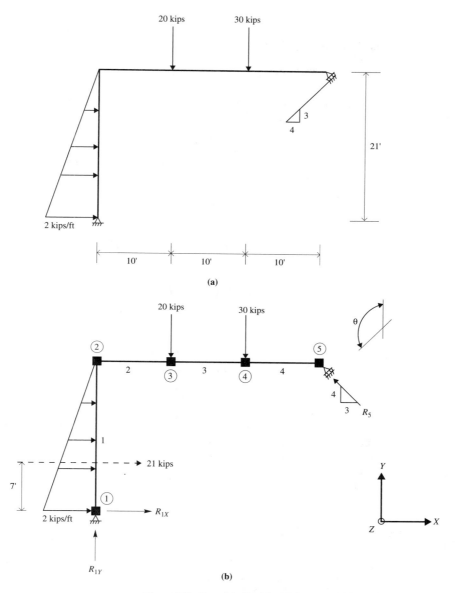

Figure 2.8 Example Problem 2.3.

will be more convenient to work with the two components R_{5X} and R_{5Y}, which can be computed from Eqs. (2.15) and (2.16) using the second quadrant angle θ, as

$$R_{5X} = R_5 \cos \theta = -0.6 R_5$$
$$R_{5Y} = R_5 \sin \theta = 0.8 R_5$$

from which we can determine that

$$R_{5X} = -0.75 R_{5Y}$$

The quantity R_{5X} is negative since it will act in the negative X direction for the assumed positive direction of R_5 shown in Figure 2.8b. We can now compute the reactive forces by applying the equilibrium conditions as follows:

Summing moments about Z at joint 1 (assuming both R_{5X} and R_{5Y} are positive) gives

$$\Sigma_1 M_Z \rightarrow -(21)(7) - (20)(10) - (30)(20) - (R_{5X})(21) + (R_{5Y})(30) = 0$$

which can be solved for R_{5Y} after substituting the preceding relationship for R_{5X} in terms of R_{5Y}

$$-(21)(7) - (20)(10) - (30)(20) - (-0.75 R_{5Y})(21) + (R_{5Y})(30) = 0$$

$$\rightarrow R_{5Y} = 20.699 \text{ kips} \uparrow$$

With R_{5Y} known, we can now compute R_{5X} as

$$R_{5X} = -0.75 R_{5Y} = 15.524 \text{ kips} \leftarrow$$

after which the value of the force R_5 can be computed as

$$R_5 = \sqrt{R_{5X}^2 + R_{5Y}^2} = 25.874 \text{ kips} \nwarrow$$

Finally the reactive force components R_{1X} and R_{1Y} can be computed by summing forces in the X and Y directions.

$$\Sigma F_X \rightarrow R_{1X} = 5.476 \text{ kips} \leftarrow$$

$$\Sigma F_Y \rightarrow R_{1Y} = 29.301 \text{ kips} \uparrow$$

Note that the inclination of the roller support has a significant effect upon the magnitudes of the reactive forces when compared to the results of the previous example problem with a horizontal roller. The orientation of both roller and slide supports must be accurately represented in the mathematical model of any structure since even a small inclination can have a significant effect on the analysis results.

EQUATIONS OF CONDITION

In the previous example problems, it was only possible to determine three independent reactive quantities for a plane structure since there were only three independent equations of equilibrium. However, it is possible to have more than three independent equations if the structure is constructed in a specific manner. If these additional independent equations exist, it will be possible to compute one additional reactive quantity for each new equation. The additional equations are known as *equations of construction* or *equations of condition*. Both terms are equally used in the engineering literature. These equations can be generated if one or more segments of the structure can be isolated from the total structure in

Equations of Condition

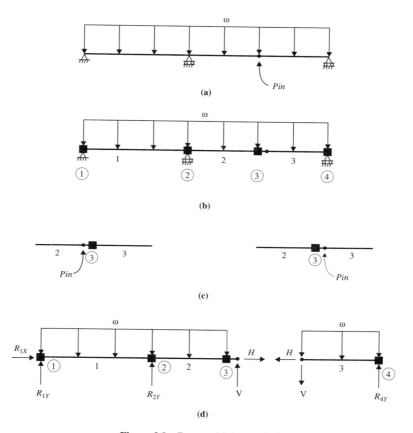

Figure 2.9 Beam with internal pin.

such a way that it is possible to develop new equilibrium equations without introducing any additional unknown quantities into the analysis.

As an example, consider the two span beam shown in Figure 2.9a. The beam is continuous over the interior roller support so that a moment can be transmitted across the support, but no moment can be transmitted into the support. The pinned support and the two roller supports will result in four independent reactive force components at the support joints. The special construction condition for this beam, which will permit us to compute these four reactive force components, is the interior pin in the right span. This pin can transmit both horizontal and vertical force components between the segments of the beam on either side, but it cannot transmit a moment.

The mathematical model for the beam must contain one joint at each of the support points. In addition, a joint must be included at the location of the interior pin, which results in a mathematical model with four joints, as shown in Figure 2.9b. The effect of the interior pin can be simulated in the mathematical model by either of the two conditions shown in Figure 2.9c, in which the member to either the left or the right of joint 3 is considered to be pinned to the joint. Both members cannot be pinned to the joint, since one of the members must be rigidly attached to restrain the joint against rotation. For this particular analysis, we will assume that member 3 is pinned to joint 3. The only way that this choice could affect the final results of the analysis is if an external concentrated moment were applied to joint 3. For that situation, the moment would be transmitted into the member which was rigidly attached to the joint. This is not a factor in this problem.

We can generate three independent equilibrium equations in terms of the four unknown reactive forces R_{1X}, R_{1Y}, R_{2Y} and R_{4Y} by using any of the sets of equilibrium conditions described previously for a plane structure. The fourth independent equation can be obtained by isolating the segments of the beam to the left and right of the pin as free bodies, as shown in Figure 2.9d. Since only a horizontal force component H and a vertical force component V can be transmitted through the pin between joint 3 and the left end of member 3, it is possible to sum moments about Z at the pin for either the left segment or the right segment of the beam without introducing any additional unknown quantities into the equation. Using this additional equation, we will now have four equations and four unknowns which can be solved for the four reactive forces.

It might seem at first that we actually have two additional independent equations since we can sum moments for both the left segment and the right segment at the pin. However, this is not the case since the combination of these two equations corresponds to the single equation which would be obtained by summing moments about joint 3 for the total structure. This would not be an independent equation since we have already generated our allotment of three global equations for the total structure. The general rule is:

> If there are N members connected to an internal pin in a frame, only $N - 1$ independent equations of condition can be generated by summing moments about the pin for various segments of the structure. One of the members in the mathematical model must be rigidly connected to the joint at the pin location.

The equations of condition are local equations for those particular segments of the structure for which they were generated rather than global equations for the total structure.

Note that if the pin were not present in the beam, there also would be an unknown moment M at the point where the two segments are cut apart. This moment would enter into the moment equation for either segment of the beam. We then would have four equations with five unknowns and the extra equation would not have helped. The structure must have this special construction condition in order to generate the additional equation to solve for the reactions.

Equations of Condition

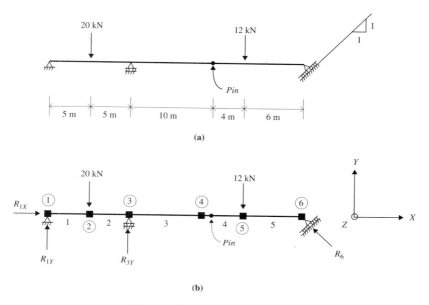

Figure 2.10 Example Problem 2.4.

Example Problem 2.4

The use of an equation of condition can be demonstrated by computing the reactions for the beam shown in Figure 2.10a. The mathematical model for the beam is shown in Figure 2.10b. In addition to the three joints at each of the support points and the two joints at the locations of the concentrated forces, a joint has also been included at the location of the interior pin. This results in a mathematical model with a total of six joints and five members. All members are assumed to be rigidly attached to their end joints except member 4, which is pinned where it is attached to joint 4. This will simulate the action of the interior pin in the beam. The four independent reaction quantities consist of the three force components R_{1X}, R_{1Y}, R_{3Y} and the inclined force R_6 at the roller support. The analysis procedure which can be used to compute these reactive forces is:

Summing moments about Z for the segment of the beam to the right of the pin gives

$$\Sigma_{\text{pin/right}} M_Z \to -(12)(4) + (R_{6Y})(10) = 0 \to R_{6Y} = 4.8 \text{ kN} \uparrow$$

Since the roller is at a one to one slope we can now compute R_{6X} as

$$R_{6X} = -R_{6Y} = 4.8 \text{ kN} \leftarrow$$

from which R_6 can be determined to be

$$R_6 = \sqrt{R_{6X}^2 + R_{6Y}^2} = 6.788 \text{ kN} \nwarrow$$

Next summing moments about Z at joint 1 for the total structure gives

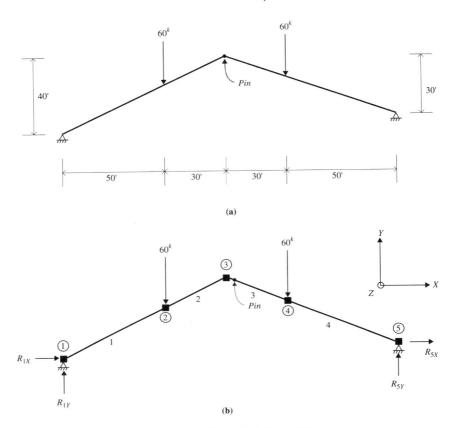

Figure 2.11 Example Problem 2.5.

$$\Sigma_1 M_Z \rightarrow -(20)(5) + (R_{3Y})(10) - (12)(24) + (R_{6Y})(30) = 0$$
$$\rightarrow R_{3Y} = 24.4 \text{ kN} \uparrow$$

The values of R_{1X} and R_{1Y} can now be computed by summing forces in the X and Y directions for the total structure

$$\Sigma F_X \rightarrow R_{1X} = 4.8 \text{ kN} \rightarrow$$
$$\Sigma F_Y \rightarrow R_{1Y} = 2.8 \text{ kN} \uparrow$$

This solution now can be verified by summing moments about some point which is not on the line of action of any of the reactive forces. This will be left as an exercise for the reader.

Example Problem 2.5

Compute the reactions for the plane frame shown in Figure 2.11a. The mathematical model for the frame is shown in Figure 2.11b. Member 3 is assumed to be pinned to joint 3 to simulate the interior pin. This problem is somewhat different than the

Equations of Condition

previous beam problem since it will not be possible to solve for any of the reactive force components independently. At least one set of simultaneous equations must be solved at some point in the analysis.

The analysis can be performed as follows:

Summing moments about Z at joint 1 for the total structure gives

$$\Sigma_1 M_Z \rightarrow -(60)(50) - (60)(110) - (R_{5X})(10) + (R_{5Y})(160) = 0$$

which reduces to

$$-10R_{5X} + 160R_{5Y} = 9600$$

Summing moments about the pin for the right segment of the structure gives

$$\Sigma_{\text{pin/right}} M_Z \rightarrow -(60)(30) + (R_{5X})(30) + (R_{5Y})(80) = 0$$

which reduces to

$$30R_{5X} + 80R_{5Y} = 1800$$

We now have two simultaneous equations with R_{5X} and R_{5Y} as the unknowns. Solving these equations gives

$$R_{5X} = 85.712 \text{ kips} \leftarrow$$
$$R_{5Y} = 54.643 \text{ kips} \uparrow$$

The values of R_{1X} and R_{1Y} now can be computed by summing forces in the X and Y directions for the total structure.

$$\Sigma F_X \rightarrow R_{1X} = 85.712 \text{ kips} \rightarrow$$
$$\Sigma F_Y \rightarrow R_{1Y} = 65.357 \text{ kips} \uparrow$$

The accuracy of this solution can be verified by summing moments about Z for the total structure at some convenient point, such as joint 3. This check is left to the reader.

Stability, Determinacy and Indeterminacy

In the previous example problems, we only considered the case where the equation of condition existed because of an internal pin which resulted in a moment release between two segments of the structure. Other types of internal releases also can exist which will lead to other forms of the equations of condition. For example, Figure 2.12a shows a statically determinate beam with five independent reactive force components due to the restraints supplied by the two pinned supports and the roller support. The left and right segments of the beam are joined by an internal roller connection which results in both a moment release and a horizontal force release between the segments. Since only vertical force can be transmitted between the two segments of the beam, two additional independent equilibrium equations can be obtained by isolating the two segments and summing forces in the

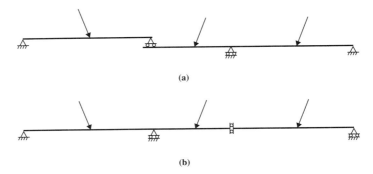

Figure 2.12 Other internal releases.

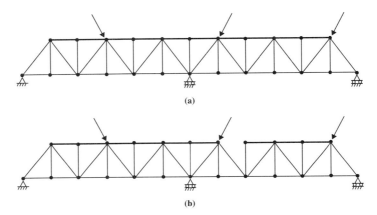

Figure 2.13 Truss equation of condition.

X direction and summing moments about Z at the connection point for either segment. Another type of internal release which will lead to an equation of condition is shown in Figure 2.12b. The left and right segments of this beam are joined by a vertical slide connection which can transmit both moment and horizontal force but no vertical force. An independent equation of condition can be obtained by summing forces in the Y direction for either the left or right segments of the beam. A connection of this type could be constructed by connecting two vertical plates, which are welded to the ends of the beam segments, by bolts which pass through vertical slotted holes. Both of the internal connections shown in Figure 2.12 might be used to permit thermal expansion in a structure.

It is also possible to have equations of condition in plane trusses. Figure 2.13a shows a truss with four independent reactive force components which is statically indeterminate. Figure 2.13b shows a truss with the same support conditions which is statically determinate. The equation of condition can be generated for the second truss by summing moments about Z for either the right or left segment of the truss which are connected by the single pin. This same process cannot be used in the first truss due to the horizontal member which connects the two segments. If the truss is divided into segments by making a vertical cut through the pin, it also will be necessary to cut through the member. This member force then would enter into the moment equation, thus introducing an additional unknown in the problem.

A structural engineer must be very careful when analyzing any type of structure to ensure that any internal force or moment releases which exist in the structure are considered properly during the analysis. It is important that the engineer have the experience and knowledge to develop a mathematical model for a structure which represents its actual behavior within acceptable bounds. This is true whether the analysis is to be performed manually or on the computer.

Since we now see that the number of independent equilibrium equations can exceed three for a plane structure and six for a space structure, it is necessary to revise the previously described conditions for determining whether a structure is statically determinate and stable. For any plane structure, the number of independent equilibrium equations will be

$$NEQ = NC + 3 \qquad (2.20)$$

where NC is the number of independent equations of condition. Therefore, by using the same reasoning that was used previously, the stability and static determinacy conditions for any plane structure are:

if $NR < NC + 3$, the structure is unstable;

if $NR = NC + 3$, the structure is statically determinate;

and

if $NR > NC + 3$, the structure is statically indeterminate.

A similar set of conditions exist for a space structure except that the number of independent equilibrium equations will be $NC + 6$. Of course, it is also possible for a structure to be geometrically unstable even though the number of independent reactive quantities is equal to the number of independent equilibrium equations. The stability of any structure can be verified by computing the determinant of the equation coefficients. If the determinant is nonzero, the structure is stable, while if it is zero, the structure is unstable.

COMPUTER PROGRAMS SDPTRUSS AND SDPFRAME

The computer programs SDPTRUSS and SDPFRAME, which are supplied for this book, can be used by the readers to verify their manual solutions for the reactions for plane trusses and plane frames. The program SDPTRUSS will analyze a statically determinate plane truss while the program SDPFRAME will analyze a statically determinate plane frame. For analysis purposes, a beam can be considered to be a plane frame. The instructions for using SDPTRUSS are in the disk file

SDPTRUSS.DOC and the instructions for SDPFRAME are in the disk file SDPFRAME.DOC. These files are in ASCII form. A hard copy can be obtained by listing them on the printer using the MS DOS command COPY NAME.EXT PRN:, where NAME is the name of the file and .EXT is its extension. The files are already formatted to give a 1-inch left margin when printed at 10 characters per inch. The printer should be set to give 1-inch top and bottom margins. The instructions for setting the printer format should be contained in the instruction manual for the particular printer which is being used.

The input data for each program must be supplied in an ASCII disk file. The required format for this data file for each program is shown in the program instructions along with a listing of a sample data file. For now, we will merely treat these programs as "black boxes" in which we can insert input and receive output in order to check our manual calculations for the reactions for a plane truss or a plane frame. The executable file, which has a .EXE extension, for either SDPTRUSS or SDPFRAME can be executed by merely typing the program name, without the extension, and pressing the Enter key. These programs will be discussed in more detail in Chapters 3 and 6.

The input data files for SDPTRUSS and SDPFRAME can be created in the format described in the program instructions using any ASCII editor, such as the program EDIT, which is supplied with Version 5 and newer versions of the MS DOS operating system. The geometry of the mathematical model for a plane truss or plane frame is described in the input data for each program by specifying the position of the joints as a set of X and Y coordinates in the global coordinate system for the structure. The requirements for locating the joints in the mathematical model are the same as those used in the previous example problems. These requirements are described in the program instructions. The origin of the global coordinate system may be at any convenient point. The locations of the members are defined by specifying the joint numbers at their two ends. All members in a plane truss, which is to be analyzed by SDPTRUSS, are assumed to be pinned to their end joints, while all members in a plane frame, which is to be analyzed by SDPFRAME, are initially assumed to be rigidly attached to the joints. However, it is also possible to modify the end connections for specific members in a plane frame by additional input data after the locations of the members have been defined.

All active and reactive concentrated forces are expressed as X and Y components in the global coordinate system for both plane trusses and plane frames, while the positive direction of all active and reactive moments about the Z axis for plane frames are defined by the right hand rotation rule. All distributed member forces on plane frame members are expressed as components q_x and q_y in a local right hand orthogonal coordinate system, which is unique to each member in the frame, as shown in Figure 2.14. The three axes for the local member coordinate system are designated as x, y and z, with the local x axis extending along the longitudinal axis of the member from the first joint toward the second joint, and the local z axis extending out of the plane of the structure parallel to the global Z ax-

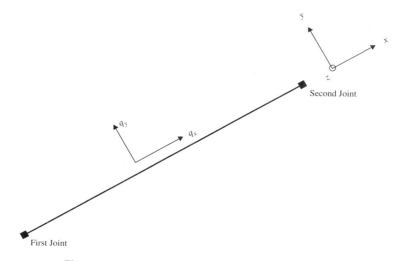

Figure 2.14 Frame member distributed force components.

is. Therefore, the x component of the distributed force will be along the longitudinal axis of the member and the y component will be perpendicular to the member axis. The user must be careful to define the correct signs for these components with respect to the directions of the local axes. The designations for the first joint and the second joint for a member are defined in the input data as explained in the program instructions.

In addition to the analysis programs SDPTRUSS and SDPFRAME, four companion programs named SDPTDATA, SDPFDATA, PLOTSDPT and PLOTSDPF are also available. The program SDPTDATA can be used to interactively create the input data file for SDPTRUSS while the program SDPFDATA can be used to create the input data file for SDPFRAME. The data files will be written to the disk in the correct format for use in the analysis. This eliminates the need for the program user to remember the exact format requirements for the data file. However, the program instructions still should be consulted to determine what input information will be required to describe any structure. Although it first might seem that it will be faster and easier to create the data files for the analysis of any structure using the programs SDPTDATA or SDPFDATA, the majority of the students who have used these programs in the author's structural analysis courses found that it was faster to create the files manually using EDIT after they became familiar with the required file format. They claimed that answering the questions from SDPTDATA and SDPFDATA during the interactive creation of the files slowed them down. The readers can make their own judgements as to which mode of operation they prefer.

After the data files have been created, the programs PLOTSDPT and PLOTSDPF can be used to plot the geometry and show the support restraints for the mathematical models for the structures before they are analyzed by

```
Example Problem 2.1 - Analysis by Program SDPTRUSS
8,13,2,2
Joint Coordinates
1,0.0,0.0
2,8.0,0.0
3,8.0,6.0
4,16.0,0.0
5,16.0,6.0
6,24.0,0.0
7,24.0,6.0
8,32.0,0.0
Member Data
1,1,2
2,1,3
3,2,3
4,2,4
5,3,4
6,3,5
7,4,5
8,4,6
9,4,7
10,5,7
11,6,7
12,6,8
13,7,8
Support Restraints
1,1,1
8,0,1
Joint Loads
3,10.0,0.0
4,0.0,-25.0
```

Figure 2.15 Example Problem 2.1, SDPTRUSS input.

SDPTRUSS or SDPFRAME. The input to these programs will be the same data files which are used by the analysis programs. Errors in the input data often can be seen in these plots which are not obvious from a listing of the input data file.

The following computer solutions demonstrate how the programs SDPTRUSS and SDPFRAME can be used to verify the manual analyses for several of the previous example problems. The reader should carefully study these computer solutions to gain an understanding of the input requirements for the programs.

Computer Solution for Example Problem 2.1

Figure 2.15 shows a listing of the input data file for the analysis of Example Problem 2.1 by the program SDPTRUSS. The joint coordinates are expressed in meters and the joint forces are expressed in kilonewtons. The origin for the global coordinate system is located at joint 1. A plot of the geometry of the mathematical model used in this analysis is shown in Figure 2.16. This plot was generated by the program PLOTSDPT. The results of the analysis are shown in Figure 2.17. These results consist of the member axial forces and the X and Y reactive force components at the support joints. The computed values for the reactions at joint 1 and joint 8 agree with those obtained from the manual solution. We will not be

Example Problem 2.1 - Analysis by Program SDPTRUSS

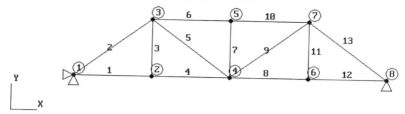

Figure 2.16 Example Problem 2.1, geometry plot from PLOTSDPT.

```
Example Problem 2.1 - Analysis by Program SDPTRUSS

Member Forces (Tension Positive)

Member          Force

  1            24.167
  2           -17.708
  3             0.000
  4            24.167
  5            17.708
  6           -38.333
  7             0.000
  8            19.167
  9            23.958
 10           -38.333
 11             0.000
 12            19.167
 13           -23.958

Reactions

Joint          RX              RY

  1          -10.000          10.625
  8            0.000          14.375
```

Figure 2.17 Example Problem 2.1, SDPTRUSS output.

concerned with the member forces at this time. The computation of the member forces for a plane truss will be discussed in the next chapter along with a more detailed description of the program SDPTRUSS.

Computer Solution for Example Problem 2.2

Figure 2.18 shows a listing of the input data file for the program SDPFRAME for Example Problem 2.2, with the joint coordinates expressed in feet, the joint forces expressed in kips and the distributed member forces expressed in kips per foot. A plot of the geometry of the mathematical model for the structure, which was generated by the program PLOTSDPF, is shown in Figure 2.19. The results of

```
Example Problem 2.2 - Analysis by Program SDPFRAME
5,4,0,2,2,1
Joint Coordinates
1,0.0,0.0
2,0.0,21.0
3,10.0,21.0
4,20.0,21.0
5,30.0,21.0
Member Data
1,1,2
2,2,3
3,3,4
4,4,5
Support Restraints
1,1,1,0
5,0,1,0
Joint Loads
3,0.0,-20.0,0.0
4,0.0,-30.0,0.0
Distributed Member Loads
1,0.0,0.0,-2.0,0.0
```

Figure 2.18 Example Problem 2.2, SDPFRAME input.

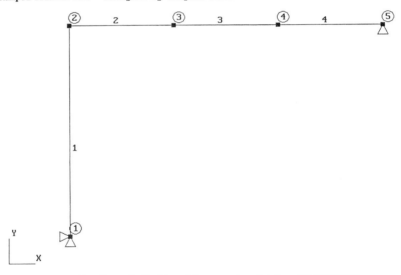

Figure 2.19 Example Problem 2.2, geometry plot from PLOTSDPF.

the analysis are shown in Figure 2.20. The computer solution for the reactive force components at joints 1 and 5 agree exactly with those obtained in the manual analysis. The definitions for the member end loads, which are listed in the results, will be discussed in Chapter 6 where the analysis procedure which is used in the program SDPFRAME will be described.

The local y components of the distributed force on member 1 at the first joint and at the second joint are equal to -2.0 and 0.0 respectively since joint 1 is de-

Example Problem 2.2 - Analysis by Program SDPFRAME

Member End Loads

Member	Joint	Sx	Vy	Mz
1	1	18.433	21.000	0.00
	2	-18.433	0.000	147.00
2	2	-0.000	18.433	-147.00
	3	0.000	-18.433	331.33
3	3	-0.000	-1.567	-331.33
	4	0.000	1.567	315.67
4	4	0.000	-31.567	-315.67
	5	0.000	31.567	0.00

Reactions

Joint	RX	RY	MZ
1	-21.000	18.433	0.00
5	0.000	31.567	0.00

Figure 2.20 Example Problem 2.2, SDPFRAME output.

fined as the first joint and joint 2 is defined as the second joint for the member in the input data. This causes the local x axis for the member to be vertical and positive up and the local y axis to be horizontal and positive to the left. The distributed force is assumed to have a linear variation between the values at the two ends of the member during the analysis. If the first joint and second joint definitions were reversed for member 1 in the input data, then the y distributed force components at the first joint and the second joint would be 0.0 and 2.0, respectively, since the local x axis would now be positive down and the local y axis would be positive to the right. Either of these two forms of the input data for the distributed force is acceptable and will lead to the same analysis results for the reactions. The verification of this is left as an exercise for the reader.

Computer Solution for Example Problem 2.4

Figure 2.21 shows a listing of the input data file for program SDPFRAME for Example Problem 2.4. The primary differences in creating this input file, compared to the input file shown in Figure 2.18 for the analysis of Example Problem 2.2, is that the input must include information to describe the internal pin and the inclined support. The internal pin is represented in the same manner that was used in the manual analysis by inserting joint 4 in the mathematical model of the beam at the location of the pin and then specifying that member 4, which is to the right of joint 4, is attached to the joint by a pin. A plot of the mathematical model by the program PLOTSDPF is shown in Figure 2.22. Note that the mathematical model has a total of seven joints and six members, since a dummy joint and a dummy member also have been included in the computer input data to represent the inclined roller support as described in the user instructions for SDPFRAME.

```
Example Problem 2.4 - Analysis by Program SDPFRAME
7,6,2,3,2,0
Joint Coordinates
1,0.0,0.0
2,5.0,0.0
3,10.0,0.0
4,20.0,0.
5,24.0,0.0
6,30.0,0.0
7,35.0,-5.0
Member Data
1,1,2
2,2,3
3,3,4
4,4,5
5,5,6
6,6,7
Pinned Members
4,0,1
6,0,1
Support Restraints
1,1,1,0
3,0,1,0
7,1,1,0
Joint Loads
2,0.0,-20.0,0.0
5,0.0,-12.0,0.0
```

Figure 2.21 Example Problem 2.4, SDPFRAME input.

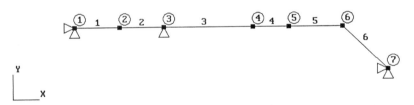

Figure 2.22 Example Problem 2.4, geometry plot from PLOTSDPF.

Dummy member 6 is pinned to joint 6 and is rigidly attached to dummy joint 7, which is restrained by a pinned support. The internal pin at the end of member 6 where it connects to joint 6 supplies an additional equation of condition for the mathematical model. This fifth equation is needed since this mathematical model has five independent reactive force components, as indicated by the five translation restraints at its supports, while the actual structure only has four support translation restraints. The results of the computer analysis are shown in Figure 2.23. The reactive force components at dummy joint 7 are equal to the reactive force components at the roller support in the actual structure. This can be easily verified by considering the free body diagram of the dummy member shown in Figure 2.24.

Example Problem 2.4 - Analysis by Program SDPFRAME

Member End Loads

Member	Joint	Sx	Vy	Mz
1	1	4.800	2.800	0.00
	2	-4.800	-2.800	14.00
2	2	4.800	-17.200	-14.00
	3	-4.800	17.200	-72.00
3	3	4.800	7.200	72.00
	4	-4.800	-7.200	0.00
4	4	4.800	7.200	0.00
	5	-4.800	-7.200	28.80
5	5	4.800	-4.800	-28.80
	6	-4.800	4.800	0.00
6	6	6.788	0.000	0.00
	7	-6.788	0.000	0.00

Reactions

Joint	RX	RY	MZ
1	4.800	2.800	0.00
3	0.000	24.400	0.00
7	-4.800	4.800	0.00

Figure 2.23 Example Problem 2.4, SDPFRAME output.

Summing moments about joint 6 shows that the moments of the two reactive force components at joint 7 exactly balance. Therefore, the resultant reactive force must have a line of action along the length of the dummy member and its magnitude must be equal to the reactive force, which would occur at the inclined roller support at joint 6 in the actual structure.

In many cases during an analysis, it is possible to use several variations of the mathematical model for a structure to arrive at the same analysis results. For this structure, it would also be acceptable to consider dummy joint 7 to be restrained by a fixed support, while considering dummy member 6 to be pinned to this joint. This would add an additional reactive moment at dummy joint 7, but the internal pin connection for the dummy member to this joint would provide another equation of condition. The reactive force components at joint 7 would be the same for this mathematical model as for the mathematical model which was just analyzed. The reactive moment at joint 7 would be zero. These two support variations are shown in Figure 2.25.

A similar technique to that used in this computer solution can be used when analyzing any structure with inclined roller supports with the programs SDPTRUSS or SDPFRAME. However, it is not necessary to insert a pin at the end of the dummy member in SDPTRUSS since all plane truss members are automatically assumed to be pinned to their end joints. This is explained in the user instructions for the programs.

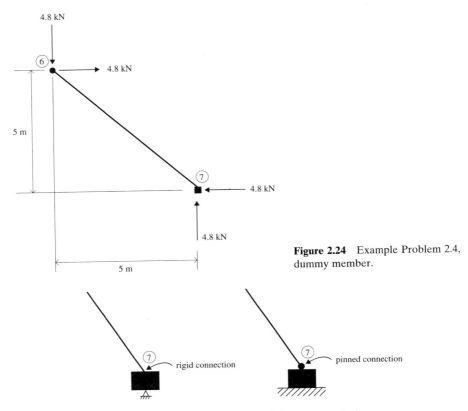

Figure 2.24 Example Problem 2.4, dummy member.

Figure 2.25 Example Problem 2.4, support variations.

The reader might now want to gain some experience with the computer by analyzing Example Problems 2.3 and 2.5 with the program SDPFRAME.

INFLUENCE LINES FOR BEAM REACTIONS

In all of the previous analyses, the loads were stationary on the structure. However, in many structures, such as highway bridges or crane runways, the loads can move to any position. Since it is usually necessary to determine the maximum values for the reactions during the design of a structure, a procedure must be developed for determining the position of the moving loads which will maximize the reactive loads. A very useful tool for determining the effect of the position of the loads upon the magnitudes of the reactions is an influence line. The following discussion will be concerned only with influence lines for beam reactions. Influence lines for other structural quantities and other types of structures will be considered in later chapters

Influence Lines For Beam Reactions

Basic Definition of an Influence Line

An influence line for a beam reaction is a curve, which extends over the length of the beam, where the ordinate at any point is equal to the magnitude of the reaction due to a unit downward force at that point on the beam. It might seem strange that an influence line is developed for a downward force since the sign convention which has been used up to now considers an upward force to be positive. A downward force is being used since influence lines are primarily used to determine the effects of gravity loads on structures, such as a truck loading on a bridge. These loads act downward. Essentially all structural analysis reference books and handbooks use this convention. Rather than attempting to fight tradition, we will also use a downward force here.

An influence line can be developed for any particular reactive quantity for a beam by performing a series of analyses to compute the magnitude of the quantity for a unit downward force at various positions on the beam. The shape of the influence line then can be determined by plotting these computed values versus the position of the load and drawing a curve through the points over the length of the beam. This curve will be the influence line for that particular reactive quantity, and its ordinate at any point will be equal to the value of the reactive quantity due to a unit downward force on the beam at that point.

Example Problem 2.6

Compute the influence lines for the vertical reactive force components at each support for the beam whose mathematical model is shown in Figure 2.26a. A joint is not shown at the location of the downward unit concentrated force since it will continuously change positions as the force moves along the beam.

The *influence table* in Figure 2.26b contains the computed values for the left reactive force R_{2Y} and the right reactive force R_{3Y} for a unit concentrated downward force at various values of the position variable X corresponding to 2-meter intervals along the beam. These reactive forces can be easily computed by summing moments about Z at each support for each position of the unit force. Figure 2.26c shows the curves which were obtained by plotting these computed values versus the position of the unit force. These curves are the influence lines for the vertical reactive forces. An influence line was not developed for the horizontal reactive force component at the left support, since it is obvious that it will be zero for any position of a downward force on the beam. Since the units of the reactive force will be the same as the units of the unit force, the units of the influence line ordinates will be kN/kN (i.e.; kN of reactive force per kN of active force). The ordinates are actually dimensionless. This is very important, as we will see when we discuss the computation of beam reactive forces using influence lines, since the same influence line can be used for forces with any units. The negative ordinates of the influence lines for specific zones along the beam indicate that the beam would try to lift off the support when a downward force is located in these zones. The reactive force must be down to restrain the beam against upward translation. It has been assumed during the analysis that the roller support at joint 3 is constructed such that it can supply a downward reactive force.

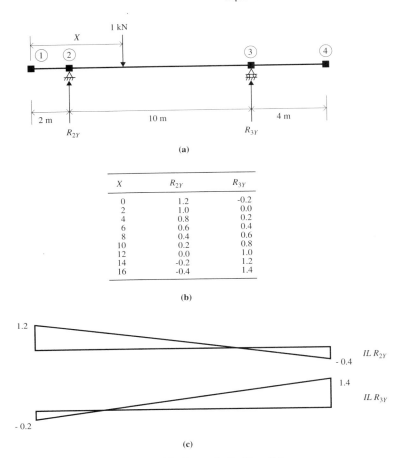

Figure 2.26 Example Problem 2.6.

Example Problem 2.7

Compute the influence lines for the vertical reactive forces for the statically determinate beam whose mathematical model is shown in Figure 2.27a. The computed values for the left reactive force R_{1Y}, the interior reactive force R_{2Y} and the right reactive force R_{4Y} for a unit force at selected points along the beam are shown in the influence table in Figure 2.27b. These values were computed by performing the following set of operations for each position of the unit force:

1. Summing moments about Z for the segment of the beam to the right of the pin for any position of the unit force will result in an equation with R_{4Y} as the only unknown. This equation can be solved for R_{4Y}.
2. Next, summing moments about Z at joint 1 will result in an equation containing R_{2Y} and R_{4Y}. Since R_{4Y} is now known, this equation can be solved for R_{2Y}.
3. Finally, summing forces in the Y direction will result in an equation which can be solved for R_{1Y}, since R_{2Y} and R_{4Y} are known.

Influence Lines For Beam Reactions

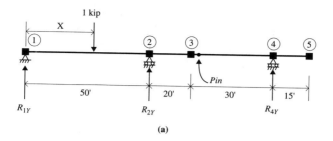

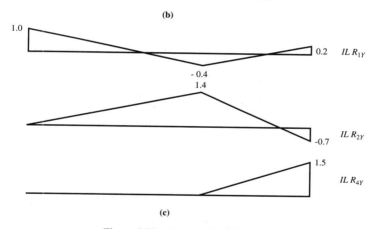

Figure 2.27 Example Problem 2.7.

It is also possible to perform these calculations in other ways, but this is probably the most efficient. The influence lines which are obtained by plotting these values are shown in Figure 2.27c. These curves show that the variation of the reactive forces with the load position is much more complicated for this beam than for the previous beam. Not only do these influence lines have both positive and negative ordinates, but they also have kinks at the location of the interior pin.

A careful study of these influence lines will give us some insight into what happens to the reactive forces as the loads change position on the beam. For example, the influence line for the reactive force R_{2Y} shows that this reactive force is positive for a downward force at any location to the left of the pin and increases in magnitude as the force moves from the left end of the beam toward the right. However, after the force moves to the right of the pin the reactive force starts to decrease and becomes negative when the force is to the right of the right support. Similar information can be obtained concerning the variation of the left and right reactive forces from their influence lines. Information of this type can be very beneficial to a structural engineer during the design of a beam and its supports.

If the solution of the equilibrium equations for either of the previous examples are performed algebraically to determine expressions for the reactive forces in terms of the position variable X, it will be found that X only occurs to the first power in each equation. Therefore, the reactions vary linearly with X and the influence lines must be composed of straight line segments. In fact, it can be stated as a general rule that the influence lines for the reactions for any statically determinate straight beam will always be composed of straight line segments, although they may be kinked or have vertical discontinuities in the ordinates at specific points. By recognizing this beforehand, it is usually possible to draw the complete influence line for any reactive quantity with a small number of computed points.

Note that in both of these example problems, the sum of the ordinates of the influence lines for the vertical reactive force components is exactly equal to 1.0 at any point on the beam since the influence lines correspond to the effect of a unit force on the beam. This simple relationship can be very useful for checking that the influence lines were drawn properly for the computed reactions.

Properties of Influence Lines

In order to use influence lines effectively, it is necessary to understand their properties. Although the ordinates of an influence line represent the effect of a unit downward concentrated force acting on the beam, an influence line also can be used to consider the effects of other types of loads. The following sections will show that in addition to the ordinates of the influence line, the area under the influence line and its slope also have physical significance.

Concentrated Forces. Since the ordinate ϕ of the influence line, for the reactive quantity R, at any specific point on a beam, is equal to the value of R due to a unit concentrated downward force at the point, the value of R due to a force of magnitude W at the point can be obtained by the product of W and ϕ. If the beam is loaded by a series of n concentrated forces, as shown in Figure 2.28, the value of R can be determined by adding the contribution of each individual force

$$R = W_1\phi_1 + W_2\phi_2 + \cdots + W_i\phi_i + \cdots + W_n\phi_n \quad (2.21)$$

where ϕ_i is the influence line ordinate at the location of force W_i.

Distributed Forces. An influence line can also be used to determine the effect of a distributed force on a beam, such as the load q which acts over the portion of the beam defined by the position variables X_1 and X_2 shown in Figure 2.29. The portion of the distributed force over an infinitesimal length dX, at a point de-

Influence Lines For Beam Reactions

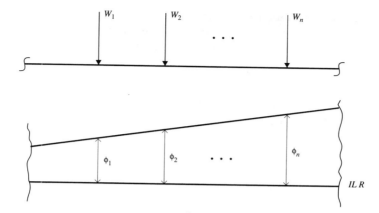

Figure 2.28 Concentrated forces.

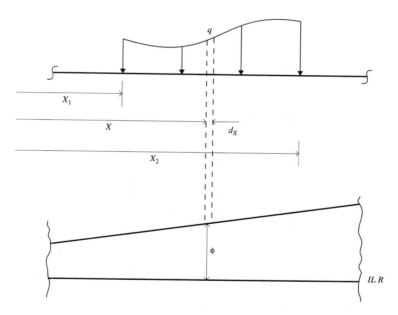

Figure 2.29 Distributed force.

fined by the position variable X, will be equivalent to an infinitesimal concentrated force dW, where

$$dW = q\, dX \tag{2.22}$$

and the value of the reactive quantity dR due to this infinitesimal force will be

$$dR = dW\, \phi \tag{2.23}$$

where ϕ is the influence line ordinate at X. The value of R due to the total distributed force now can be computed by summing the contributions of all of the infinitesimal forces over the loaded length of the beam.

$$R = \int_{X_1}^{X_2} dR \qquad (2.24)$$

from which, on substituting the expressions in Eqs. (2.23) and (2.22) in turn, we obtain

$$R = \int_{X_1}^{X_2} q\phi \, dX \qquad (2.25)$$

If the distributed force is uniform over the loaded length, with a magnitude ω, it can be moved outside the integral sign to give

$$R = \omega \int_{X_1}^{X_2} \phi \, dX \qquad (2.26)$$

which is equivalent to

$$R = \omega A \qquad (2.27)$$

where A is the area under the influence line over the length of the uniform distributed force. If the load is not uniform the integral of $q\phi$ must be evaluated.

Concentrated Moments. The effect of one or more concentrated moments acting on a beam also can be determined by using an influence line. The procedure which can be used is to first replace the moment by an equivalent couple which can be represented by a pair of forces W, with a moment arm $2a$, as shown in Figure 2.30, where

$$M = 2aW \qquad (2.28)$$

The moment is considered to be positive if it is clockwise. This is consistent with considering the positive force for an influence line to be down, since it is also in the opposite direction of a right hand system. The reactive quantity R can now be computed by multiplying each of these forces by the corresponding ordinates of the influence line. The ordinates of the influence line where the two forces are applied can be computed in terms of the ordinate and the slope of the influence line at the point of application of the concentrated moment. The resulting value for R is

$$R = W\left(\phi + a\frac{d\phi}{dX}\right) - W\left(\phi - a\frac{d\phi}{dX}\right) \qquad (2.29)$$

which reduces to

Influence Lines For Beam Reactions

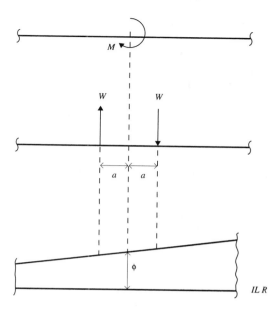

Figure 2.30 Concentrated moment.

$$R = 2a\, W\, \frac{d\phi}{dX} \tag{2.30}$$

or

$$R = M\, \frac{d\phi}{dX} \tag{2.31}$$

Therefore, the effect of a concentrated moment can be determined by multiplying the magnitude of the moment by the slope of the influence line at its point of application. The slope of the influence line is positive if the ordinates increase in the positive X direction. If more than one concentrated moment exists on the beam, then the value of R can be determined by summing the effect of each individual moment by multiplying its magnitude by the slope of the influence line at its location

$$R = M_1\, \frac{d\phi_1}{dX} + M_2\, \frac{d\phi_2}{dX} + \cdots + M_i\, \frac{d\phi_i}{dX} + \cdots + M_n\, \frac{d\phi_n}{dX} \tag{2.32}$$

where $d\phi_i/dX$ is the slope of the influence line at the point of application of the moment M_i.

COMPUTATION OF REACTIONS USING INFLUENCE LINES

Now that we are familiar with the definition of an influence line and its properties, we can consider how an influence line can be used to compute the reactions for a beam under various conditions. The influence lines might be considered to be a general solution for the reactions for a beam. The following example problems show how we can obtain a particular solution for a specific set of loads on the beam.

Example Problem 2.8

Compute the vertical reactive forces at each end of the simply supported beam subjected to the loads shown in Figure 2.31a. The first step in the analysis is to transform the inclined force acting on the top of the vertical strut into a moment and a horizontal and vertical force component acting directly on the beam, as shown in the mathematical model of the beam in Figure 2.31b. We now can use this loading to compute the reactive forces using the influence lines for R_{1Y} and R_{4Y} shown in Figure 2.31c. The verification of the shape of these influence lines is left as an exercise for the reader.

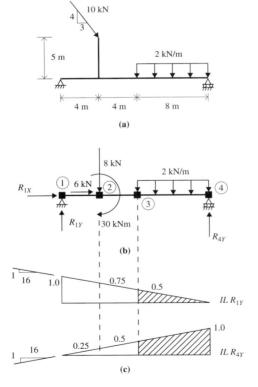

Figure 2.31 Example Problem 2.8.

Computation of Reactions Using Influence Lines

To compute either reactive force, the vertical concentrated force component will be multiplied by the ordinate of the influence line at its location, the moment will be multiplied by the slope of the influence line, and the uniform distributed force will be multiplied by the area under the influence line over its length. The area under the influence line for R_{1Y} over the length of the distributed force will be triangular, while the area for R_{4Y} will be trapezoidal, as indicated by the shaded zones in Figure 2.31c.

$$R_{1Y} = (8)(0.75) + (30)\left(-\frac{1}{16}\right) + (2)\left[\frac{(8)(0.5)}{2}\right] = 8.125 \text{ kN}$$

$$R_{4Y} = (8)(0.25) + (30)\left(\frac{1}{16}\right) + (2)\left[\frac{(8)(0.5 + 1.0)}{2}\right] = 15.875 \text{ kN}$$

Note that in these calculations the downward concentrated force and the downward distributed force are treated as positive quantities since a positive influence line ordinate corresponds to a downward force. The concentrated moment is positive since it is clockwise, while the slope of the influence line for R_{1Y} at the point of application of the concentrated moment is negative and the slope of the influence line for R_{4Y} is positive. The reader can easily verify that these are the correct vertical reactive forces for this beam by a simple equilibrium check. The horizontal reactive force at the left end of the beam can be found by summing forces in the X direction.

Example Problem 2.9

Compute the maximum values for the vertical reactive forces for a bridge with the geometry shown previously in Figure 2.27a, due to a HS20-44 truck loading. The truck can move to any position on the bridge and it can face in either direction.

The computation of the maximum reactions is a two step process. The first step is to determine the positions of the truck which will result in the maximum magnitudes for each reaction. In many situations, it will be immediately obvious how the loading must be situated to produce the maximum effect, but in some cases, it might be necessary to try several trial positions to arrive at the maximum position. In addition, if the influence line has both positive and negative ordinates, it will be necessary to determine two load positions for each reaction since both the maximum positive and maximum negative values will be needed for the design of the supports. The second step is to compute the value of the reactions using the maximum load positions. These computations can be performed directly using the ordinates of the influence lines at the truck axle positions.

Figure 2.32 shows the truck positions for the maximum positive and negative values for each vertical reactive force component for this bridge.

Note that the maximum effects are not always obtained for the truck facing in the same direction. Both directions must be considered to ensure that the maximum effect has been obtained. For each case, the maximum effect is obtained with the variable rear axle spacing of the truck set at its minimum value of 14 feet. Of course,

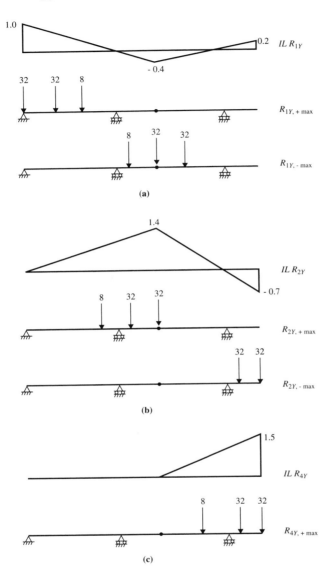

Figure 2.32 Example Problem 2.9.

for another bridge with a different geometry, a different rear axle spacing might be used. Any spacing within the limits of 14 feet and 30 feet for an HS20-44 loading is permissible.

The maximum positive and negative values for the left reactive force R_{1Y} and interior reactive force R_{2Y} can be determined by using the truck locations shown in Figures 2.32a and 2.32b

$$R_{1Y,+\max} = 32(1.0) + 32(0.72) + 8(0.44) = 58.56 \text{ kips} = 58.56 \text{ kips} \uparrow$$

$$R_{1Y,-\max} = 8(-0.12) + 32(-0.4) + 32(-0.213) = -20.58 \text{ kips} = 20.58 \text{ kips} \downarrow$$

$$R_{2Y,+\max} = 8(0.84) + 32(1.12) + 32(1.4) = 87.36 \text{ kips} = 87.36 \text{ kips} \uparrow$$

$$R_{2Y,-\max} = 32(-0.047) + 32(-0.7) = -23.90 \text{ kips} = 23.90 \text{ kips} \downarrow$$

Since the influence line for R_{4Y} only has positive ordinates, only one value for R_{4Y} will be computed using the truck location shown in Figure 2.32c

$$R_{4Y,+\max} = 8(0.567) + 32(1.033) + 32(1.5) = 85.58 \text{ kips} = 85.58 \text{ kips} \uparrow$$

These reactive forces are the maximum forces which would be applied to the beam by the supports. The maximum forces which would occur on the supports due to the truck loading would have the same magnitudes but in the opposite directions. These support forces would be used in the design of the foundations for the bridge. Of course, the dead weight of the bridge also should be considered in the design process. This is a simple task since the dead weight usually can be assumed to be uniform over the length of the bridge. The contribution of this downward uniform distributed force can be determined by multiplying its magnitude by the total area under the influence line for each reactive force. The dead load reactive forces should be added to the maximum reactive forces for the moving loads to obtain the final design values.

A similar process can be used for any type of moving loads. It is usually not very difficult to determine the loading position to produce the maximum effect for any reactive quantity. If the loading were a distributed force which could act over any portions of the beam, such as the loading on the floor system in a warehouse, the maximum positive reaction would be obtained by placing the distributed force over those portions of the beam where the influence line ordinates are positive. The distributed force would be situated where the ordinates are negative to obtain the maximum negative reaction. These values would then be used for design, since these loading patterns might occur at some time during the life of the structure as the loads are moved around on the floor system. It is not important when the maximum reactions occur. The only important consideration is the maximum values that could occur during the life of the structure.

COMPUTER PROGRAM BEAMIL

The disk file BEAMIL.DOC contains the instructions for the computer program BEAMIL. This program can be used to generate influence lines for the reactive forces, and also for the shear force and bending moment at interior points, in a statically determinate or statically indeterminate beam. For the present, we will only be concerned with using this program to generate influence lines for the reactive forces for statically determinate beams. It will be used in later chapters for other purposes. The analysis procedure which is used in this program will not be discussed here since it is not important for the use of the program. The primary purpose for supplying this program is to give the readers a means to verify their manual solutions for the suggested problems at the ends of the chapters which deal with beam influence lines. The program also can be used as a tool to investigate the behavior of structures with variations in the geometry or support conditions. By using this program to investigate a variety of structures, the reader can learn a great deal about structural behavior without the need to perform a large number of tedious and time consuming manual calculations.

The mathematical model of any beam which is to be analyzed by BEAMIL will consist of a set of joints connected by straight horizontal members. Joints are required at each support point, at each free end, at each interior pin and also at each point where there is a change in the cross section or material in a statically indeterminate beam. If the beam is statically determinate, the cross section properties and material properties are not important since they will not affect the influence line ordinates. The joints must be numbered sequentially from left to right from 1 to NJ, where NJ is the total number of joints. The locations of the joints are defined by their X global coordinate, with the X axis extending toward the right and the Y axis up. The origin of the global coordinate system must be at joint 1. The members will be automatically numbered from 1 to NJ $-$ 1 from left to right by the program. The required format for the input data file for the program is described in BEAMIL.DOC. The input data file may be created interactively using the program BMILDATA if desired.

Figure 2.33 shows a listing of the input data file for the analysis of the beam shown previously in Figure 2.27. The mathematical model has five joints and four members, with one joint at each of the three support points, one joint at the location of the interior pin and one joint at the free end of member 4. Figure 2.34 shows the plot of the influence line and a listing of the ordinates at selected points along the beam, for the vertical reactive force at the left support. The listed ordinates are at the same locations as the ordinates in the influence table in Figure 2.27b. This plot and the listed values agree with the manual solution obtained previously. Note that the plot indicates the location of the interior pin by inserting a lowercase p over joint 3.

When performing this analysis, the program was instructed to use one interior computation point for each member. The influence line ordinates are comput-

```
Sample Beam for Program BEAMIL
5,1,1,1,3
Joint Coordinates
1,0.0
2,50.0
3,70.0
4,100.0
5,115.0
Interior Pins
3
Support Restraints
1,1,1,0
2,0,1,0
4,0,1,0
```

Figure 2.33 Sample data file for program BEAMIL.

Suggested Problems

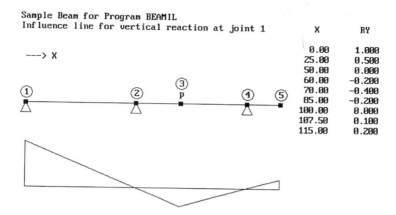

Figure 2.34 Sample reaction influence line from BEAMIL.

ed and listed for each joint and for each interior computation point, as explained in the program instructions.

SUGGESTED PROBLEMS

SP2.1 Classify the structures in Figure SP2.1 as unstable, statically determinate or statically indeterminate on the basis of the number of unknown reactive quantities and the number of independent equations of equilibrium.

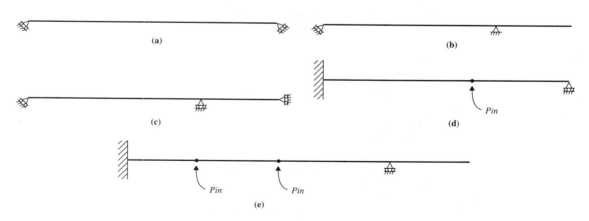

Figure SP2.1

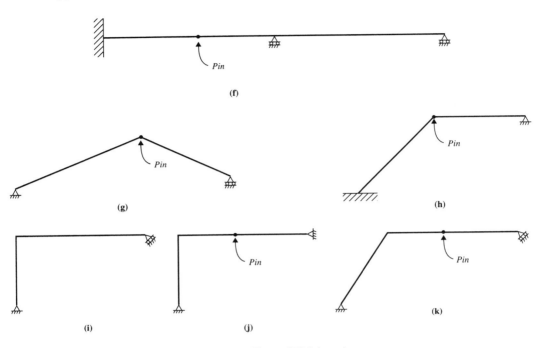

Figure SP2.1 *(cont.)*

SP2.2 Generate a set of equilibrium equations for each of the two structures shown in Figure SP2.2 by summing forces in the global X and Y directions and summing moments about the center support. Compute the determinant of the coefficient matrix for each set of equations. What conclusions can you draw concerning the stability of the two structures? Compute the reactive forces for the structure which is stable.

SP2.3–SP2.10 Compute the reactions for each of the structures shown in Figures SP2.3 to SP2.10. Verify your solutions using the programs SDPTRUSS and SDPFRAME.

SP2.11–SP2.14 Draw the influence lines for all of the vertical reactive forces for each of the beams shown in Figures SP2.11 to SP2.14. Verify your solutions using the program BEAMIL.

SP2.15 Draw the influence lines for the vertical reactive forces for the beam shown in Figure SP2.15. Use the influence lines to compute the magnitude of the reactive forces due to the active loads on the beam.

SP2.16 Draw the influence lines for the vertical reactive forces for the beam shown in Figure SP2.16. Compute the maximum magnitude for each of these reactive forces due to a HS20-44 truck which can move in either direction along the beam.

Suggested Problems

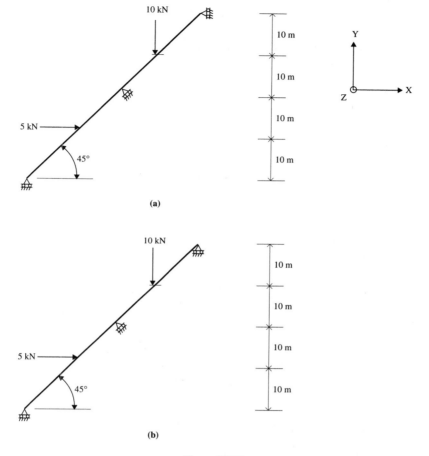

Figure SP2.2

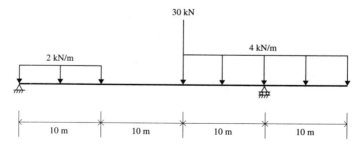

Figure SP2.3

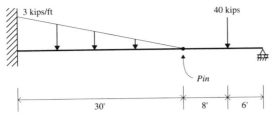

Figure SP2.4

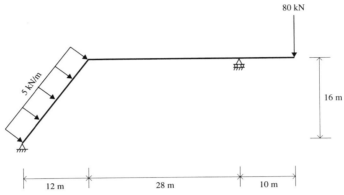

Figure SP2.5

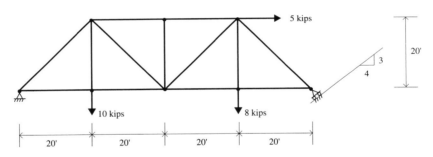

Figure SP2.6

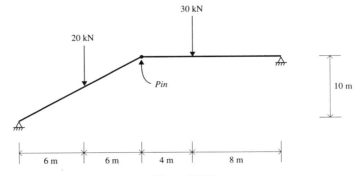

Figure SP2.7

Suggested Problems

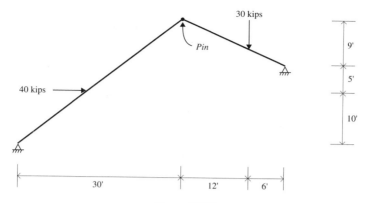

Figure SP2.8

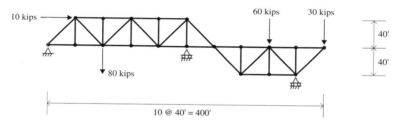

Figure SP2.9

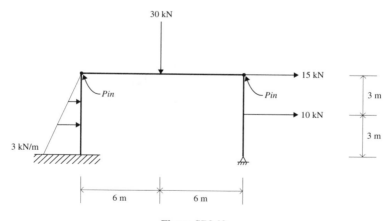

Figure SP2.10

Figure SP2.11

68   Equilibrium and Reactions   Chap. 2

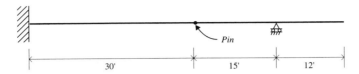

Figure SP2.12

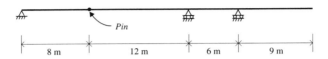

Figure SP2.13

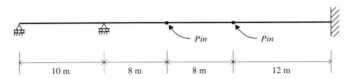

Figure SP2.14

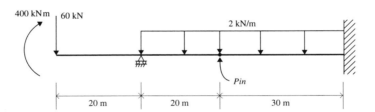

Figure SP2.15

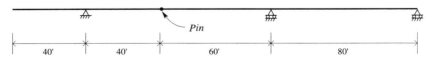

Figure SP2.16

3

Analysis of Statically Determinate Plane Trusses

MATHEMATICAL MODEL FOR A PLANE TRUSS

In this and the following chapters, we will go beyond the computation of the reactions for a structure and develop procedures for performing the complete analysis for the member end loads, the member deformations and the joint displacements. The type of analysis which must be performed for any structure will depend upon the properties of its mathematical model. Decisions must be made by the engineer whether the structure must be treated as a two dimensional or a three dimensional system and whether it should be analyzed as a truss or a frame. If the structure can be assumed to act as truss, the analysis usually will be significantly simplified compared to that which would be required if it were to be analyzed as a frame. In this chapter, we will be concerned with the computation of the reactions and the member forces for statically determinate plane trusses. The analysis of statically determinate space trusses, plane frames and space frames will be considered in later chapters.

The mathematical model for a plane truss consists of a set of joints which are connected by straight members with all of the joints and all of the members located in a common plane. To qualify as a plane truss, the mathematical model for the structure must have the following specific properties:

   1. All members must be connected to their end joints such that no moment is transmitted between the joints and the members. Therefore, the connection between the member and the joint at each end must be equivalent to a fric-

tionless pin which permits unrestricted rotation of the end of the member about an axis perpendicular to the plane of the structure.
2. All loads on the structure must consist only of concentrated forces acting directly on the joints in the plane of the structure. Moments acting on the joints or intermediate loads acting directly on the members are not permitted.
3. Only translation restraints may exist at the support joints. Therefore, only pinned supports and roller supports, which translate in the plane of the structure, are permitted.

If all of the previous conditions are satisfied, only axial forces will be transmitted between the joints and the members, and only reactive forces in the plane of the structure will exist at the support joints. There will be no shear forces or bending moments in the members.

STABILITY AND STATIC DETERMINACY OF PLANE TRUSSES

The unknown quantities which must be determined during the analysis of a plane truss will consist of the axial forces in the members and the reactive forces at the support joints. The unknowns for each member will consist of the axial forces transmitted between the member and the joints at its two ends. Figure 3.1 shows these two end forces for any typical member n. The force acting on the end connected to the first joint is designated as S_{1n} and the force acting on the end connected to the second joint is designated as S_{2n}. Either end joint may be designat-

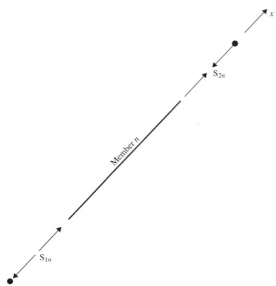

Figure 3.1 Plane truss member end loads.

Stability and Static Determinacy of Plane Trusses

ed as the first joint during the analysis. This designation must be made by the analyst before the analysis is started. The positive directions for both of these end forces are in the positive direction of the local member x axis, which extends along the member from the first joint toward the second joint. The orientation of the local member y and z axes are not important for a plane truss member. There will also be one unknown quantity for each translation restraint at the support joints. These unknowns will consist of a reactive force component in the direction of each restraint. This will result in two unknown reactive force components for each pinned support and one unknown reactive force component for each roller support. Therefore, the total number of unknowns NUN for the truss will be

$$NUN = 2NM + NR \quad (3.1)$$

where NM is the number of members and NR is the number of support restraints.

If a structure is in equilibrium, then each individual part of that structure also must be in equilibrium. Therefore, if we can ensure that each joint and each member in a plane truss is in equilibrium, under a given set of joint loads, the total truss must be in equilibrium. The independent equilibrium equations will consist of two equations for each joint, corresponding to the sum of forces in any two directions in the plane of the truss at the joint, and one equation for each member, corresponding to the sum of forces along the local x axis for the member. The total number of independent equilibrium equations NEQ will be

$$NEQ = 2NJ + NM \quad (3.2)$$

where NJ is the number of joints in the truss.

Finally, for the truss to be statically determinate and stable, the number of independent equilibrium equations must be equal to the number of unknowns:

$$2NJ + NM = 2NM + NR \quad (3.3)$$

or

$$2NJ = NM + NR \quad (3.4)$$

Of course, as shown in the previous chapter, when generating the global equilibrium equations for computing the reactions for a structure, the requirement that the number of equations is equal to the number of unknowns is only a necessary condition for stability. The truss could still be geometrically unstable due to the arrangement of the members or the orientation of the support restraints. This condition will be discussed in more detail later.

In addition to the previously required relationship for static determinacy and stability of a plane truss, we can also define two other possible situations:

if $2NJ > NM + NR$, the plane truss is unstable;

and

if $2NJ < NM + NR$, the plane truss is statically indeterminate.

In this chapter, we will only be concerned with the analysis of statically determinate plane trusses. The analysis of statically indeterminate trusses will be considered in later chapters.

ANALYSIS OF A PLANE TRUSS BY JOINT EQUILIBRIUM

Although there are two end forces for each member in a truss, these end forces are not independent. Summing forces along the local member x axis for any member n gives

$$S_{1n} + S_{2n} = 0 \tag{3.5}$$

from which

$$S_{1n} = -S_{2n} \tag{3.6}$$

The two end forces have equal magnitudes and act in opposite directions. Therefore, to simplify the analysis, the axial force in any member n in a truss will merely be designated by the single variable S_n, with a positive value indicating tension in the member and a negative value indicating compression. The value of S_n corresponds to the end force S_{2n}, since a positive value for this force represents tension in the member. The total number of unknowns for the truss has now been reduced to NM + NR and the number of remaining independent equilibrium equations has been reduced to 2NJ since one equation has been used for each member to establish the relationship shown in Eq. (3.6)

The NM member forces and the NR reactive force components can be computed for a statically determinate and stable plane truss by solving a set of 2NJ simultaneous joint equilibrium equations. These equations can be generated by summing forces in the global X and Y directions at each joint. The global force components acting on the first joint, for any member n, can be expressed in terms of the member axial force S_n and the member orientation angles θ_{nX} and θ_{nY}, as defined in Figure 3.2, as

$$S_{nX} = S_n \cos \theta_{nX} = S_n C_{nX} \tag{3.7}$$

$$S_{nY} = S_n \cos \theta_{nY} = S_n C_{nY} \tag{3.8}$$

where the quantities C_{nX} and C_{nY} are known as the *direction cosines* for the member. The force components acting on the second joint will have these same magnitudes but opposite signs. The member orientation angle θ_{nX} is measured counterclockwise from the global X axis to the local member x axis, with the local x axis directed away from the global origin, while the angle θ_{nY} is measured clockwise from the global Y axis.

The direction cosines for member n can be easily computed from the geometry of the truss by using the global X and Y coordinates of the end joints, as defined in Figure 3.3. The X and Y length projections of the member can be expressed as

Analysis of a Plane Truss by Joint Equilibrium

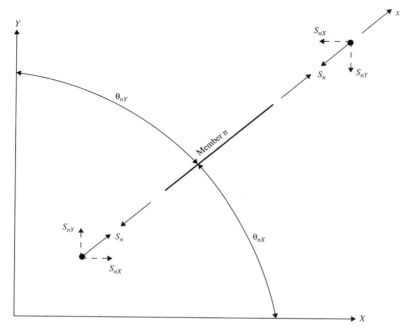

Figure 3.2 Member force components.

$$L_{nX} = X_{n2} - X_{n1} \tag{3.9}$$

$$L_{nY} = Y_{n2} - Y_{n1} \tag{3.10}$$

from which the length of the member can be computed as

$$L_n = \sqrt{L_{nX}^2 + L_{nY}^2} \tag{3.11}$$

The direction cosines can now be expressed as

$$C_{nX} = \frac{L_{nX}}{L_n} \tag{3.12}$$

and

$$C_{nY} = \frac{L_{nY}}{L_n} \tag{3.13}$$

The signs of the projections of the members on the global axes will automatically result in the correct signs for the direction cosines, while the signs of the direction cosines will generate the correct signs for the member force components in Eqs. (3.7) and (3.8) for any member orientation. If the preceding process is used to compute the direction cosines, there is no need to determine numerical values in degrees or radians for the angles θ_{nX} and θ_{nY} for any member.

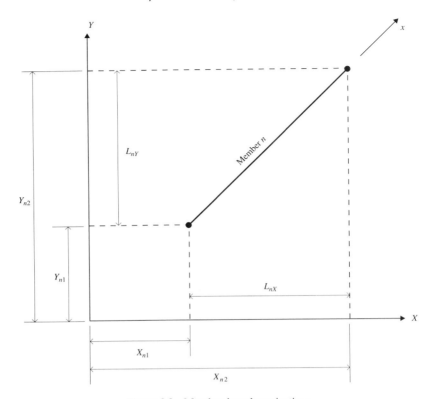

Figure 3.3 Member length projections.

Example Problem 3.1

Compute the member forces and the reactive force components for the plane truss whose mathematical model is shown in Figure 3.4a. The truss has five joints, seven members and three support restraints. The reactive forces correspond to a horizontal and vertical force component at joint 1 and a vertical force component at joint 4. If we now check the relationship between the number of joints, the number of members and the number of support restraints, we have

$$NJ = 5$$
$$NM = 7 \quad \rightarrow \quad 2NJ = NM + NR$$
$$NR = 3$$

Therefore, the truss is statically determinate. It also appears that this truss is stable, but we cannot verify this until we determine whether we are able to determine a set of finite member forces and reactive force components which will satisfy the joint equilibrium equations.

The direction cosines for the members can be computed by using Eqs. (3.9) through (3.13). We must only decide which end joint for each member is to be designated as the first joint. One possible approach is to always consider the end joint

Analysis of a Plane Truss by Joint Equilibrium

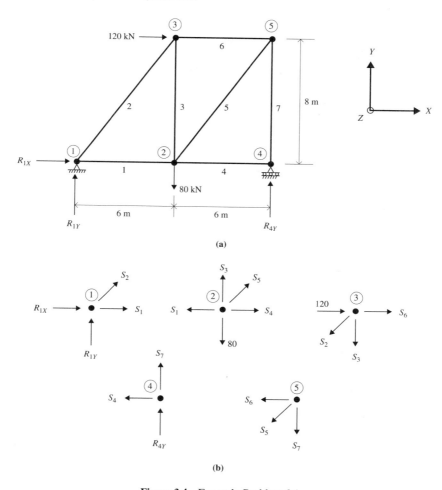

Figure 3.4 Example Problem 3.1.

with the lowest joint number to be the first joint regardless of the orientation of the individual members. Using a consistent and simple scheme of this type often can help to reduce errors in the analysis and also make later checking of the calculations easier. This scheme will be used in all of the example problems in this book for both manual and computer analyses.

Member 1 (joint 1 to joint 2):

$$L_{1X} = 6 \qquad\qquad C_{1X} = 1.0$$
$$\rightarrow L_1 = 6 \rightarrow$$
$$L_{1Y} = 0 \qquad\qquad C_{1Y} = 0$$

Member 2 (joint 1 to joint 3):

$$L_{2X} = 6 \qquad\qquad C_{2X} = 0.6$$
$$\rightarrow \quad L_2 = 10 \quad \rightarrow$$
$$L_{2Y} = 8 \qquad\qquad C_{2Y} = 0.8$$

Member 3 (joint 2 to joint 3):

$$L_{3X} = 0 \qquad\qquad C_{3X} = 0$$
$$\rightarrow \quad L_3 = 8 \quad \rightarrow$$
$$L_{3Y} = 8 \qquad\qquad C_{3Y} = 1.0$$

Member 4 (joint 2 to joint 4):

$$L_{4X} = 6 \qquad\qquad C_{4X} = 1.0$$
$$\rightarrow \quad L_4 = 6 \quad \rightarrow$$
$$L_{4Y} = 0 \qquad\qquad C_{4Y} = 0$$

Member 5 (joint 2 to joint 5):

$$L_{5X} = 6 \qquad\qquad C_{5X} = 0.6$$
$$\rightarrow \quad L_5 = 10 \quad \rightarrow$$
$$L_{5Y} = 8 \qquad\qquad C_{5Y} = 0.8$$

Member 6 (joint 3 to joint 5):

$$L_{6X} = 6 \qquad\qquad C_{6X} = 1.0$$
$$\rightarrow \quad L_6 = 6 \quad \rightarrow$$
$$L_{6Y} = 0 \qquad\qquad C_{6Y} = 0$$

Member 7 (joint 4 to joint 5):

$$L_{7X} = 0 \qquad\qquad C_{7X} = 0$$
$$\rightarrow \quad L_7 = 8 \quad \rightarrow$$
$$L_{7Y} = 8 \qquad\qquad C_{7Y} = 1.0$$

Figure 3.4b shows the free body diagrams for each of the five joints in the truss, assuming that all members are in tension and that the reactive force components act in the positive directions of the global X and Y axes. The 2NJ equilibrium equations, which can be obtained by summing forces in the X and Y directions for joints 1 though 5, can be expressed in terms of the member forces S_1 through S_7 and the reactive force components R_{1X}, R_{1Y} and R_{4Y} as

$$\Sigma_1 F_X \rightarrow S_1 C_{1X} + S_2 C_{2X} + R_{1X} = 0$$
$$\Sigma_1 F_Y \rightarrow S_1 C_{1Y} + S_2 C_{2Y} + R_{1Y} = 0$$
$$\Sigma_2 F_X \rightarrow -S_1 C_{1X} + S_3 C_{3X} + S_4 C_{4X} + S_5 C_{5X} = 0$$
$$\Sigma_2 F_Y \rightarrow -S_1 C_{1Y} + S_3 C_{3Y} + S_4 C_{4Y} + S_5 C_{5Y} - 80 = 0$$
$$\Sigma_3 F_X \rightarrow -S_2 C_{2X} - S_3 C_{3X} + S_6 C_{6X} + 120 = 0$$

Analysis of a Plane Truss by Joint Equilibrium

$$\Sigma_3 F_Y \rightarrow -S_2 C_{2Y} - S_3 C_{3Y} + S_6 C_{6Y} = 0$$

$$\Sigma_4 F_X \rightarrow -S_4 C_{4X} + S_7 C_{7X} = 0$$

$$\Sigma_4 F_Y \rightarrow -S_4 C_{4Y} + S_7 C_{7Y} + R_{4Y} = 0$$

$$\Sigma_5 F_X \rightarrow -S_5 C_{5X} - S_6 C_{6X} - S_7 C_{7X} = 0$$

$$\Sigma_5 F_Y \rightarrow -S_5 C_{5Y} - S_6 C_{6Y} - S_7 C_{7Y} = 0$$

The member force components, which are applied to the joint by the members, are considered to be positive at the first joint and negative at the second joint for each member. The signs of the direction cosines will result in the correct signs for the equation coefficients.

After substituting the numerical values for the direction cosines, the equilibrium equations for the truss can be expressed in matrix form as

$$\begin{bmatrix} 1.0 & 0.6 & 0 & 0 & 0 & 0 & 0 & 1.0 & 0 & 0 \\ 0 & 0.8 & 0 & 0 & 0 & 0 & 0 & 0 & 1.0 & 0 \\ -1.0 & 0 & 0 & 1.0 & 0.6 & 0 & 0 & 0 & 0 & 0 \\ 0 & 0 & 1.0 & 0 & 0.8 & 0 & 0 & 0 & 0 & 0 \\ 0 & -0.6 & 0 & 0 & 0 & 1.0 & 0 & 0 & 0 & 0 \\ 0 & -0.8 & -1.0 & 0 & 0 & 0 & 0 & 0 & 0 & 0 \\ 0 & 0 & 0 & -1.0 & 0 & 0 & 0 & 0 & 0 & 0 \\ 0 & 0 & 0 & 0 & 0 & 0 & 1.0 & 0 & 0 & 1.0 \\ 0 & 0 & 0 & 0 & -0.6 & -1.0 & 0 & 0 & 0 & 0 \\ 0 & 0 & 0 & 0 & -0.8 & 0 & -1.0 & 0 & 0 & 0 \end{bmatrix} \begin{Bmatrix} S_1 \\ S_2 \\ S_3 \\ S_4 \\ S_5 \\ S_6 \\ S_7 \\ R_{1X} \\ R_{1Y} \\ R_{4Y} \end{Bmatrix} = \begin{Bmatrix} 0 \\ 0 \\ 0 \\ 80 \\ -120 \\ 0 \\ 0 \\ 0 \\ 0 \\ 0 \end{Bmatrix}$$

We now have a set of 10 linear simultaneous equations with the 7 member forces and the 3 reactive force components as the unknowns. Although it is possible to solve this set of 10 equations manually, it is obviously not practical. Fortunately, this is a task for which the computer is ideally suited. Essentially every commercial spreadsheet program has the capability to solve a set of linear simultaneous equations, either by a direct program command or by an indirect process using a combination of matrix inversion and matrix multiplication. There are also many specialized numerical analysis programs which can be used to solve these equations. To meet our specific needs here, the program SEQSOLVE has been supplied with this book. This program can solve up to 127 linear simultaneous equations using the Gauss-Jordan Elimination Method. The instructions for using SEQSOLVE are given in the disk file SEQSOLVE.DOC.

Figure 3.5 shows a listing of the input data file for SEQSOLVE for the preceding equations. The output from the program is shown in Figure 3.6. Since the program was able to solve the equilibrium equations, to obtain a finite solution for all of the member forces and reactive force components, the truss is stable. Note that the computed values for several of the member forces are negative. This means that these members are in compression. All other members are in tension as originally assumed. A quick check will show that the computed values for the reactive force components are the same as would be obtained using the analysis techniques described in Chapter 2.

```
Example Problem 3.1 - Plane Truss Analysis by Simultaneous Equations
10
Non Zero Equation Coefficients
1,1,1.0,0
1,2,0.6,0
1,8,1.0,0
2,2,0.8,0
2,9,1.0,0
3,1,-1.0,0
3,4,1.0,0
3,5,0.6,0
4,3,1.0,0
4,5,0.8,0
5,2,-0.6,0
5,6,1.0,0
6,2,-0.8,0
6,3,-1.0,0
7,4,-1.0,0
8,7,1.0,0
8,10,1.0,0
9,5,-0.6,0
9,6,-1.0,0
10,5,-0.8,0
10,7,-1.0,1
Non Zero Equation Constants
4,80.0,0
5,-120.0,1
```

Figure 3.5 Example Problem 3.1, SEQSOLVE input.

```
Example Problem 3.1 - Plane Truss Analysis by Simultaneous Equations

Solution of Equations
     1         90.00000
     2         50.00000
     3        -40.00000
     4          0.00000
     5        150.00000
     6        -90.00000
     7       -120.00000
     8       -120.00000
     9        -40.00000
    10        120.00000
```

Figure 3.6 Example Problem 3.1, SEQSOLVE output.

This same approach can be used to analyze any statically determinate truss. If the program SEQSOLVE is not able to obtain a solution to the equations, the message "Coefficient matrix appears to be singular" will be displayed and the solution process will be terminated. This will mean either that the truss is unstable or that there was an error in either generating the joint equilibrium equations or in creating the input data file for the program. Since it is very easy for errors to occur while generating the equilibrium equations, particularly in the signs of the coefficients, the equations and the computer input should be carefully checked before stating that the truss is actually unstable.

Analysis of a Plane Truss by Joint Equilibrium

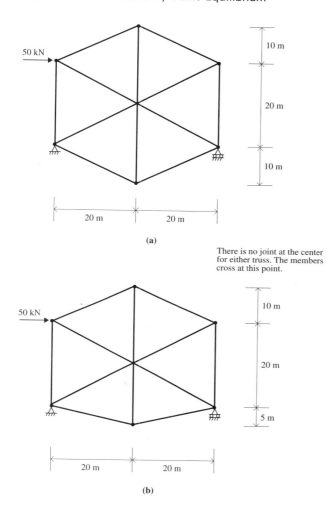

Figure 3.7 Unstable and stable trusses.

It is not always obvious when a truss is unstable. In some cases, a small change in the geometry can mean the difference between a stable and an unstable condition. For example, Figure 3.7 shows two trusses with very similar geometries. Each truss has six joints, nine members and three support restraints. It appears that they each meet the requirements for a statically determinate and stable plane truss. However, if the 2NJ equilibrium equations are generated for each truss, it will be found that it is not possible to find a solution for the equations for the truss in Figure 3.7a, while a solution can be found for the equations for the truss in Figure 3.7b. The arrangement of the members in the first truss creates a mechanism which cannot support the applied loads, while the arrangement of the members in the second truss forms a stable structure. The form of this mechanism in the first truss is not obvious.

It might be interesting to perform a series of analyses in which the bottom vertical dimension is changed incrementally between the two limits to see what happens to the magnitudes of the member forces and the reactive forces as the truss geometry approaches the unstable condition. This is left as an exercise for the reader.

METHOD OF JOINTS

It appears at first glance from the previous example problem that a manual analysis for the member forces and reactive forces in a plane truss is not practical. The manual solution of a large set of linear simultaneous equations is a tedious, time consuming and error prone operation. However, for many trusses, it is possible to perform the analysis manually if the calculations are performed in a specific order. In fact, it is possible to analyze the truss in Example Problem 3.1 without solving more than two equilibrium equations simultaneously at any step in the analysis.

First, solve for the reactive force components at the support joints using the global equilibrium equations presented in Chapter 2. The specific operations which can be used are

$$\Sigma_1 M_Z \rightarrow R_{4Y} = 120 \text{ kN} = 120 \text{ kN} \uparrow$$

$$\Sigma F_Y \rightarrow R_{1Y} = -40 \text{ kN} = 40 \text{ kN} \downarrow$$

$$\Sigma F_X \rightarrow R_{1X} = -120 \text{ kN} = 120 \text{ kN} \leftarrow$$

Next, substitute these values into the joint equilibrium equations, which were developed in Example Problem 3.1, to obtain a set of 10 linear equations with only the 7 member forces S_1 through S_7 as the unknowns. The resulting equations, in matrix form, are

$$\begin{bmatrix} 1.0 & 0.6 & 0 & 0 & 0 & 0 & 0 \\ 0 & 0.8 & 0 & 0 & 0 & 0 & 0 \\ -1.0 & 0 & 0 & 1.0 & 0.6 & 0 & 0 \\ 0 & 0 & 1.0 & 0 & 0.8 & 0 & 0 \\ 0 & -0.6 & 0 & 0 & 0 & 1.0 & 0 \\ 0 & -0.8 & -1.0 & 0 & 0 & 0 & 0 \\ 0 & 0 & 0 & -1.0 & 0 & 0 & 0 \\ 0 & 0 & 0 & 0 & 0 & 0 & 1.0 \\ 0 & 0 & 0 & 0 & -0.6 & -1.0 & 0 \\ 0 & 0 & 0 & 0 & -0.8 & 0 & -1.0 \end{bmatrix} \begin{Bmatrix} S_1 \\ S_2 \\ S_3 \\ S_4 \\ S_5 \\ S_6 \\ S_7 \end{Bmatrix} = \begin{Bmatrix} 120 \\ 40 \\ 0 \\ 80 \\ -120 \\ 0 \\ 0 \\ -120 \\ 0 \\ 0 \end{Bmatrix}$$

These equations now can be solved in pairs to determine the member forces.

Solving equations 1 and 2 for S_1 and S_2 gives

Method of Joints

$$S_1 = 90 \text{ kN}$$
$$S_2 = 50 \text{ kN}$$

Solving equations 5 and 6 for S_3 and S_6 gives

$$S_3 = -40 \text{ kN}$$
$$S_6 = -90 \text{ kN}$$

Solving equations 3 and 4 for S_4 and S_5 gives

$$S_4 = 0$$
$$S_5 = 150 \text{ kN}$$

Finally, solving equation 10 for S_7 gives

$$S_7 = -120 \text{ kN}$$

The preceding values agree with those obtained from the computer solution of the 10 equations.

If we now check off the numbers of the equations which were used in the preceding solution, we will see that equations 7, 8 and 9 were not needed to determine the member forces. These three equations are left over since three global equilibrium equations were used to solve for the reactive forces. Since there only can be 10 independent equilibrium equations for this truss, these three equations must be some combination of the seven joint equations which were used to obtain the solution. This does not mean that these equations are worthless, since they can be used to check our solution. If they are not satisfied, then something is wrong with our computed values for the member forces or the reactions.

The method of analysis which was just used is usually called the *Method of Joints*. This name is a very good description of the analysis procedure since the member forces were determined by solving the equations corresponding to ΣF_X and ΣF_Y at joints 1, 3 and 2 in turn and then the single equation corresponding to ΣF_Y at joint 5. The equations which were left over corresponded to ΣF_X and ΣF_Y at joint 4 and ΣF_X at joint 5. Although this procedure cannot be used for all statically determinate trusses, it does give a very convenient method for computing the member forces for those trusses for which it is applicable.

For many trusses, the complete analysis for the member forces can be carried out, after the reactive forces have been computed, without the need to solve any simultaneous equations. The trick is to work with the X and Y components of the member forces during the analysis rather than with the actual member forces. If one of the force components for a member can be determined from the joint equilibrium requirements, the other force component can be determined by using the ratio of the member projected lengths on the global X and Y axes, as defined previously in Eqs. (3.9) and (3.10).

$$S_{nX} = S_{nY} \frac{L_{nX}}{L_{nY}} \tag{3.14}$$

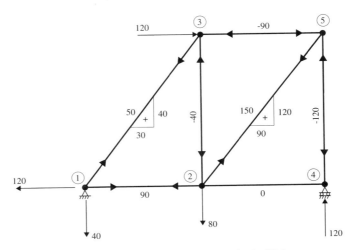

Figure 3.8 Example Problem 3.1, Method of Joints.

and

$$S_{nY} = S_{nX} \frac{L_{nY}}{L_{nX}} \tag{3.15}$$

With both force components known, the member force can be computed as

$$S_n = \sqrt{S_{nX}^2 + S_{nY}^2} \tag{3.16}$$

The sign for S_n must be assigned to correspond to whether the member is in tension or compression. Figure 3.8 shows a sketch which can be used to keep track of the member force components as they are computed. The arrows at the ends of the members were drawn as each member was analyzed to represent the direction of the forces acting on the joints from the members. These arrows are very helpful in keeping track of the signs of the force components during the calculations.

The order of the computations, after the reactive forces have been computed, would be as follows:

$\Sigma_1 F_Y \to S_{2Y} = 40 \text{ kN} \to S_{2X} = (3/4) S_{2Y} = 30 \text{ kN} \to S_2 = 50 \text{ kN}$

$\Sigma_1 F_X \to S_{1X} = 90 \text{ kN} \to S_1 = 90 \text{ kN}$

$\Sigma_3 F_X \to S_{6X} = -90 \text{ kN} \to S_6 = -90 \text{ kN}$

$\Sigma_3 F_Y \to S_{3Y} = -40 \text{ kN} \to S_3 = -40 \text{ kN}$

$\Sigma_2 F_Y \to S_{5Y} = 120 \text{ kN} \to S_{5X} = (3/4) S_{5Y} = 90 \text{ kN} \to S_5 = 150 \text{ kN}$

$\Sigma_2 F_X \to S_{4X} = 0 \to S_4 = 0$

$\Sigma_5 F_Y \to S_{7Y} = -120 \text{ kN} \to S_7 = -120 \text{ kN}$

Method of Sections

For this particular truss, it should be possible for most engineers to mentally perform all of the required calculations for the member forces without even writing out an equation. Of course, all trusses are not this simple, but by a combination of a sketch of this type and a few written equations, it is often possible to perform the complete analysis. Since the member force components are used in the calculations, it is usually not necessary to compute the actual member forces until all of the calculations have been completed.

METHOD OF SECTIONS

In some cases, it might not be necessary for an engineer to compute all of the member forces in a particular truss in order to obtain the information which is needed. As an example, consider the scenario in which it has been discovered during a safety inspection that member 5 in the truss in Example Problem 3.1 is very heavily corroded, which has resulted in a significant reduction in its effective cross section area. We want to perform a quick calculation to compute the force in this member to determine whether it is overstressed without going through the process of computing any other member forces. The process which can be used, after the reactive forces have been computed, is:

Assume that members 4, 5 and 6 have been cut to divide the truss into two sections, as shown by the two free body diagrams in Figure 3.9. It is assumed in drawing these diagrams that the cut members are in tension.

The force in member 5 can now be computed by summing forces in the vertical direction for either the left or the right section of the truss. Performing this operation for the left section gives

$$\Sigma_{\text{left}} F_Y \rightarrow S_{5Y} - 40 - 80 = 0 \rightarrow S_{5Y} = 120 \text{ kN} \rightarrow S_5 = 150 \text{ kN}$$

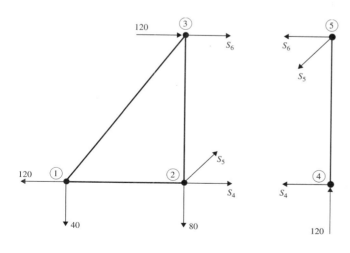

Figure 3.9 Example Problem 3.1, Method of Sections.

We now have the desired member force and can proceed with the safety check for the truss.

If desired, it is also possible to compute the forces in members 4 and 6 by summing moments about appropriate points. Summing moments about joint 5 for the right segment gives

$$\Sigma_{5/\text{right}} M_Z \rightarrow S_4 = 0$$

while summing moments about joint 2 for the left segment gives

$$\Sigma_{2/\text{left}} M_Z \rightarrow S_6 = -90 \text{ kN}$$

This analysis procedure is known as the *Method of Sections*. The reason for this name is obvious.

The Method of Sections can also be used in conjunction with the Method of Joints to analyze some trusses which cannot be analyzed entirely by the Method of Joints. This is demonstrated in the following example problem.

Example Problem 3.2

Compute the reactions and the member forces for the plane truss whose mathematical model is shown in Figure 3.10a. The truss has 9 joints, 15 members and 3 reactive force components. It is statically determinate if it is stable. As we will see later, the truss is stable.

The reactive force components can be computed using the global equations of equilibrium.

$$\Sigma_1 M_Z \rightarrow R_{8Y} = 18.56 \text{ kips} \uparrow$$

$$\Sigma F_Y \rightarrow R_{1Y} = 37.44 \text{ kips} \uparrow$$

$$\Sigma F_X \rightarrow R_{1X} = 0$$

If we now try to compute the member forces using the Method of Joints, we will find that we encounter difficulty since there are three or more members attached to each joint in the truss. Only two unknown member forces can be computed at each joint by using the joint equilibrium equations. The only way in which we can proceed at this time without solving a large number of simultaneous equations is to use the Method of Sections to compute some of the member forces. Figure 3.10b shows the free body diagram of the left section of the truss if members 5, 7 and 8 are assumed to be cut. The forces in the cut members can be computed by the following operations on this section

$$\Sigma_5 M_Z \rightarrow S_8 = 29.00 \text{ kips}$$

$$\Sigma F_Y \rightarrow S_{7Y} = -13.44 \text{ kips} \rightarrow S_{7X} = -10.08 \text{ kips} \rightarrow S_7 = -16.80 \text{ kips}$$

$$\Sigma F_X \rightarrow S_5 = -18.92 \text{ kips}$$

The remaining member forces can now be determined using only the Method of Joints. Figure 3.10c shows the final results of these calculations. The specific order in which the member forces were determined is

$$\Sigma_2 F_X \rightarrow S_{4X} = 18.92 \text{ kips} \rightarrow S_{4Y} = 14.19 \text{ kips} \rightarrow S_4 = 23.65 \text{ kips}$$

Method of Sections

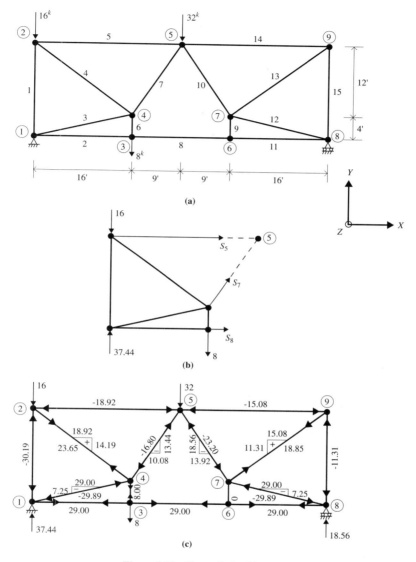

Figure 3.10 Example Problem 3.2.

$\Sigma_2 F_Y \rightarrow S_1 = -30.19$ kips

$\Sigma_1 F_Y \rightarrow S_{3Y} = -7.25$ kips $\rightarrow S_{3X} = -29.00$ kips $\rightarrow S_3 = -29.89$ kips

$\Sigma_1 F_X \rightarrow S_2 = 29.00$ kips

$\Sigma_3 F_Y \rightarrow S_6 = 8.00$ kips

$\Sigma_6 F_X \rightarrow S_{11} = 29.00$ kips

$\Sigma_6 F_Y \to S_9 = 0$

$\Sigma_8 F_X \to S_{12X} = -29.00$ kips $\to S_{12Y} = -7.25$ kips $\to S_{12} = -29.89$ kips

$\Sigma_8 F_Y \to S_{15} = -11.31$ kips

$\Sigma_9 F_Y \to S_{13Y} = 11.31$ kips $\to S_{13X} = 15.08$ kips $\to S_{13} = 18.85$ kips

$\Sigma_9 F_X \to S_{14} = -15.08$ kips

$\Sigma_7 F_Y \to S_{10Y} = -18.56$ kips $\to S_{10X} = -13.92$ kips $\to S_{10} = -23.20$ kips

The joint equations which were not used are $\Sigma_3 F_X, \Sigma_4 F_X, \Sigma_4 F_Y, \Sigma_5 F_X, \Sigma_5 F_Y$ and $\Sigma_7 F_X$. These equations can be used to check the solution. There are six check equations since three global equilibrium equations for the total truss were used to compute the reactive forces and three equilibrium equations for the left section were used to compute member forces S_5, S_7 and S_8. Only 12 equations were used from the total set of equations to compute the remaining member forces.

CLASSIFICATION OF PLANE TRUSSES

From the previous examples, we can see that it is possible to compute all of the member forces for some statically determinate trusses by using only the Method of Joints, while for other trusses a combination of the Method of Joints and the Method of Sections must be used. However, for some statically determinate and stable trusses, such as the hexagonal truss shown previously in Figure 3.7b, neither of these procedures is useful for computing the member forces. If an analysis is attempted by the Method of Joints, after the three reactive forces have been computed, it will be found that there are three members connected to every joint. If an analysis is attempted by the Method of Sections, it will be found that there are no cuts which can be used to isolate sections of the truss for computation of the forces in only three cut members. It would be very helpful if we could easily determine, before the analysis is attempted, which analysis procedure might be successful for a particular truss. This not only could save time, but also help to eliminate the frustration of false starts in the analysis process. Fortunately, this is possible, to some extent, since all plane trusses can be grouped into three distinct classifications.

Simple Trusses

A *simple truss* has a very distinct property which can be recognized by the process which can be used to construct the truss from the basic joint and member components. This process is to start with a triangle composed of three members and three joints and then to add one new joint and two new members in sequence to form the truss. The two new members may be connected to any two existing joints and the new joint may be located at any point in the plane with one restriction. This restriction is that the new joint may not be located such that the axes of the

Classification of Plane Trusses

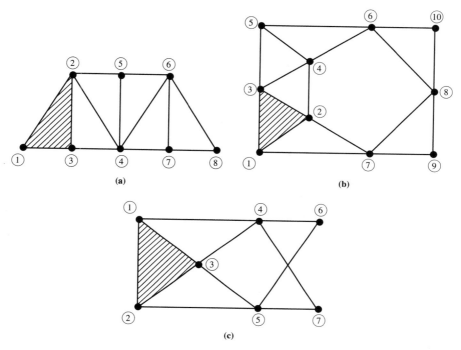

Figure 3.11 Simple trusses.

two new members form a straight line, since the new joint would be unstable. If this procedure is followed, the resulting framework will be rigid since the original triangle of members is rigid, and each pair of new members will form a rigid addition. Figure 3.11 shows several trusses which meet this requirement. The original triangle for each of these trusses is bounded by joints 1, 2 and 3. The additional joints then can be added in the order in which they are numbered along with their two connecting members to create the simple truss. The internal areas in the truss may all be triangular, as in the truss in Figure 3.11a, or any other shape, as in the truss in Figure 3.11b, as long as the basic rules for forming a simple truss are followed. It is even permissible for two or more members to cross each other, as demonstrated in the truss in Figure 3.11c. Each of the trusses is a rigid framework since it is not possible to change its shape without deforming the members. There is no limit on the number of members or the number of joints for a simple truss.

The next step is to determine how a simple truss must be supported to form a statically determinate and stable structure. From the construction process, we can see that the total number of members in the truss will be equal to the original three members plus two additional members for each additional joint beyond the original three joints. This leads to the following relationship between the number of members and the number of joints:

$$NM = 3 + 2(NJ - 3) \tag{3.17}$$

or

$$NM = 2NJ - 3 \tag{3.18}$$

If we now compare this relationship to the previously developed relationship in Eq. (3.4) for a statically determinate and stable plane truss, we can conclude that a simple truss must have three support restraints to be statically determinate and stable. These restraints either may be supplied by a combination of one pinned support and one roller support or by a combination of three roller supports. These restraints may occur at any joints and the roller supports may be at any orientation as long as the line of action of the roller support force does not pass through the pin support for the first combination or the lines of action of the three roller support forces do not meet at a common point for the second combination.

An important property of a simple truss is that all of the member forces can be determined by the Method of Joints after the three reactive forces have been computed using the global equations of equilibrium. The analysis process can be started at the last joint which was added in the construction sequence since there will only be two members connected to this joint. The analysis can then proceed through the joints in the reverse order in which they were added to determine the remaining member forces. There will never be more than two unknown member forces at any joint if the joints are analyzed in this reverse order. In many cases, the individual global member force components can be computed independently, as demonstrated previously. The worst condition which can ever occur during the analysis is the need to solve two linear simultaneous equations.

Compound Trusses

If we try to generate the statically determinate truss which was analyzed in Example Problem 3.2 by the simple truss construction process, we will find that we will encounter difficulty in trying to form the five sided area in the center of the truss. Since we cannot form this truss using the rules for a simple truss, it must fall into one of the other classifications. It is also obvious that this is not a simple truss since it was not possible to compute all of the member forces using the Method of Joints. This method of analysis always can be used for a simple truss. This truss falls into the classification known as a *compound truss*.

A compound truss is formed by joining two simple trusses by the equivalent of three nonparallel and nonconcurrent forces. One possible connecting scheme could consist of a common pin and one connecting member. The common pin would supply two connecting force components while the member would supply a single connecting force along its longitudinal axis. Another connecting scheme could consist of three nonparallel members whose axes do not cross at a common point. The truss in Example Problem 3.2 can be considered to correspond to either of these two connecting schemes, as shown in Figure 3.12. Figure 3.12a shows

Classification of Plane Trusses

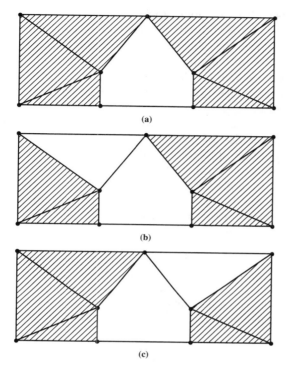

Figure 3.12 Compound trusses.

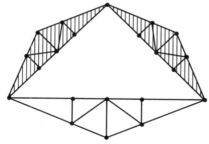

Figure 3.13 Compound truss.

this truss as two simple trusses connected by a common pin and a single member, while Figures 3.12b and 3.12c show two variations of two simple trusses connected by three members.

There are many other possible arrangements for a compound truss. Figure 3.13 shows two simple trusses which are connected by a common pin and another simple truss. The connecting truss is equivalent to a single connecting force with a line of action through the two joints at its ends. It is also acceptable to connect a simple truss to a compound truss or even to connect two compound trusses. The only restriction is that the connections must be equivalent to three nonparallel nonconcurrent forces. Theoretically, this process could go on forever.

If the number of joints and the number of members are counted in any compound truss, it will be found that the supports must supply three reactive forces to form a statically determinate structure. As with a simple truss, these supports may consist of either one pinned support and one roller support or three roller supports. The same restrictions that were stated for the roller orientations for a simple truss also apply to a compound truss. The three reactive forces can always be computed using the global equations of equilibrium for the truss.

Although it is possible to determine all of the member forces for some compound trusses using only the Method of Joints, the majority of compound trusses require a combination of the Method of Joints and the Method of Sections. The Method of Sections can be used to determine the three connecting forces between the simple trusses. After each simple truss has been isolated and all forces acting on it are known, its member forces can be determined by the Method of Joints. This was demonstrated previously in Example Problem 3.2. The following example also demonstrates this analysis process.

Example Problem 3.3

Compute the member forces for the compound truss whose mathematical model is shown in Figure 3.14a. The truss has 11 joints, 19 members and 3 independent reactive force components. Therefore, it is statically determinate if it is stable. The reactive forces can be computed using the global equilibrium equations by the following operations:

$$\Sigma_1 M_Z \rightarrow R_{11Y} = 6.5 \text{ kN} \uparrow$$

$$\Sigma F_Y \rightarrow R_{1Y} = 3.5 \text{ kN} \uparrow$$

$$\Sigma F_X \rightarrow R_{1X} = 12 \text{ kN} \leftarrow$$

The Method of Joints only can be used at joints 1 and 11 to compute member forces S_1, S_2, S_{18} and S_{19}. All other joints have three or more unknown member forces. Therefore, we must resort to the Method of Sections to complete the analysis.

One of several possible forms for this compound truss is to consider it to be composed of two simple trusses connected by member 10 and a common pin at joint 6. Two other possible forms are to consider two simple trusses to be connected by members 8, 9 and 10 or two simple trusses connected by members 10, 11 and 12. For the present analysis, we will use the form consisting of a common pin and a single connecting member.

The compound truss can be separated into two simple trusses by passing a cutting plane through member 10 and joint 6. The connecting forces for the two sections will consist of a force S_{10} along the axis of the connecting member and a horizontal force H_6 and a vertical force V_6 at the common pin, as shown in the free body diagrams of the two simple trusses in Figure 3.14b. Note that the connecting forces are equal but in opposite directions on each truss. These connected forces can be computed by using the equilibrium conditions for either simple truss.

Using the global equilibrium equations for the left truss gives

$$\Sigma_6 M_Z \rightarrow S_{10} = 2.25 \text{ kN} \rightarrow$$

Classification of Plane Trusses

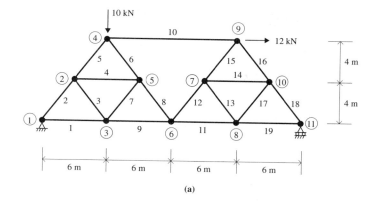

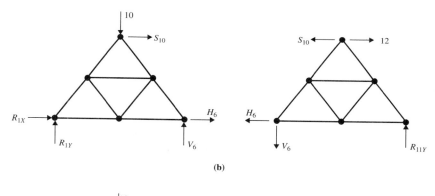

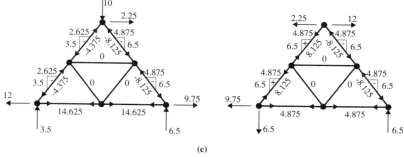

Figure 3.14 Example Problem 3.3.

$$\Sigma F_Y \rightarrow V_6 = 6.5 \text{ kN} \uparrow$$
$$\Sigma F_X \rightarrow H_6 = 9.75 \text{ kN} \rightarrow$$

The same results will be obtained by using the global equilibrium equations for the right truss, except that the connecting forces will be in the opposite directions. Since the connecting forces are equal and opposite on each section, it is not necessary to perform both analyses.

The member forces for each simple truss now can be found by the Method of Joints. We can keep track of the results of the various operations on the joints by recording the various member force components on the sketches shown in Figure 3.14c as they are computed.

The following operations can be performed for the left simple truss

$$\Sigma_1 F_Y \rightarrow S_{2Y} = -3.5 \text{ kN} \rightarrow S_{2X} = -2.625 \text{ kN} \rightarrow S_2 = -4.375 \text{ kN}$$

$$\Sigma_1 F_X \rightarrow S_1 = 14.625 \text{ kN}$$

$$\Sigma_6 F_Y \rightarrow S_{8Y} = -6.5 \text{ kN} \rightarrow S_{8X} = -4.875 \text{ kN} \rightarrow S_8 = -8.125 \text{ kN}$$

$$\Sigma_6 F_X \rightarrow S_9 = 14.625$$

It is not possible to independently compute the components of the forces in members 3 and 7 by summing forces in the X and Y directions at joint 3 due to the orientation of the two members. However, it is possible to develop two simultaneous equations for the member forces:

$$\Sigma_3 F_X \rightarrow -S_1 - 0.6S_3 + 0.6S_7 + S_9 = 0$$

$$\Sigma_3 F_Y \rightarrow 0.8S_3 + 0.8S_7 = 0$$

which, on substituting the known values for S_1 and S_9 and solving the set of equations gives $S_3 = 0$ and $S_7 = 0$.

We can now proceed to solve for the remaining member forces in the left simple truss by the following operations:

$$\Sigma_2 F_Y \rightarrow S_{5Y} = -3.5 \text{ kN} \rightarrow S_{5X} = -2.625 \text{ kN} \rightarrow S_5 = -4.375 \text{ kN}$$

$$\Sigma_2 F_X \rightarrow S_4 = 0$$

$$\Sigma_4 F_X \rightarrow S_{6X} = -4.875 \text{ kN} \rightarrow S_{6Y} = -6.5 \text{ kN} \rightarrow S_6 = -8.125 \text{ kN}$$

The three equations which were not used for this truss are $\Sigma_4 F_Y$, $\Sigma_5 F_X$ and $\Sigma_5 F_Y$.

The following operations can be performed for the right simple truss:

$$\Sigma_{11} F_Y \rightarrow S_{18Y} = -6.5 \text{ kN} \rightarrow S_{18X} = -4.875 \text{ kN} \rightarrow S_{18} = -8.125 \text{ kN}$$

$$\Sigma_{11} F_X \rightarrow S_{19} = 4.875 \text{ kN}$$

$$\Sigma_6 F_Y \rightarrow S_{12Y} = 6.5 \text{ kN} \rightarrow S_{12X} = 4.875 \text{ kN} \rightarrow S_{12} = 8.125 \text{ kN}$$

$$\Sigma_6 F_X \rightarrow S_{11} = 4.875 \text{ kN}$$

As in the left truss, we must solve two simultaneous equations to determine the forces in members 13 and 17:

$$\Sigma_8 F_X \rightarrow -S_{11} - 0.6S_{13} + 0.6S_{17} + S_{19} = 0$$

$$\Sigma_8 F_Y \rightarrow 0.8S_{13} + 0.8S_{17} = 0$$

which, on substituting the known values for S_{11} and S_{19} and solving gives $S_{13} = 0$ and $S_{17} = 0$.

The remaining member forces for the right simple truss can now be computed by the following operations

Classification of Plane Trusses

$$\Sigma_7 F_Y \to S_{15Y} = 6.5 \text{ kN} \to S_{15X} = 4.875 \text{ kN} \to S_{15} = 8.125 \text{ kN}$$

$$\Sigma_7 F_X \to S_{14} = 0$$

$$\Sigma_9 F_X \to S_{16X} = -4.875 \text{ kN} \to S_{16Y} = -6.5 \text{ kN} \to S_{16} = -8.125 \text{ kN}$$

The three equations which were not used for this truss are $\Sigma_9 F_Y$, $\Sigma_{10} F_X$ and $\Sigma_{10} F_Y$. The equations which were not used in the determination of the member forces for each truss can be used to check the solution.

Complex Trusses

Any plane truss which is statically determinate and stable and does not fit into the classification for a simple truss or a compound truss will be classified as a *complex truss*. An example of a truss which falls into this classification is the hexagonal truss in Figure 3.7b. Several other examples are shown in Figure 3.15. It is obvious that each of these trusses is a complex truss since they each have more than

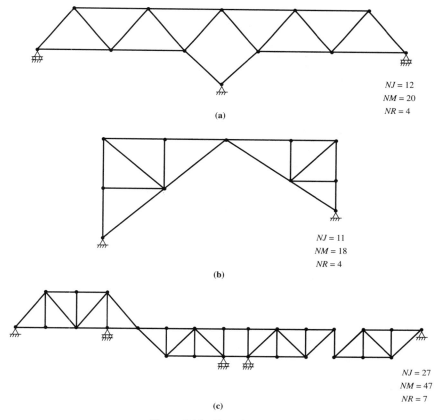

Figure 3.15 Complex trusses.

three independent reactive force components. The truss configuration shown in Figure 3.15a was developed and patented by an engineer named E. M. Wichert. It was used primarily for railroad and highway bridges. This truss configuration was so unique that a complete book was written about it by D. B. Steinman (1932). The book describes the advantages of this type of truss and discusses various procedures which can be used in its analysis.

Complex trusses might have any member configuration and any number of independent reactive force components, equal to or greater than 3, as long as the required relationship between the number of joints, the number of members and the number of support restraints for a statically determinate truss, as defined in Eq. (3.4), is satisfied. In addition, the truss must be stable. Since essentially each complex truss is a special case, we will not try to discuss any specialized analysis procedures here. In some cases it might be obvious how the reactions and member forces can be computed using a combination of equations of condition, the Method of Joints and the Method of Sections. In other cases, no obvious simple analysis process can be found and the analyst must resort to generating and solving the complete set of 2NJ joint equilibrium equations. Since low cost personal computers are now available to solve large systems of linear simultaneous equations, this is probably the best approach to take if a simple analysis approach is not obvious.

ZERO FORCE MEMBERS

There is a general rule, known as the *Zero Force Rule*, which often can be used as an aid in the analysis of a plane truss, in which there are three or more members connected to each joint, if certain conditions are satisfied in the truss geometry. This rule can be stated as follows:

> If there are three unknown member forces at a joint and two of these forces have the same line of action, then the component of the force in the third member, perpendicular to the other two members, will be equal to the sum of the components of all other member forces and active forces acting on the joint in the perpendicular direction. If there are no other nonzero member forces or active forces acting on the joint, then the force in the third member will be zero.

It will be found that zero force members will exist in many trusses. However, this does not mean that they are unimportant in the overall behavior of the truss. These members are often inserted as bracing members, during the design of a truss, to reduce the overall length of compression members to increase their buckling strength.

Example Problem 3.4

As an example of the use of the Zero Force Rule, consider the complex truss whose mathematical model is shown in Figure 3.16. The truss has 12 joints, 20 members and

Zero Force Members

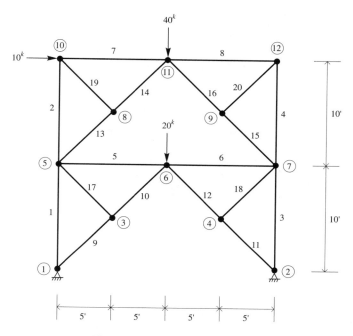

Figure 3.16 Example Problem 3.4.

4 independent reactive force components. Therefore, it is statically determinate. The vertical reactive force components at joints 1 and 2 can be computed by using the following global equilibrium operations:

$$\Sigma_1 M_Z \rightarrow R_{2Y} = 40 \text{ kips}$$

$$\Sigma_2 M_Z \rightarrow R_{1Y} = 20 \text{ kips}$$

However, it is not possible to compute R_{1X} and R_{2X} since there is only one remaining independent global equilibrium equation. Some type of combined analysis which considers the interaction of the member forces and the reactive forces must be used.

It does not appear that the Method of Joints will be useful to compute the member forces since there are two unknown member forces and an unknown reactive force component at joints 1 and 2 and three or more unknown member forces at joints 3 through 12. In addition, there is no obvious section that can be used to isolate a portion of the truss for independent analysis. However, it is possible to perform a complete analysis of this truss by using a modified form of the Method of Joints due to its particular geometry.

If we first consider the equilibrium of joints 3, 4, 8 and 9, we see that two members at each of these joints have the same line of action, while the third member is oriented in a different direction. Applying the Zero Force Rule at these joints shows that the forces in members 17, 18, 19 and 20 must be zero. By using these values, it is now possible to compute the additional member forces and reactive force components by the following joint operations:

$\Sigma_{10} F_X$ and $\Sigma_{10} F_Y \rightarrow S_2 = 0$ and $S_7 = -10$ kips

$\Sigma_{12} F_X$ and $\Sigma_{12} F_Y \rightarrow S_4 = 0$ and $S_8 = 0$

$\Sigma_{11} F_X$ and $\Sigma_{11} F_Y \rightarrow S_{14} = -21.213$ kips and $S_{16} = -35.355$ kips

$\Sigma_8 F_X \rightarrow S_{13} = -21.213$ kips

$\Sigma_9 F_X \rightarrow S_{15} = -35.355$ kips

$\Sigma_5 F_X$ and $\Sigma_5 F_Y \rightarrow S_1 = -15$ kips and $S_5 = 15$ kips

$\Sigma_7 F_X$ and $\Sigma_7 F_Y \rightarrow S_3 = -25$ kips and $S_6 = 25$ kips

$\Sigma_6 F_X$ and $\Sigma_6 F_Y \rightarrow S_{10} = -7.071$ kips and $S_{12} = -21.213$ kips

$\Sigma_3 F_X \rightarrow S_9 = -7.071$ kips

$\Sigma_4 F_X \rightarrow S_{11} = -21.213$ kips

$\Sigma_1 F_X \rightarrow R_{1X} = 5$ kips

$\Sigma_2 F_X \rightarrow R_{2X} = -15$ kips

Since only 2NJ-2 joint equilibrium equations were used to compute the member forces and the remaining reactive force components, the remaining two joint equations $\Sigma_1 F_Y$ and $\Sigma_2 F_Y$ can be used as check equations to verify the accuracy of the analysis.

The results of the analysis in the previous example indicate why members 17 through 20 are present in the truss even though the force in each of these members is zero. They apparently have been included to reduce the effective length of the diagonal compression members. By cutting the effective length of these members in half, their buckling strength in the plane of the truss has been increased by a factor of 4 compared to the situation if the bracing members were not present, since the buckling strength of a member varies as the inverse of the square of its length.

COMPUTER PROGRAM SDPTRUSS

In Chapter 2, we used the program SDPTRUSS to determine the reactions for statically determinate plane trusses. We also can use this program to verify the manual solutions for the member forces in the example problems in this chapter and for the suggested problems at the end of the chapter. The program computes the member forces for a statically determinate plane truss by generating a set of equilibrium equations by summing forces in the global X and Y unrestrained directions at each joint. This results in a set of NM linear simultaneous equations with the member forces as the unknowns. This approach is used, rather than generating the complete set of 2NJ equations with both the member forces and the reactive forces as the unknowns, since it simplifies keeping track of the unknowns and also reduces the number of simultaneous equations which must be solved. The simultaneous equations are solved using the Gauss-Jordan Elimination Method. The specific solution routine which is used is the same one which is used in the program SEQSOLVE. After the simultaneous equations have been solved for the member forces, the NR reactive force components are computed by summing forces in the restrained X and Y directions at each support joint.

The program SDPTRUSS was written using the QBasic programming language. This particular language is distributed with Version 5 and newer versions of the MS DOS operating system. The source code for SDPTRUSS is contained in the ASCII disk file SDPTRUSS.BAS. We will not go into the details of the source code since the purpose of this book is to teach structural engineering rather than computer programming. The source code has been included in this file so that those readers who are interested can see how the analysis procedure can be implemented in a computer program. The source code can be listed on the printer either with the MS DOS command COPY SDPTRUSS.BAS PRN: or by loading the ASCII file into a word processor such as WordPerfect and then printing it.

The program also could have been developed using any of the other popular programming languages such as FORTRAN, Pascal, or C. The QBasic language was used since it is available on most computers operating under MS DOS. The specific language is not important, since it is a fairly easy task to transform a program from one programming language to another after the program logic has been developed. Any reader who is familiar with one of these other languages should be able to make sense out of the source listing even though they have not used QBasic previously. There are many excellent books available which describe the details of the QBasic language (Arnson et al., 1991; Schneider, 1991; Schneider and Peter Norton Computing Group, 1991). Numerous REM (i.e.; REMark) statements have been included in the source code to describe the various operations which are being performed. The executable file, which is stored in the disk file SDPTRUSS.EXE, was generated using the compiler in Version 7.1 of the Microsoft Basic Professional Development System (Microsoft Corp., 1989; Aitken, 1991).

In addition to the requirements that the truss must be statically determinate and stable, there are two other restrictions on any truss which is to be analyzed by SDPTRUSS. First, the program is limited to a truss with a maximum of 127 members since the QBasic language is limited to a maximum array size of 64 kilobytes. The maximum size coefficient matrix, for the equilibrium equations, which will fit in a 64 kilobyte array is 127 rows by 127 columns. Second, all roller supports must translate in the global X or Y directions. Inclined roller supports are not permitted in order to simplify the analysis process. However, it is possible to get around this restriction by modifying the mathematical model for the truss by eliminating the inclined roller support and adding an additional member and pinned support, as demonstrated previously in the computer solution by the program SDPFRAME for Example Problem 2.4 in Chapter 2. This procedure is explained in detail in the program user instructions in the disk file SDPTRUSS.DOC. There are no other

```
Example Problem 3.4 - Analysis by Program SDPTRUSS
12,20,2,3
Joint Coordinates
1,0.0,0.0
2,20.0,0.0
3,5.0,5.0
4,15.0,5.0
5,0.0,10.0
6,10.0,10.0
7,20.0,10.0
8,5.0,15.0
9,15.0,15.0
10,0.0,20.0
11,10.0,20.0
12,20.0,20.0
Member Data
1,1,5
2,5,10
3,2,7
4,7,12
5,5,6
6,6,7
7,10,11
8,11,12
9,1,3
10,3,6
11,2,4
12,4,6
13,5,8
14,8,11
15,7,9
16,9,11
17,3,5
18,4,7
19,8,10
20,9,12
Support Restraints
1,1,1
2,1,1
Joint Loads
6,0.0,-20.0
10,10.0,0.0
11,0.0,-40.0
```

Figure 3.17 Example Problem 3.4, SDPTRUSS input.

restrictions on the number of joints, the number of support restraints or the truss geometry.

Figure 3.17 shows a listing of the input data file for SDPTRUSS for the complex truss in Example Problem 3.4. A plot of the geometry of the mathematical model by the program PLOTSDPT is shown in Figure 3.18 and the output results from SDPTRUSS are shown in Figure 3.19. The computed values for the member forces and the reactive forces agree with those obtained in the manual solution, as should be expected.

Example Problem 3.4 - Analysis by Program SDPTRUSS

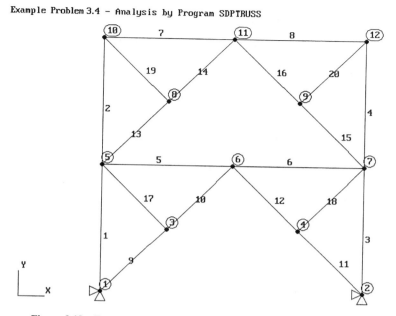

Figure 3.18 Example Problem 3.4, geometry plot from PLOTSDPT.

INFLUENCE LINES FOR PLANE TRUSS MEMBER FORCES

In the previous chapter, the influence line for a reactive force for any structure was defined as a curve which extended over the length of the structure for which the ordinate at any point was equal to the value of the reactive force due to a unit downward force acting at that location on the structure. Influence lines also can be developed for the member forces in trusses. In this case, the ordinate of the influence line at any point will be equal to the particular member force for which it has been developed due to a unit downward force at that point on the truss. These influence lines can be very useful for design since they give a convenient way to determine the maximum member forces which might occur due to moving loads on the truss. The maximum member forces are needed for design of the members.

There is actually nothing new to be learned to develop influence lines for truss member forces since the same analysis procedures discussed previously in this chapter for computing the member forces in plane trusses can be used. The only difference is that we will now be dealing with a single unit downward force which can move to any position along the load line of the truss, where the load line is defined as the path that the loads acting on the truss will take as they move across the truss. For example, Figure 3.20 shows the same truss geometry with three different load lines. In Figure 3.20a the traffic moves across the top chord of a highway bridge, while in Figure 3.20b the traffic moves across the lower chord of the bridge.

Example Problem 3.4 - Analysis by Program SDPTRUSS

Member Forces (Tension Positive)

Member	Force
1	-15.000
2	0.000
3	-25.000
4	0.000
5	15.000
6	25.000
7	-10.000
8	-0.000
9	-7.071
10	-7.071
11	-21.213
12	-21.213
13	-21.213
14	-21.213
15	-35.355
16	-35.355
17	0.000
18	0.000
19	0.000
20	0.000

Reactions

Joint	RX	RY
1	5.000	20.000
2	-15.000	40.000

Figure 3.19 Example Problem 3.4, SDPTRUSS output.

Figure 3.20c shows a third case where a conveyor belt which transports coal in a power plant might switch levels as it moves along the supporting truss. Each of these three cases could result in different influence lines for the forces in specific members in the truss. The analyst must know exactly how the loads will be applied to the truss before any type of analysis can be performed.

Example Problem 3.5

Compute the influence lines for members 4, 5 and 6 for the bridge truss whose mathematical model is shown in Figure 3.21a if the traffic loads move along the bottom chord of the truss. The first step in developing the influence lines for the member forces is to develop the influence lines for the reactive forces since the magnitude of the reactive forces for any position of the unit force will be needed when computing the member force influence lines. These influence lines can be developed using the procedure described in the previous chapter. The influence lines for R_{1Y} and R_{8Y} are shown in Figure 3.21b. The reactive force R_{1X} will be zero for any position of a unit downward force on the load line. It was assumed in developing these influence lines that when the unit force is at any position between joints on the load line, the loads

Influence Lines for Plane Truss Member Forces

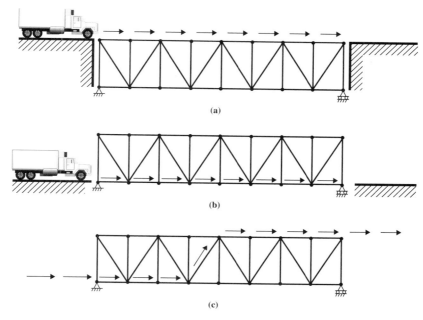

Figure 3.20 Truss load lines.

acting on the adjacent joints will be proportional to the reactions for a simply supported stringer extending between the two joints. This is a reasonable assumption since a typical floor system for a bridge of this type would consist of a set of transverse floor beams spanning between the trusses on either side of the roadway, with longitudinal stringers spanning between the floor beams, as shown in the plan view of the bridge in Figure 3.21c. It will be assumed that both lanes have the same loads and the loads in the right lane are transferred into the right truss while the loads in the left lane are transferred into the left truss.

The desired influence lines can be easily computed by cutting the truss into two sections, as shown by the free body diagrams in Figure 3.22a. The member forces can be computed for any position of the downward unit force by the following operations:

$$\Sigma_{3/\text{left}} M_Z \to S_4$$
$$\Sigma_{\text{left}} F_Y \text{ or } \Sigma_{\text{right}} F_Y \to S_5$$
$$\Sigma_{4/\text{right}} M_Z \to S_6$$

The value of R_{1Y} and R_{8Y} can be determined for use in these calculations from the reactive force influence lines. The resulting values of the member forces for each position of the unit force on the load line are shown in the influence table in Figure 3.22b and the plotted influence lines are shown in Figure 3.22c. As with the reactive force influence lines, the ordinates corresponding to each load line joint have been connected by straight lines. Note that member 4 will be in tension for any position of a concentrated force on the load line while member 6 will be in compression. The force in member 5 will change from compression to tension as the force

102 Analysis of Statically Determinate Plane Trusses   Chap. 3

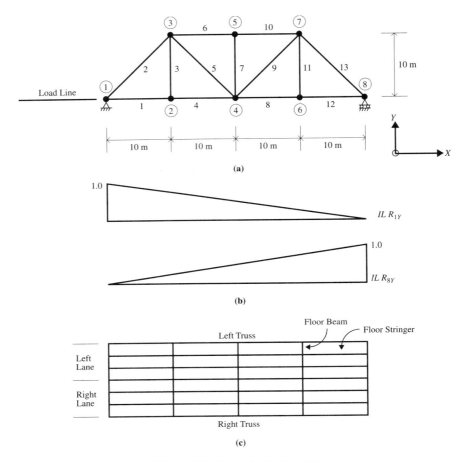

Figure 3.21 Example Problem 3.5.

moves across the bridge. Therefore, both a maximum tension force and a maximum compression force would have to be computed for member 5 for any traffic loading on the bridge. It is left as an exercise for the reader to compute the maximum forces for these members for an HS20-44 truck.

It is often possible to find some combination of the Method of Sections and the Method of Joints to compute the ordinates of the influence lines for specific members in a truss so that the number of calculations can be minimized. Of course, if neither of these procedures can be conveniently used, the analyst can always fall back on the procedure of developing and solving 2NJ joint equilibrium equations for each position of the unit force. The advantage of this procedure is that it will give information for developing the influence lines for each reactive force and each member force

Computer Program PTRUSSIL

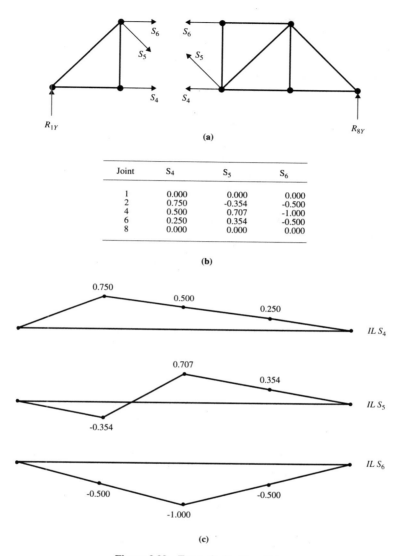

Figure 3.22 Example Problem 3.5.

in the truss. The only thing that will have to be changed for each position of the unit force is the right side constant vector for the simultaneous equations. The equation coefficients do not change with the force position.

COMPUTER PROGRAM PTRUSSIL

The computer program PTRUSSIL, which is supplied for this book, will plot the influence lines for the member forces for a statically determinate or a statically indeterminate plane truss. The analysis procedure which is used in the program will not be discussed here. The program is included so that the readers can verify their manual solutions for the suggested problems at the end of this chapter. The instructions for using PTRUSSIL are contained in the ASCII disk file PTRUSSIL.DOC.

The format for the input data file for PTRUSSIL is very similar to the file format for the program SDPTRUSS. The primary differences in the input data for the two programs are that the joint loads are specified in the input for SDPTRUSS, while the load line joints are specified in the input for PTRUSSIL. The program then computes the force in the member, for which the influence line is to be generated, corresponding to a unit downward force at each of the load line joints. In addition, since the program PTRUSSIL also can analyze statically indeterminate trusses, it is necessary to define the modulus of elasticity for each different material in the truss and the cross section area for each different member cross section. However, if there is only one material and one cross section in the truss the program will assign arbitrary values for these quantities since their specific values will

```
Example Problem 3.5 - Analysis by Program PTRUSSIL
8,13,1,1,2,5
Joint Coordinates
1,0.0,0.0
2,10.0,0.0
3,10.0,10.0
4,20.0,0.0
5,20.0,10.0
6,30.0,0.0
7,30.0,10.0
8,40.0,0.0
Member Data
1,1,2
2,1,3
3,2,3
4,2,4
5,3,4
6,3,5
7,4,5
8,4,6
9,4,7
10,5,7
11,6,7
12,6,8
13,7,8
Support Restraints
1,1,1
8,0,1
Load Line Joints
1
2
4
6
8
```

Figure 3.23 Example Problem 3.5, PTRUSSIL input.

not affect the analysis if they are constant throughout the truss. This is described in the program instructions.

Figure 3.23 shows the listing of the input data file for the analysis of the truss in Example Problem 3.5 by PTRUSSIL. Figures 3.24 through 3.26 show the plotted influence lines and a listing of the ordinates at the load line joints for members 4, 5 and 6 as they are shown on a VGA monitor. These influence lines agree with those shown in Figure 3.22, which were obtained by the manual analysis of this truss.

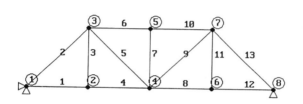

Figure 3.24 Example Problem 3.5, influence line from PTRUSSIL.

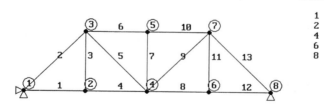

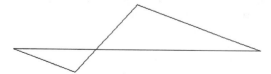

Figure 3.25 Example Problem 3.5, influence line from PTRUSSIL.

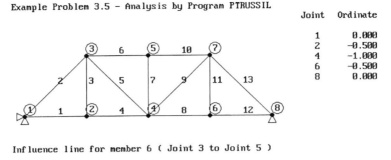

Figure 3.26 Example Problem 3.5, influence line from PTRUSSIL.

REFERENCES

AITKEN, PETER G. (1991). *Microsoft Basic 7.1—A Programmer's reference*. New York: John Wiley.

ARNSON, ROBERT, CHRISTY GEMMELL, and HARRY HENDERSON (1991). *The Waite Group's MS-DOS QBasic Programmer's Reference*. Redmond, Washington: Microsoft Press.

MICROSOFT CORP. (1989a). *Microsoft BASIC Professional De elopment System—Programmer's Guide*. Redmond, Washington: Microsoft Press.

MICROSOFT CORP. (1989b). *Microsoft BASIC Professional De elopment System—BASIC Language Reference*. Redmond, Washington: Microsoft Press.

SCHNEIDER, DAVID L. (1991). *Microsoft QBasic: An Introduction to Structured Programming for Engineering, Mathematics, and the Sciences*. New York: Macmillan.

SCHNEIDER, DAVID L., and the PETER NORTON COMPUTING GROUP (1991). *QBasic Programming*. New York: Brady.

STEINMAN, D. B. (1932). *The Wichert Truss*. New York: D. Van Nostrand.

SUGGESTED PROBLEMS

SP3.1 Determine whether each of the trusses shown in Figure SP3.1 is unstable, statically determinate or statically indeterminate by counting the number of joints, members and support restraints. Classify those trusses which are statically determinate as simple, compound or complex.

SP3.2 Develop a set of 2NJ simultaneous equations for each of the plane trusses in Figure 3.7 with the member forces and the reactive forces as the unknowns. Solve the equations using the program SEQSOLVE. Verify the results of the analysis with the program SDPTRUSS.

Suggested Problems

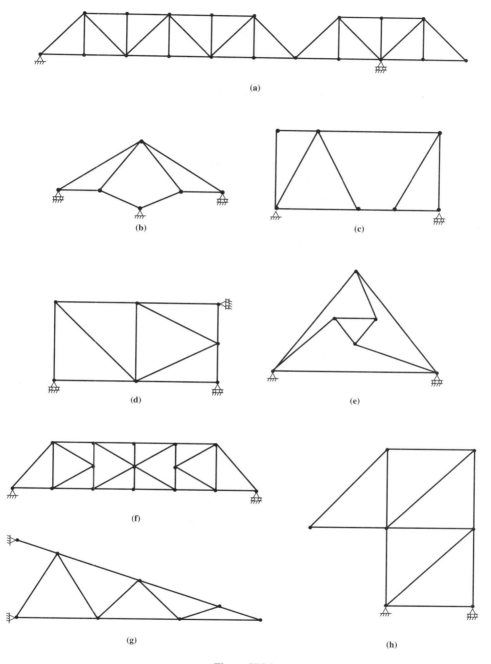

Figure SP3.1

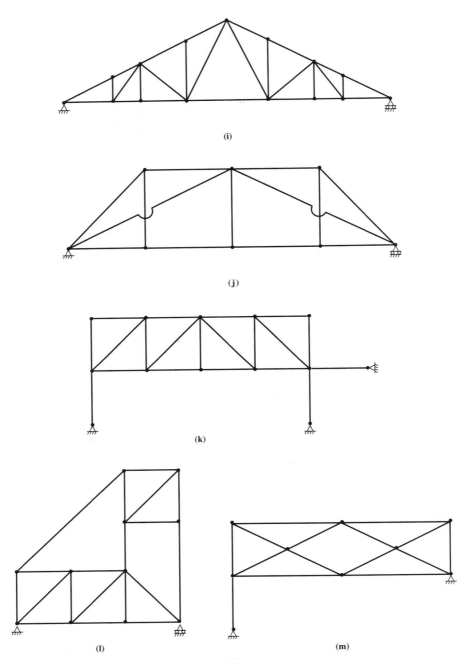

Figure SP3.1 *(cont.)*

Suggested Problems

SP3.3–SP3.4 Develop a set of 2NJ simultaneous equations for each of the plane trusses shown in Figures SP3.3 and SP3.4 with the member forces and the reactive forces as the unknowns. Solve the equations using the program SEQSOLVE. Verify the results of the analysis with the program SDPTRUSS.

SP3.5–SP3.14 Compute all of the member forces in the plane trusses shown in Figures SP3.5 to SP3.14 using the Method of Joints. Verify the results of the analysis with the program SDPTRUSS.

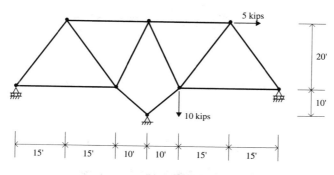

Figure SP3.3

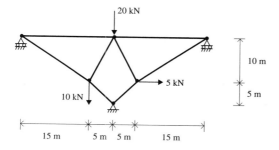

Figure SP3.4

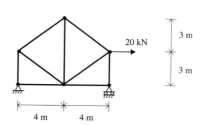

Figure SP3.5

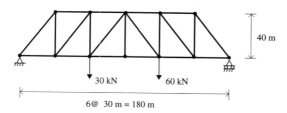

Figure SP3.6

110   Analysis of Statically Determinate Plane Trusses   Chap. 3

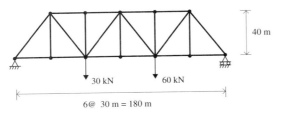

Figure SP3.7

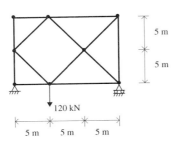

Figure SP3.8

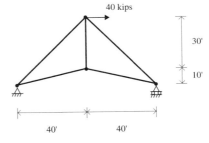

Figure SP3.9

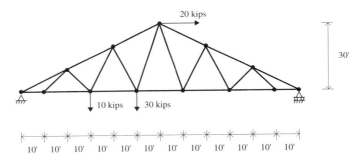

Figure SP3.10

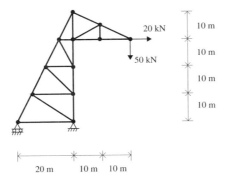

Figure SP3.11

Suggested Problems

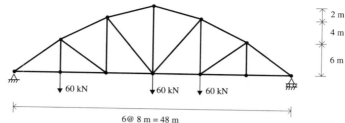

Figure SP3.12

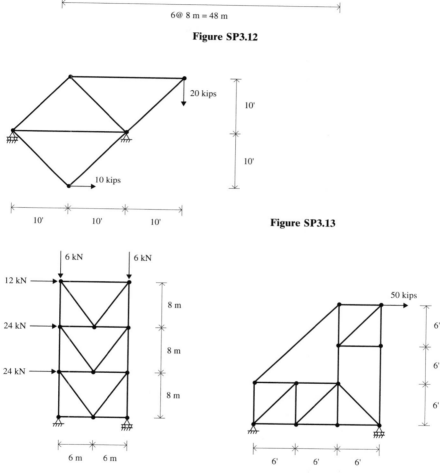

Figure SP3.13

Figure SP3.14

Figure SP3.15

SP3.15–SP3.18 Compute all of the member forces in the plane trusses shown in Figures SP3.15 to SP3.18 by using a combination of the Method of Sections and the Method of Joints. Verify the results of the analysis with the program SDPTRUSS.

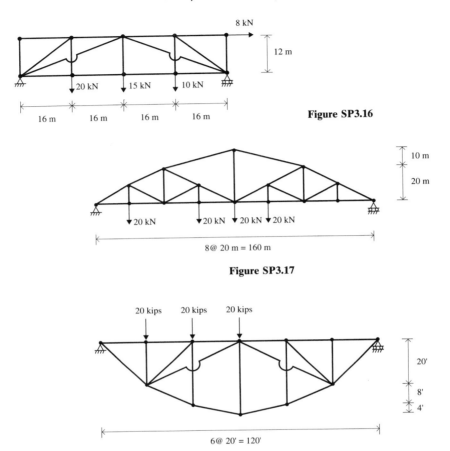

Figure SP3.16

Figure SP3.17

Figure SP3.18

SP3.19–SP3.20 Compute the reactive forces for the plane trusses shown in Figures SP3.19 and SP3.20 by utilizing an equation of condition. Compute all of the member forces by the Method of Joints after the reactive forces have been determined. Verify the results of the analysis with the program SDPTRUSS.

SP3.21–SP3.24 Draw the influence lines for the forces in the indicated members in the plane trusses shown in Figures SP3.21 to SP3.24. Verify the results with the program PTRUSSIL.

SP3.25 Draw the influence lines for the forces in the indicated members in the highway bridge truss shown in Figure SP3.25. Verify the results with the program PTRUSSIL. Compute the maximum tension and compression force for each of these members due to a HS20-44 truck which can move to any position along the load line of the bridge.

Suggested Problems

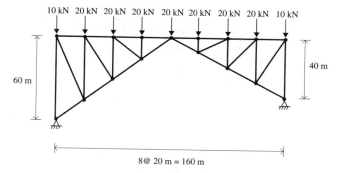

Figure SP3.19

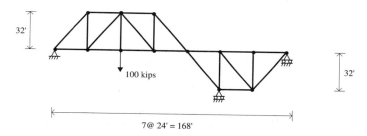

Figure SP3.20

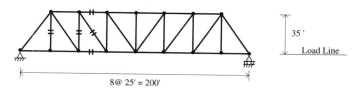

Figure SP3.21

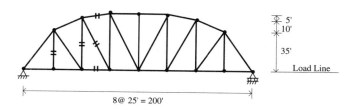

Figure SP3.22

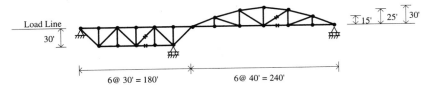

Figure SP3.23

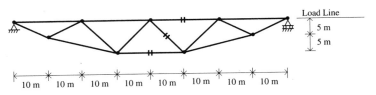

Figure SP3.24

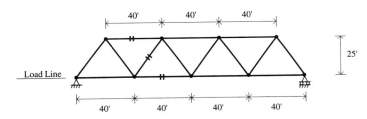

Figure SP3.25

4

Analysis of Statically Determinate Space Trusses

MATHEMATICAL MODEL FOR A SPACE TRUSS

The mathematical model for a space truss consists of a set of joints which are connected by straight members. The difference between a space truss and a plane truss is that the joints in a space truss may be located at any positions in three dimensional space, thus permitting the members to be inclined at any orientation. Otherwise, the properties of the mathematical models of the two types of trusses are very similar:

1. All members must be attached to their end joints such that no moment is transmitted between the joints and the members. The connection at each end joint will be equivalent to a ball and socket which permits rotation of each end of the member about any axis while restraining translation of the end with respect to the joint.
2. All loads acting on the structure must consist only of concentrated forces acting on the joints. The forces may act in any direction in three dimensional space.
3. All support joints are only restrained against translation. They are free to rotate about any axis. This will result in three possible types of support joints: a joint which is restrained against translation in any direction, which will supply three independent reactive force components; a joint which is restrained against translation in two perpendicular directions and free to translate in the third direction, which will supply two independent reactive force components; and a joint which is free to translate in two perpendicular di-

rections and restrained against translation in the third direction, which will supply one reactive force component.

As in a plane truss, only axial forces will be transmitted between the joints and the members in a space truss. There will be no shear forces and no bending or twisting moments at any interior points in the members.

STABILITY AND STATIC DETERMINACY OF SPACE TRUSSES

The primary difference in the analysis of a space truss, compared to a plane truss, is that there will be three independent equations of equilibrium for each joint, corresponding to the sum of the force components in three directions, while there will be only two equations per joint for a plane truss. However, there will still be two equal and opposite axial forces at the end of each member and one reactive force corresponding to each support restraint. Therefore, the stability and static determinacy conditions for a space truss can be easily determined by a slight modification of the conditions which were established in the previous chapter for a plane truss. The resulting conditions are:

$$\text{if } 3NJ > NM + NR, \text{ the space truss is unstable;}$$

$$\text{if } 3NJ = NM + NR, \text{ the space truss is statically determinate;}$$

and

$$\text{if } 3NJ < NM + NR, \text{ the space truss is statically indeterminate,}$$

where NJ is the number of joints, NM is the number of members and NR is the number of independent support restraints. A minimum of six support restraints is required. Of course, as with a plane truss, it is also possible for a space truss which satisfies either of the latter two conditions to be still geometrically unstable due to the arrangement of its members or support restraints. Three of the many possible situations which would cause a geometrically unstable condition in a space truss which meets these conditions are: an unrestrained joint with less than two members connected to it; an unrestrained joint with three or more members connected to it which all lie in the same plane; and six or more support restraints which are oriented such that they do not prevent rigid body rotation of the truss about all axes or rigid body translation in all directions. An unstable condition will become obvious during the analysis of the truss since a point will be reached in the analysis where equilibrium of a joint, or of a member or of the overall structure cannot be satisfied. In this chapter, we will only be concerned with the analysis of statically determinate and stable space trusses.

ANALYSIS OF A SPACE TRUSS BY JOINT EQUILIBRIUM

The NM member forces and the NR reactive force components can be computed for a statically determinate and stable space truss by solving a set of 3NJ simultaneous joint equilibrium equations which can be obtained by summing forces in the global X, Y and Z directions at each joint in the truss. The three global force components which are applied to the first joint by any member n in the truss, as shown in Figure 4.1, can be expressed in terms the global orientation angles θ_{nX}, θ_{nY} and θ_{nZ} of the member by the relationships

$$S_{nX} = S_n \cos \theta_{nX} = S_n C_{nX} \tag{4.1}$$

$$S_{nY} = S_n \cos \theta_{nY} = S_n C_{nY} \tag{4.2}$$

$$S_{nZ} = S_n \cos \theta_{nZ} = S_n C_{nZ} \tag{4.3}$$

where C_{nX}, C_{nY} and C_{nZ} are the direction cosines for the member. The direction cosines can be computed using the projections of the member on the global axes as

$$C_{nX} = \frac{L_{nX}}{L_n} \tag{4.4}$$

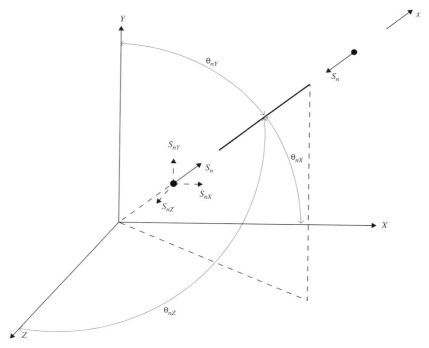

Figure 4.1 Space truss member orientation.

$$C_{nY} = \frac{L_{nY}}{L_n} \quad (4.5)$$

$$C_{nZ} = \frac{L_{nZ}}{L_n} \quad (4.6)$$

and the projected lengths and the length of the member can be determined from the global coordinates at the first joint and the second joint for the member as

$$L_{nX} = X_{n2} - X_{n1} \quad (4.7)$$
$$L_{nY} = Y_{n2} - Y_{n1} \quad (4.8)$$
$$L_{nZ} = Z_{n2} - Z_{n1} \quad (4.9)$$

and

$$L_n = \sqrt{L_{nX}^2 + L_{nY}^2 + L_{nZ}^2} \quad (4.10)$$

The signs of the direction cosines will automatically result in the correct signs for the member force components for any orientation of the member in space. The three global force components on the second joint will have the same magnitudes as those on the first joint but they will have opposite signs.

Example Problem 4.1

Compute the member forces and the reactions for the space truss whose mathematical model is shown in Figure 4.2 by solving a set of 3NJ simultaneous joint equilibrium equations. The truss has six members, four joints and six independent support translation restraints. The reactive force components correspond to R_Y and R_Z at joint 1, R_Y and R_Z at joint 2 and R_X and R_Y at joint 3.

$$NJ = 4$$
$$NM = 6 \quad \rightarrow \quad 3NJ = NM + NR$$
$$NR = 6$$

Therefore, the truss is statically determinate. We will assume that it is stable. This will be checked later when we attempt to solve the joint equilibrium equations.

The direction cosines for the members can be computed by Eqs. (4.4) through (4.10) as follows:

Member 1 (joint 1 to joint 2):

$L_{1X} = 10$ $\qquad$ $C_{1X} = 1.0$
$L_{1Y} = 0 \quad \rightarrow \quad L_1 = 10 \quad \rightarrow \quad C_{1Y} = 0$
$L_{1Z} = 0$ $\qquad$ $C_{1Z} = 0$

Member 2 (joint 2 to joint 3):

$L_{2X} = -5$ $\qquad$ $C_{2X} = -0.48564$
$L_{2Y} = 0 \quad \rightarrow \quad L_2 = 10.29563 \quad \rightarrow \quad C_{2Y} = 0$
$L_{2Z} = -9$ $\qquad$ $C_{2Z} = -0.87416$

Analysis of a Space Truss by Joint Equilibrium

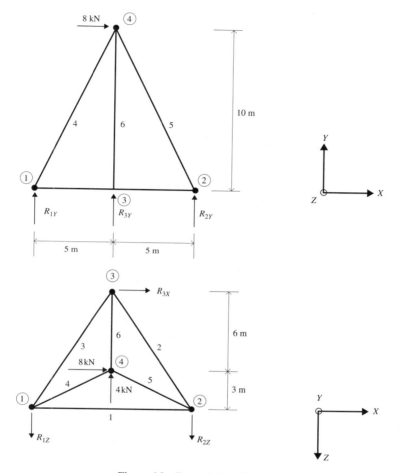

Figure 4.2 Example Problem 4.1.

Member 3 (joint 1 to joint 3):

$L_{3X} = 5$ $C_{3X} = 0.48564$

$L_{3Y} = 0$ → $L_3 = 10.29563$ → $C_{3Y} = 0$

$L_{3Z} = -9$ $C_{3Z} = -0.87416$

Member 4 (joint 1 to joint 4):

$L_{4X} = 5$ $C_{4X} = 0.43193$

$L_{4Y} = 10$ → $L_4 = 11.57584$ → $C_{4Y} = 0.86387$

$L_{4Z} = -3$ $C_{4Z} = -0.25916$

Member 5 (joint 2 to joint 4):

$L_{5X} = -5$ $\qquad$ $C_{5X} = -0.43193$

$L_{5Y} = 10$ $\quad \rightarrow \quad$ $L_5 = 11.57584$ $\quad \rightarrow \quad$ $C_{5Y} = 0.86387$

$L_{5Z} = -3$ $\qquad$ $C_{5Z} = -0.25916$

Member 6 (joint 3 to joint 4):

$L_{6X} = 0$ $\qquad$ $C_{6X} = 0$

$L_{6Y} = 10$ $\quad \rightarrow \quad$ $L_6 = 11.66190$ $\quad \rightarrow \quad$ $C_{6Y} = 0.85749$

$L_{6Z} = 6$ $\qquad$ $C_{6Z} = 0.51450$

The 12 equilibrium equations now can be obtained by summing forces in the X, Y and Z directions at each joint:

$$\Sigma_1 F_X \rightarrow S_1 C_{1X} + S_3 C_{3X} + S_4 C_{4X} = 0$$

$$\Sigma_1 F_Y \rightarrow S_1 C_{1Y} + S_3 C_{3Y} + S_4 C_{4Y} + R_{1Y} = 0$$

$$\Sigma_1 F_Z \rightarrow S_1 C_{1Z} + S_3 C_{3Z} + S_4 C_{4Z} + R_{1Z} = 0$$

$$\Sigma_2 F_X \rightarrow -S_1 C_{1X} + S_2 C_{2X} + S_5 C_{5X} = 0$$

$$\Sigma_2 F_Y \rightarrow -S_1 C_{1Y} + S_2 C_{2Y} + S_5 C_{5Y} + R_{2Y} = 0$$

$$\Sigma_2 F_Z \rightarrow -S_1 C_{1Z} + S_2 C_{2Z} + S_5 C_{5Z} + R_{2Z} = 0$$

$$\Sigma_3 F_X \rightarrow -S_2 C_{2X} - S_3 C_{3X} + S_6 C_{6X} + R_{3X} = 0$$

$$\Sigma_3 F_Y \rightarrow -S_2 C_{2Y} - S_3 C_{3Y} + S_6 C_{6Y} + R_{3Y} = 0$$

$$\Sigma_3 F_Z \rightarrow -S_2 C_{2Z} - S_3 C_{3Z} + S_6 C_{6Z} = 0$$

$$\Sigma_4 F_X \rightarrow -S_4 C_{4X} - S_5 C_{5X} - S_6 C_{6X} + 8 = 0$$

$$\Sigma_4 F_Y \rightarrow -S_4 C_{4Y} - S_5 C_{5Y} - S_6 C_{6Y} = 0$$

$$\Sigma_4 F_Z \rightarrow -S_4 C_{4Z} - S_5 C_{5Z} - S_6 C_{6Z} - 4 = 0$$

These linear equations can be solved using the computer program SEQSOLVE, as demonstrated in Chapter 3 for the analysis of a plane truss. Figure 4.3 shows a listing of the input data file for SEQSOLVE for this set of equations while Figure 4.4 shows the output from the program. Unknowns 1 through 6 are the member forces S_1 through S_6 while unknowns 7 through 12 are the reactive forces R_{1Y}, R_{1Z}, R_{2Y}, R_{2Z}, R_{3X} and R_{3Y} respectively. A quick check will show that the computed values for the reactions do satisfy the global equilibrium requirements for the truss. Therefore, the truss is stable since a set of member forces and reactive forces could be found which satisfied equilibrium.

METHOD OF JOINTS

It is also possible to analyze the previous truss by first solving for the reactive force components using the global equilibrium conditions for the truss and then work-

Method of Joints

```
Example Problem 4.1 - Space Truss Analysis by Simultaneous Equations
12
Non Zero Equation Coefficients
1,1,1.0,0
1,3,0.48564,0
1,4,0.43193,0
2,4,0.86387,0
2,7,1.0,0
3,3,-0.87416,0
3,4,-0.25916,0
3,8,1.0,0
4,1,-1.0,0
4,2,-0.48564,0
4,5,-0.43193,0
5,5,0.86387,0
5,9,1.0,0
6,2,-0.87416,0
6,5,-0.25916,0
6,10,1.0,0
7,2,0.48564,0
7,3,-0.48564,0
7,11,1.0,0
8,6,0.85749,0
8,12,1.0,0
9,2,0.87416,0
9,3,0.87416,0
9,6,0.51450,0
10,4,-0.43193,0
10,5,0.43193,0
10,6,0.0,0
11,4,-0.86387,0
11,5,-0.86387,0
11,6,-0.85749,0
12,4,0.25916,0
12,5,0.25916,0
12,6,-0.51450,1
Non Zero Equation Constants
10,-8.0,0
12,4.0,1
```

Figure 4.3 Example Problem 4.1, SEQSOLVE input.


```
Example Problem 4.1 - Space Truss Analysis by Simultaneous Equations
Solution of Equations

       1          -1.85183
       2           9.76184
       3          -6.71127
       4          11.83315
       5          -6.68837
       6          -5.18305
       7         -10.22230
       8          -2.80005
       9           5.77788
      10           6.80005
      11          -8.00000
      12           4.44442
```

Figure 4.4 Example Problem 4.1, SEQSOLVE output.

ing through the joints to solve for the member forces one at a time. This is similar to the Method of Joints described in the previous chapter for plane trusses, except that it is now being applied in three dimensions. The following operations can be used to analyze this truss by this procedure:

Summing moments about a line parallel to the X axis through joints 1 and 2 will result in an equation with R_{3Y} as the only unknown, which can be solved to give

$$\Sigma_{1\text{-}2} M_X \rightarrow R_{3Y} = 4.444 \text{ kN}$$

Next, summing moments about a line parallel to the Z axis through joint 1 will result in an equation in terms of R_{2Y} and R_{3Y}. Substituting the preceding value for R_{3Y} and solving for R_{2Y} gives

$$\Sigma_1 M_Z \rightarrow R_{2Y} = 5.778 \text{ kN}$$

The value of R_{1Y} now can be determined by summing forces in the Y direction:

$$\Sigma F_Y \rightarrow R_{1Y} = -10.222 \text{ kN}$$

Summing moments about a line parallel to the Y axis through the point at the intersection of the lines of action of the forces R_{2Z} and R_{3X} will result in an equation which can be solved for R_{1Z}. This point will be designated as point 0.

$$\Sigma_0 M_Y \rightarrow R_{1Z} = -2.8 \text{ kN}$$

Finally, the remaining two reactive forces can be determined by summing forces in the X and Z directions:

$$\Sigma F_X \rightarrow R_{3X} = -8.0 \text{ kN}$$

$$\Sigma F_Z \rightarrow R_{2Z} = 6.8 \text{ kN}$$

Note that these are the same values obtained from the solution of the complete set of simultaneous equations.

The member forces can now be determined by the following joint operations:

$$\Sigma_1 F_Y \rightarrow S_{4Y} = 10.222 \text{ kN} \rightarrow S_4 = S_{4Y}/C_{4Y} = 11.833 \text{ kN}$$

$$\Sigma_2 F_Y \rightarrow S_{5Y} = -5.778 \text{ kN} \rightarrow S_5 = S_{5Y}/C_{5Y} = -6.688 \text{ kN}$$

$$\Sigma_3 F_Y \rightarrow S_{6Y} = -4.444 \text{ kN} \rightarrow S_6 = S_{6Y}/C_{6Y} = -5.183 \text{ kN}$$

$$\Sigma_1 F_Z \rightarrow S_{3Z} = 5.867 \text{ kN} \rightarrow S_3 = S_{3Z}/C_{3Z} = -6.711 \text{ kN}$$

$$\Sigma_1 F_X \rightarrow S_1 = -1.852 \text{ kN}$$

$$\Sigma_2 F_Z \rightarrow S_{2Z} = -8.533 \text{ kN} \rightarrow S_2 = S_{2Z}/C_{2Z} = 9.762 \text{ kN}$$

The equations which have not been used to compute these member forces are $\Sigma_2 F_X$, $\Sigma_3 F_X$, $\Sigma_3 F_Z$, $\Sigma_4 F_X$, $\Sigma_4 F_Y$ and $\Sigma_4 F_Z$. These equations can be used to check the computed member forces. There are six unused joint equilibrium equations since six global equilibrium equations were used to compute the reactive force components. There are only twelve independent equilibrium equations for this truss.

Method of Joints

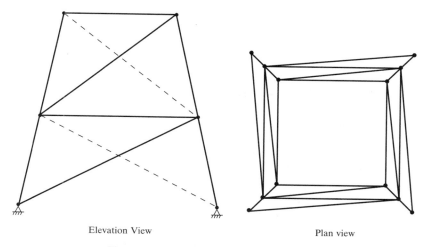

Figure 4.5 Statically determinate space truss.

Unfortunately, the majority of statically determinate space trusses cannot be analyzed by the Method of Joints. In addition, the Method of Sections is usually not of much value due to the difficulty of working in three dimensions. In many cases, the truss will have more than six independent reactive force components, with no obvious equations of condition which can be used in computing these values, or the truss will have more than three members attached to each joint so that it is not possible to solve for the member forces at each joint in turn. Figure 4.5 shows a four sided space truss which has both of these conditions. It has 12 joints, 24 members and 12 independent reactive force components since each of the four support joints is restrained against translation in all directions. According to this count, it is statically determinate. It is also stable, although this might not be obvious. However, any attempt to compute the reactive forces or the member forces will immediately run into difficulty. The 12 reactive force components cannot be computed using the six global equilibrium equations for the structure. The best that can be accomplished is to express six of the forces in terms of the other six. In addition, there are more than three members attached to each unrestrained joint. Although there are a number of specialized procedures which could be discussed here for solving various types of statically determinate space trusses which cannot be analyzed by the Method of Joints, it is not worth the effort due to the availability of low cost computers. The easiest approach for a truss of this type is to generate and solve the set of 3NJ joint equilibrium equations using an equation solving program such as SEQSOLVE. This process will always work as long as the truss is statically determinate and stable.

There are many structures which can be considered to act as a space truss. Two examples are electric transmission towers and long span roof structures in many large auditoriums and sports arenas. Unfortunately, most of these structures are statically indeterminate. The actual number of statically determinate space trusses which a structural engineer will encounter is probably very small.

COMPUTER PROGRAM SDSTRUSS

The computer program SDSTRUSS, which is supplied for this book, will analyze a statically determinate space truss of arbitrary geometry for the member forces and the reactive force components. The instructions for using the program are contained in the ASCII disk file SDSTRUSS.DOC. The program user can either generate the input file with an ASCII Editor using the format described in the program instructions, or the program SDSTDATA can be used to interactively create the file. The QBasic source code for the program is contained in the ASCII disk file SDSTRUSS.BAS for those readers who are interested in seeing how the joint equilibrium equations for a space truss can be generated and solved in a simple computer program.

The program SDSTRUSS is very similar to the program SDPTRUSS, which was discussed in the previous chapter. The analysis procedures in the two programs are essentially identical, except that in SDSTRUSS the three dimensional nature of the truss is considered. The locations of the individual joints in the mathematical model of the space truss are defined by their X, Y and Z coordinates in the global coordinate system for the structure. The locations of the members are defined by specifying their two end joints. The members forces are computed by solving a set of NM joint equilibrium equations which are generated by summing forces in the unrestrained global directions at each joint. After the member forces have been determined, the NR reactive force components are computed by summing forces in the restrained global directions at each support joint. The program is limited to the analysis of a space truss with 127 members for the same reason described in the previous chapter for the program SDPTRUSS. This should meet most educational requirements. The program can be used by the reader to verify the manual solutions for any of the suggested problems at the end of this chapter.

Figure 4.6 shows the listing of the input data file for SDSTRUSS for the space truss in Example Problem 4.1. A plot of the geometry of the mathematical model of the truss by the program PLOTSDST is shown in Figure 4.7, while the results of the analysis from SDSTRUSS are shown in Figure 4.8. The member forces and the reactive forces which are listed in the program output agree with the values computed previously.

The program PLOTSDST has the capability to generate a plot of the truss from essentially any viewing position in space. The view orientation is defined by the horizontal and vertical view angles, θ_H and θ_V, as defined in Figure 4.9. The positive direction of θ_H is measured by the right hand rotation rule about the global Y axis from the positive global Z axis, while the positive direction of θ_V is up-

Computer Program SDSTRUSS

```
Example Problem 4.1 - Analysis by SDSTRUSS
4,6,3,1
Joint Coordinates
1,0.0,0.0,0.0
2,10.0,0.0,0.0
3,5.0,0.0,-9.0
4,5.0,10.0,-3.0
Member Data
1,1,2
2,2,3
3,1,3
4,1,4
5,2,4
6,3,4
Support Restraints
1,0,1,1
2,0,1,1
3,1,1,0
Joint Loads
4,8.0,0.0,-4.0
```

Figure 4.6 Example Problem 4.1, SDSTRUSS input.

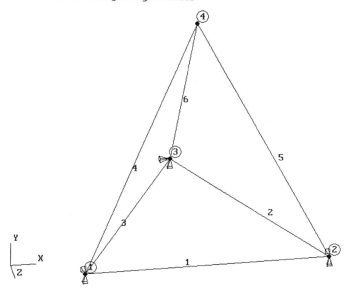

Figure 4.7 Example Problem 4.1, geometry plot from PLOTSDST.

ward from the global XZ plane. If both θ_H and θ_V are zero, the view corresponds to the projection of the truss on the global XY plane as viewed from the positive Z direction. Values of θ_H and θ_V of 90° and zero correspond to the projection on the YZ plane viewed from a positive X position while values of zero and 90° correspond to the projection on the XZ plane viewed from a positive Y position. The view along the Z axis might be considered to be a front view while the other two

Example Problem 4.1 - Analysis by SDSTRUSS

Member Forces (Tension Positive)

Member	Force
1	-1.852
2	9.762
3	-6.711
4	11.833
5	-6.688
6	-5.183

Reactions

Joint	RX	RY	RZ
1	0.000	-10.222	-2.800
2	0.000	5.778	6.800
3	-8.000	4.444	0.000

Figure 4.8 Example Problem 4.1, SDSTRUSS output.

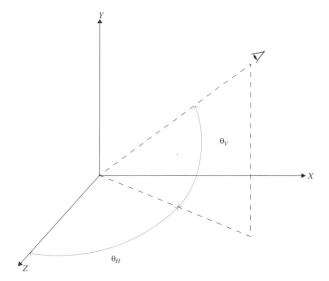

Figure 4.9 View angles.

are the side view and top view, respectively. The specific view shown in Figure 4.7 corresponds to a value of θ_H of $-10°$ and a value of θ_V of $30°$. The orientation of the global X, Y and Z axes for any view are shown in the lower left corner of the plot as a reference. The joint translation restraints are shown in the plot as small three dimensional wedges oriented in the positive X, Y and Z directions at the support joints. It might be necessary to try several different sets of orientation angles before a view is obtained which gives a good representation of the truss geometry.

SUGGESTED PROBLEMS

SP4.1 The joint coordinate in meters for the mathematical model of the truss shown in Figure SP4.1 are

Joint 1:	$X = 0$	$Y = 0$	$Z = 0$
Joint 2:	$X = 10$	$Y = 0$	$Z = -15$
Joint 3:	$X = 20$	$Y = 0$	$Z = 0$
Joint 4:	$X = 10$	$Y = 40$	$Z = -5$

The support joints are restrained against translation in all directions. The truss is subjected to the following active force components on joint 4.

$$W_{4X} = 15 \text{ kN}$$
$$W_{4Y} = -40 \text{ kN}$$
$$W_{4Z} = 0$$

Develop a set of 3NJ simultaneous equations with the member forces and the reactive forces as the unknowns. Solve the equations using the program SEQSOLVE. Verify the results of the analysis with the program SDSTRUSS.

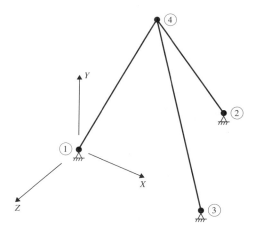

Figure SP4.1

SP4.2 The disk file SDSTRUSS.DST, which is supplied for this book, is a sample data file for a statically determinate space truss for the program SDSTRUSS. A plot of the geometry of the truss can be obtained with the program PLOTSDST. The coordinates of the joints in inches, and the support restraints and the joint forces in kips are listed in the data file.

(a) Develop a set of 3NJ simultaneous equations for this truss with the member forces and the reactive forces as the unknowns. Solve the equations using the program SEQSOLVE.

(b) Solve for the reactive forces using the global equations of equilibrium of the truss. Compute the member forces using the Method of Joints after the reactive forces have been computed.
Hint: Start with either joint 3 or joint 4. Work with the global components of the member forces during the computations.

(c) Verify the analysis with the program SDSTRUSS.

SP4.3 and **SP4.4**

(a) Determine the member forces and the reactive forces for the statically determinate space trusses shown in Figures SP4.3 and SP4.4.

(b) Verify each analysis with the program SDSTRUSS.

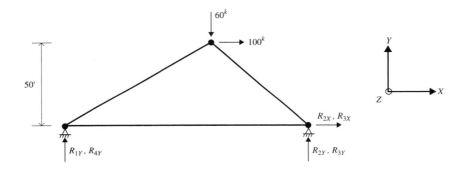

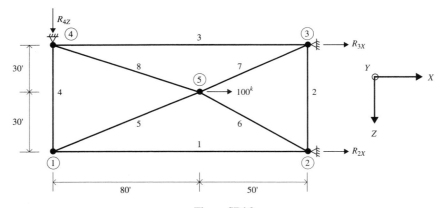

Figure SP4.3

Suggested Problems

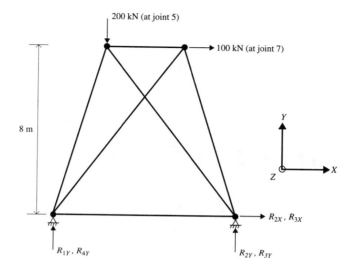

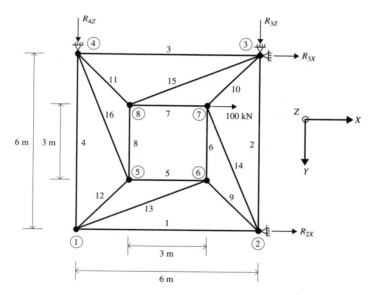

Figure SP4.4

5

Analysis of Statically Determinate Beams

DEFINITION OF A BEAM

In the previous two chapters, the discussion was limited to the analysis of two and three dimensional trusses. The mathematical models for these structures consisted of a set of joints connected by straight members, which had the specific property that, if they were subjected only to concentrated active and reactive forces at the joints, only longitudinal forces would be transmitted to the ends of the members by the joints. On the other hand, if the active and reactive loads acting on a structure are not restricted to concentrated forces on the joints, and if the members are attached to the joints such that transverse forces and moments can be transmitted to the ends of the members by the joints, then the structure is defined to be a frame. In this chapter, we will consider a specific type of frame in which all of the joints and members lie in a straight line. This type of structure is usually called a *beam*. The individual members in the mathematical model of the structure will be referred to as *beam members*. For discussion purposes, this type of structure might be considered to be a *one dimensional frame*, even though the loads on the structure may act in more than one dimension, since only one dimension is required to define any position along the length of the structure. We will consider the analysis of two and three dimensional frames in later chapters. As in the previous chapters, we will only be concerned in this chapter with the analysis of beams which are statically determinate and stable. The static determinacy and stability of the beam can be determined by comparing the number of independent reactive quantities to the number of independent equations of equilibrium, as described previously in Chapter 2.

AXIAL FORCE, SHEAR FORCE AND BENDING MOMENT

One of the primary differences between a beam member and a truss member is the type of loads which can be transmitted to any cross section of the member by the external loads acting on the structure. In a truss member, the cross section at any point along the length of the member is subjected only to a resultant axial force acting through the centroid of the cross section. The resultant loads acting on any cross section in a beam member are much more complex due to the type of active and reactive loads which can act on this type of structure. To demonstrate the nature of the loads which can act on a cross section, consider the typical simply supported beam shown in Figure 5.1a which is subjected to a general set of active and reactive loads. The global right hand XYZ coordinate system will be oriented with the positive X axis extending along the length of the beam toward the right with the positive Y axis up. The local right hand xyz coordinate system for any cross section of any member in the mathematical model of the beam will be located with the x axis passing through the centroid of the cross section and the x, y and z axes parallel and in the same positive directions as the global X, Y and Z axes. We will assume that the depth of the beam is small compared to its length, so that all active and reactive forces can be assumed to act on the longitudinal cen-

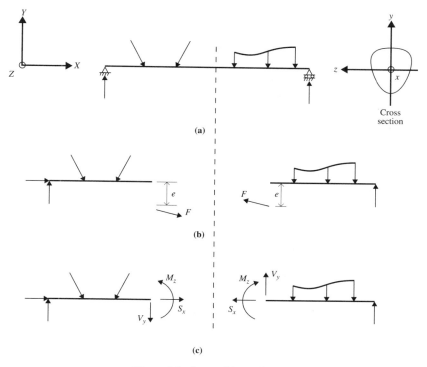

Figure 5.1 Internal beam loads.

ter line of the beam. Therefore, for illustrative purposes, the beam will be shown by a single horizontal line. If we cut the beam into two segments and consider the free body diagrams of each segment, we will find that a resultant inclined force F at an eccentricity e from the center line of the beam will be required on the end cross sections at the cut to hold the two segments in equilibrium, as shown in Figure 5.1b. The directions of the inclined forces on the ends of the left and right segments will be equal and opposite. These forces represent the actions that are transmitted through the cross section which joins the left and right segments in the beam. Although it would be possible to work directly with this eccentric inclined force, the analysis of the beam can be greatly simplified by transforming it into an equivalent set of forces, S_x and V_y, which act through the centroid of the cross section along the local x and y axes, and a moment M_z which acts about the local z axis, as shown in Figure 5.1c. The force S_x corresponds to the *axial force* on the cross section, while the force V_y and the moment M_z are known as the *shear force* and the *bending moment* respectively.

The magnitudes and the directions of S_x, V_y and M_z can be determined by using the three global equations of equilibrium for either the left segment or the right segment of the beam. When these calculations are performed, the signs for all external active and reactive loads on the beam will be defined by their directions with respect to the global XYZ coordinate system for the beam. According to the right hand rule sign convention, which was described in Chapter 1, this will mean that active and reactive global X and Y force components will be positive if they act to the right and up respectively, while moments about the global Z axis will be positive if they are counterclockwise. However, if we try to apply this sign convention to the internal cross section forces S_x and V_y and the internal cross section moment M_z we encounter difficulty since the forces and moments are in opposite directions on each segment of the beam. For analysis purposes, it is convenient to establish a unique sign convention for these internal section loads, as shown by the directions of the forces and moments on the small segment of a beam in Figure 5.2. The pair of longitudinal forces, the pair of transverse forces and the pair of moments in this figure are all considered to be positive. Therefore, for this sign convention: the axial force S_x is positive if it acts toward the right on the right end of a left segment or toward the left on the left end of a right segment; the shear force V_y is positive if it acts down on the right end of a left segment or up on the left end of a right segment; and the bending moment M_z is positive if it acts counterclockwise on the right end of a left segment or clockwise on the left end of a right segment. This is usually known as the *designer sign con ention*. Note that a positive axial force on either a left or a right segment corresponds to tension in the beam, which is consistent with the sign convention established previously for the axial forces in truss members. By using this sign convention, it is possible to work with either the left or the right segments of a beam to compute S_x, V_y and M_z at any cross section without any confusion as to the signs to be assigned to the computed quantities. Of course, when applying the global equilibrium equations to a segment, the signs used for S_x, V_y and M_z on either segment

Axial Force, Shear Force and Bending Moment

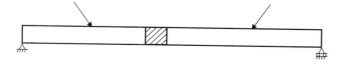

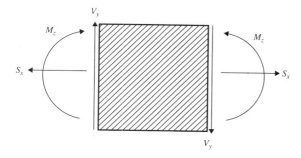

Figure 5.2 S, V, and M sign convention.

must correspond to the global sign convention for the structure. The designer sign convention is only valid for expressing the directions of S_x, V_y and M_z on an internal cross section of a beam.

Example Problem 5.1

Compute the axial force, the shear force and the bending moment on the cross section at the point which is 12 meters from the left end of the beam shown in Figure 5.3a. The mathematical model for the beam, which consists of four joints and three members, is shown in Figure 5.3b. The rules for developing this mathematical model are the same as those discussed in Chapter 2. The origin of the global coordinate system is at the left end of the beam. The reactions for the beam can be computed by resolving the inclined force at joint 2 into its X and Y components and then applying the global equilibrium equations to the overall mathematical model. During these calculations, the distributed force can be considered to be equivalent to a 10 kN downward force acting at 10 meters from the right end of the beam. The specific operations which can be used to compute the reactions are:

$$\Sigma_1 M_Z \rightarrow R_{4Y} = 11.6 \text{ kN} \uparrow$$

$$\Sigma F_Y \rightarrow R_{1Y} = 10.4 \text{ kN} \uparrow$$

$$\Sigma F_X \rightarrow R_{1X} = 9 \text{ kN} \leftarrow$$

We can now cut the beam into left and right segments at the point 12 meters from the left end. The free body diagrams for each segment are shown in Figure 5.3c. The quantities S_x, V_y and M_z on the end cross sections for each segment are shown as positive quantities, according to the designer sign convention The computations for S_x, V_y and M_z can be performed using either the left or the right segments of the beam by summing forces in the X direction, summing forces in the Y direction and summing moments about Z at the point 12 meters from the left end. We will perform the analysis for both segments here to illustrate the analysis process.

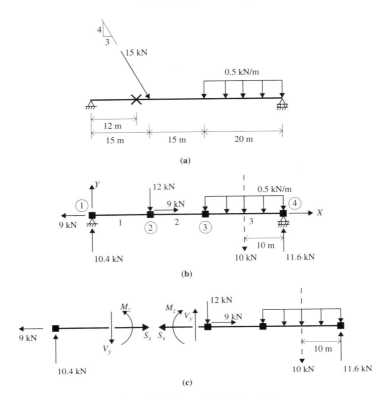

Figure 5.3 Example Problem 5.1.

By using the free body diagram for the left segment:

$\Sigma F_X \rightarrow -9 + S_x = 0 \rightarrow S_x = 9$ kN

$\Sigma F_Y \rightarrow 10.4 - V_y = 0 \rightarrow V_y = 10.4$ kN

$\Sigma_{X=12} M_Z \rightarrow -(10.4)(12) + M_z = 0 \rightarrow M_z = 124.8$ kN-m

By using the free body diagram for the right segment:

$\Sigma F_X \rightarrow -S_x + 9 = 0 \rightarrow S_x = 9$ kN

$\Sigma F_Y \rightarrow V_y - 12 - 10 + 11.6 = 0 \rightarrow V_y = 10.4$ kN

$\Sigma_{X=12} M_Z \rightarrow -M_z - (12)(3) - (10)(28) + (11.6)(38) = 0 \rightarrow M_z = 124.8$ kN-m

As expected, the same results are obtained for either segment. Since the computed values of S_x, V_y and M_z are all positive, they act in the directions shown on the free body diagrams for each segment. If negative values had been obtained for any of these quantities, it would mean that that particular quantity would act in the opposite direction. It is suggested that S_x, V_y and M_z be assumed to act in the positive directions during the analysis rather than trying to guess their actual directions beforehand.

Axial Force, Shear Force and Bending Moment

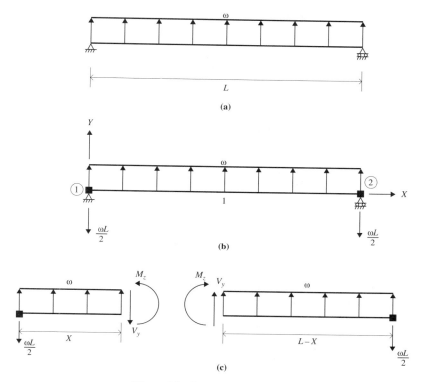

Figure 5.4 Example Problem 5.2.

This will eliminate any confusion concerning the meaning of the signs for the computed quantities.

Example Problem 5.2

Derive an algebraic expression for the shear force and the bending moment at any cross section for a simply supported beam subjected to a positive uniform distributed force ω, as shown in Figure 5.4a. The mathematical model for the beam is shown in Figure 5.4b. The origin of the global coordinate system is located at the left end of the beam. A simple set of calculations, using the global equations of equilibrium, will show that the vertical reactive force components at each end of the beam are equal to $\omega L/2$ acting down and the horizontal reactive force component at the left pinned support is zero.

The free body diagrams for the left and right segments of the beam, for a cut at a position X along its length, are shown in Figure 5.4c. The axial force S_x is not shown since it is obvious that it will be zero at any point along the beam. As in the previous example problem, we can develop the desired expressions for V_y and M_z using either the left or right segments by summing forces in the Y direction and summing moments about the point at a distance X from the left end.

By using the free body diagram of the left segment:

$$\Sigma F_Y \rightarrow -\frac{\omega L}{2} + (\omega)(X) - V_y = 0 \rightarrow V_y = \omega X - \frac{\omega L}{2}$$

$$\Sigma_X M_Z \rightarrow \left(\frac{\omega L}{2}\right)(X) - (\omega)(X)\left(\frac{X}{2}\right) + M_z = 0 \rightarrow M_z = \frac{\omega X^2}{2} - \frac{\omega L X}{2}$$

By using the free body diagram of the right segment:

$$\Sigma F_Y \rightarrow V_y + (\omega)(L - X) - \frac{\omega L}{2} = 0 \rightarrow V_y = \omega X - \frac{\omega L}{2}$$

$$\Sigma_X M_Z \rightarrow -M_z + (\omega)(L - X)\left(\frac{L - X}{2}\right) - \left(\frac{\omega L}{2}\right)(L - X) = 0 \rightarrow M_z = \frac{\omega X^2}{2} - \frac{\omega L X}{2}$$

The same results are obtained for either segment as expected. It appears that the analysis for the left segment results in fewer algebraic manipulations Therefore, it might be worthwhile to spend a few minutes before blindly jumping into any problem to determine which of the segments on either side of a cut will result in the simplest calculations to arrive at the desired result. Of course, if both segments are used, the two sets of results will give an automatic check on the computations.

These expressions now can be used to compute the magnitudes of the shear force and bending moment at any point in the beam for any numerical value for the uniform distributed force ω. If ω acts up, a positive value would be used, while if ω acts down, the value would be negative. An engineer must be careful when using expressions of this type to ensure that all quantities have consistent units. That is, if ω is expressed in kN/m, then both L and X must be expressed in meters, while if ω is expressed in kips/ft, then L and X must be in feet. Of course, X can also be expressed symbolically in terms of L, such as 0.3L, in which case algebraic expressions would be obtained for V_y and M_z at that point in the beam.

Example Problem 5.3

Derive algebraic expressions for the shear force and the bending moment at any cross section for a simply supported beam which is subjected to a partial uniform force, as shown in Figure 5.5a. The mathematical model for the beam is shown in Figure 5.5b. The origin of the global coordinate system is located at the left end of the beam. The vertical reactive force components at the left and right ends are 3ωL/8 and ωL/8, respectively, acting down. The horizontal reactive force component at the left pinned support is zero.

The primary difference between the beam in this problem and the beam in the previous problem is that neither V_y or M_z can be expressed as a single algebraic expression over the entire length of the beam. Due to the nature of the loading, it is necessary to develop one set of expressions for the left zone corresponding to $0 \leq X \leq L/2$ and another set of expressions for the right zone corresponding to $L/2 \leq X \leq L$. Since we may use either the left or the right segments of the beam to compute the expressions for V_y and M_z for any location X on the beam, we will use the particular segment for each zone which results in the least amount of algebra.

Axial Force, Shear Force and Bending Moment

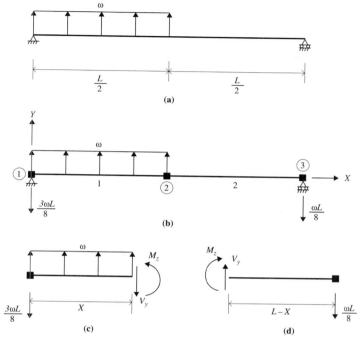

Figure 5.5 Example Problem 5.3.

Left zone: $0 \leq X \leq L/2$. Figure 5.5c shows the free body diagram of the left segment of the beam for a cut at any position X in this zone. We can now develop expressions for V_y and M_z by the following operations:

$$\Sigma F_Y \rightarrow -\frac{3\omega L}{8} + (\omega)(X) - V_y = 0 \rightarrow V_y = \omega X - \frac{3\omega L}{8}$$

$$\Sigma_X M_Z \rightarrow \left(\frac{3\omega L}{8}\right)(X) - (\omega)(X)\left(\frac{X}{2}\right) + M_z = 0 \rightarrow M_z = \frac{\omega X^2}{2} - \frac{3\omega L X}{8}$$

Right zone: $L/2 \leq X \leq L$. Figure 5.5d shows the free body diagram of the right segment of the beam for a cut at any position X in this zone. The expressions for V_y and M_z for this zone can be obtained by the following operations:

$$\Sigma F_Y \rightarrow V_y - \frac{\omega L}{8} = 0 \rightarrow V_y = \frac{\omega L}{8}$$

$$\Sigma_X M_Z \rightarrow -M_z - \left(\frac{\omega L}{8}\right)(L - X) = 0 \rightarrow M_z = \frac{\omega L X}{8} - \frac{\omega L^2}{8}$$

The reader can verify these expressions by using the free body diagram for the other segment for each zone on the beam.

We will not be concerned with the stresses which are produced in the beam by S_x, V_y and M_z in this book. The magnitudes of the stresses and their allowable values are better left for a book dealing with structural design. In many cases, the design specifications which are being used dictate how the stresses are to be computed for specific types of structures or loadings. Our only interest here will be in computing the internal loads on the cross sections for use in the design process.

RELATIONSHIP BETWEEN LOAD, SHEAR FORCE AND BENDING MOMENT

A relationship between the load acting on a beam and the shear force and bending moment on any cross section can be developed by considering the equilibrium of a small segment of a beam subjected to a positive (i.e.; upward) distributed force q_y. Figure 5.6a shows the location of the small segment of the beam while Figure 5.6b shows a free body diagram of the segment. The definitions of the various quantities shown in this figure are as follows: X defines the location of the

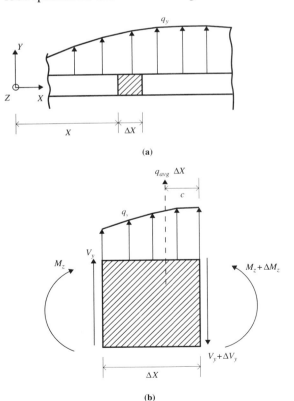

Figure 5.6 Shear force and bending moment variation.

Relationship Between Load, Shear Force and Bending Moment

left end of the segment; ΔX is the length of the segment; q_{avg} is the average magnitude of the distributed force over the length ΔX; c is the distance to the resultant of the distributed force on the segment from the right end of the segment; ΔV_y is the change in the shear force in the beam over the length ΔX; and ΔM_z is the change in the bending moment in the beam over the length ΔX. We will not consider any axial forces on the segment since there is no horizontal force component on the beam.

A relationship between q_y and V_y can be obtained by summing all of the forces acting on the segment in the vertical direction:

$$V_y + (q_{avg})(\Delta X) - (V_y + \Delta V_y) = 0 \tag{5.1}$$

which, upon canceling the positive and negative values of V_y and dividing by ΔX, gives

$$\frac{\Delta V_y}{\Delta X} = q_{avg} \tag{5.2}$$

If we now take the limit of this expression as the length of the segment approaches zero, the quantity q_{avg} will become equal to the value of q_y at the left end of the segment, while the quantity $\Delta V_y/\Delta X$ will become equal to the derivative dV_y/dX. Therefore, the resulting expression is

$$\frac{dV_y}{dX} = q_y \tag{5.3}$$

The physical interpretation of this expression is that the rate of change of the shear force V_y at any location X on the beam is equal to the magnitude of the distributed force q_y acting on the beam at that point. If q_y is acting up, the rate of change of the shear force will be positive, while if q_y is acting down, the rate of change of the shear force will be negative. Note that the units on each side of the equation are the same since both quantities have the units of force over length.

A relationship between V_y and M_z can be obtained by summing moments at the right end of the segment:

$$-M_z - (V_y)(\Delta X) - (q_{avg})(\Delta X)(c) + (M_z + \Delta M_z) = 0 \tag{5.4}$$

which, upon canceling the positive and negative values of M_z and dividing by ΔX, gives

$$\frac{\Delta M_z}{\Delta X} = V_y + q_{avg} \cdot c \tag{5.5}$$

Taking the limit of this expression as the length of the segment approaches zero, and noting that the distance c goes to zero as ΔX goes to zero, gives

$$\frac{dM_z}{dX} = V_y \tag{5.6}$$

Therefore, the rate of change of the bending moment at any point X on the beam is equal to the magnitude of the shear force at that point. If the shear force is positive, the rate of change of the bending moment will be positive, while if the shear force is negative, the rate of change of the bending moment will be negative.

Equations (5.3) and (5.6) can also be combined to obtain another expression which directly relates the distributed force at any point along the beam to the bending moment at that point. Taking the derivative of both sides of Eq. (5.6) with respect to X and then substituting Eq. (5.3) for the right side gives

$$\frac{d^2 M_z}{dX^2} = q_y \qquad (5.7)$$

Therefore, the rate of change of the rate of change of the bending moment is equal to the distributed force q_y at any point on the beam.

The relationships between the load, the shear force and the bending moment shown in Eqs. (5.3) and (5.6) can be used to compute the shear force and bending moment acting on any cross section along the length of a beam without the necessity to perform an equilibrium analysis for the segment of the beam to the right or left of the cross section. If the loading on the beam is expressed algebraically, an algebraic expression for the shear force and bending moment at any position X can be determined by direct symbolic integration. If the loading on the beam is expressed as numerical values, the shear force and bending moment can be computed at any point by performing a numerical integration of the equations.

V AND M EXPRESSIONS BY DIRECT INTEGRATION

An expression for the shear force V_y at any position in a beam can be developed by symbolically integrating Eq. (5.3). The first step is to transform the equation into the form

$$dV_y = q_y dX \qquad (5.8)$$

which, upon integrating both sides, gives

$$V_y = \int q_y dX + C_1 \qquad (5.9)$$

where C_1 is a constant. In a similar manner, an expression for the bending moment M_z can be found by transforming Eq. (5.6) into the form

$$dM_z = V_y dX \qquad (5.10)$$

which, upon integrating, and substituting the expression in Eq. (5.9), gives

$$M_z = \int V_y dX + C_2 = \iint q_y dX + C_1 X + C_2 \qquad (5.11)$$

V and M Expressions by Direct Integration

The constants C_1 and C_2 can be determined from the known boundary conditions at the ends of the portion of the beam for which the expressions for V_y and M_z are being developed. For example, at a pinned end the moment M_z will be zero, while at a free end both V_y and M_z will be zero. There will always be exactly as many known boundary conditions for a stable statically determinate beam as are required to determine the unknown constants.

Example Problem 5.4

Determine algebraic expressions for V_y and M_z for the simply supported beam shown previously in Figure 5.4a. The first step in the analysis is to develop an expression for q_y which represents the distributed force at any point on the beam. For this beam, the distributed force is constant over the length of the beam. Therefore, this expression is

$$q_y = \omega$$

Performing the first integration gives the expression for V_y as

$$V_y = \omega X + C_1$$

while performing the second integration gives the expression for M_z as

$$M_z = \frac{\omega X^2}{2} + C_1 X + C_2$$

The constants C_1 and C_2 can be determined by using the moment boundary conditions at each end of the beam:

$$M_z = 0 \text{ at } X = 0 \rightarrow C_2 = 0$$

$$M_z = 0 \text{ at } X = L \rightarrow C_1 = -\frac{\omega L}{2}$$

By using these values for C_1 and C_2, the final expressions for V_y and M_z are

$$V_y = \omega X - \frac{\omega L}{2}$$

$$M_z = \frac{\omega X^2}{2} - \frac{\omega L X}{2}$$

Note that these expressions agree with the expressions obtained in Example Problem 5.2 using an equilibrium analysis of the beam.

The location of the maximum bending moment in the beam can be determined by setting the quantity dM_z/dX to zero and solving for X. However, from Eq. 5.6 we can see that this is equivalent to setting the shear force in the beam to zero. This leads to the conclusion that the maximum bending moment in the beam will occur at the point where the shear force is zero. Performing this operation gives

$$V_y = \omega X - \frac{\omega L}{2} = 0 \rightarrow X = \frac{L}{2}$$

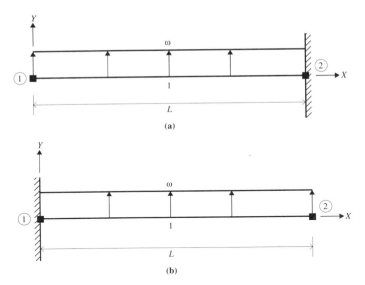

Figure 5.7 Cantilever Beams.

which shows that the maximum moment occurs at the center of the beam, as might be expected due to the symmetry of the loading and the vertical support conditions. Substituting this value for X into the expression for M_z results in

$$M_{z,\,max} = -\frac{\omega L^2}{8}$$

This is an expression which is worth remembering since a uniformly loaded, simply supported beam is very common, since the dead load of a beam usually corresponds to a downward uniform force over the length of the beam. The moment in the expression is negative due to the sign convention which is being used. If the uniform force were acting down, the moment would be positive.

It might be interesting at this point to see how the expressions for V_y and M_z will change with changes in the boundary conditions for the beam. To demonstrate the effect of the boundary conditions we will next consider the two cantilever beams whose mathematical models are shown in Figure 5.7.

Each beam is subjected to a positive uniform distributed load with the origin of the global coordinate system at the left end. The difference in the two beams is that the support conditions at the two ends are reversed. For each of these beams, the expressions for V_y and M_z, before the constants C_1 and C_2 are evaluated, will be the same as for the previous simply supported beam since the algebraic expressions for the loading on all of the beams are identical. However, the final expressions for V_y and M_z will be different due to the difference in the boundary conditions.
For the cantilever beam in Figure 5.7a:

$$V_y = 0 \text{ at } X = 0 \rightarrow C_1 = 0$$
$$M_z = 0 \text{ at } X = 0 \rightarrow C_2 = 0$$

V and M Expressions by Direct Integration

from which the final expressions for V_y and M_z are

$$V_y = \omega X$$

$$M_z = \frac{\omega X^2}{2}$$

For the cantilever beam in Figure 5.7b:

$$V_y = 0 \text{ at } X = L \rightarrow C_1 = -\omega L$$

$$M_z = 0 \text{ at } X = L \rightarrow C_2 = \frac{\omega L^2}{2}$$

from which the final expressions for V_y and M_z are

$$V_y = \omega X - \omega L$$

$$M_z = \frac{\omega X^2}{2} - \omega L X + \frac{\omega L^2}{2}$$

These different cases show that the variation of V_y and M_z over a beam is highly dependent upon the boundary conditions.

Example Problem 5.5

Determine algebraic expressions for V_y and M_z for the simply supported beam, which is subjected to a positive triangular distributed force, whose mathematical model is shown in Figure 5.8. The origin of the global coordinate system is located at the left end of the beam.

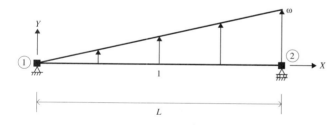

Figure 5.8 Example Problem 5.5.

The load q_y at any point on the beam can be expressed as

$$q_y = \frac{\omega X}{L}$$

which can be integrated to obtain the following expressions for V_y and M_z:

$$V_y = \frac{\omega X^2}{2L} + C_1$$

$$M_z = \frac{\omega X^3}{6L} + C_1 X + C_2$$

The constants C_1 and C_2 can be determined from the boundary conditions:

$$M_z = 0 \text{ at } X = 0 \rightarrow C_2 = 0$$

$$M_z = 0 \text{ at } X = L \rightarrow C_1 = -\frac{\omega L}{6}$$

from which the final expressions for V_y and M_z are

$$V_y = \frac{\omega X^2}{2L} - \frac{\omega L}{6}$$

$$M_z = \frac{\omega X^3}{6L} - \frac{\omega L X}{6}$$

Note that the shear force varies as the second power of X and the moment varies as the third power of X over the length of this beam, while they varied as the first power and the second power, respectively, for the beam with the uniform distributed force. The equation for the shear force V_y will always be one degree higher than the equation for the load q_y, and the equation for the bending moment M_z will always be two degrees higher. Also, note that the units of each term on the right side of the expression for V_y are identical as are the units of each term on the right side of the expression for M_z. Both of these conditions can be very useful in checking the results of the algebraic manipulations which are required to arrive at the final expressions.

Example Problem 5.6

Determine algebraic expressions for V_y and M_z for any point along the length of the beam shown previously in Figure 5.5a.

Since q_y cannot be expressed as one continuous algebraic expression over the length of the beam, it will be necessary to develop two sets of expressions for V_y and M_z for the left and right zones in the beam.

Left Zone: $0 \leq X \leq L/2$

$$q_y = \omega$$

$$V_y = \omega X + C_1$$

$$M_z = \frac{\omega X^2}{L} + C_1 X + C_2$$

Right Zone: $L/2 \leq X \leq L$

$$q_y = 0$$

$$V_y = C_3$$

$$M_z = C_3 X + C_4$$

Since we now have four constants to evaluate, we need four boundary conditions. Two of these conditions are

$$M_z = 0 \text{ at } X = 0 \text{ for the left zone} \rightarrow C_2 = 0$$

V and M Expressions by Direct Integration

and

$$M_z = 0 \text{ at } X = L \text{ for the right zone} \to C_3 L + C_4 = 0$$

The other two conditions which can be used are that the shear force and the bending moment must be continuous at the boundary between the two zones. Therefore, equating the two expressions for V_y in the left zone and the right zone at $X = L/2$ gives

$$\frac{\omega L}{2} + C_1 = C_3$$

while equating the two expressions for M_z in the left zone and the right zone at $X = L/2$ gives

$$\frac{\omega L^2}{8} + C_1 \frac{L}{2} + C_2 = C_3 \frac{L}{2} + C_4$$

We now have four algebraic equations containing the four constants C_1, C_2, C_3 and C_4. Solving these equations gives

$$C_1 = -\frac{3\omega L}{8}$$

$$C_2 = 0$$

$$C_3 = \frac{\omega L}{8}$$

$$C_4 = -\frac{\omega L^2}{8}$$

The final expressions for V_y and M_z in the two zones on the beam are as follows:
Left Zone: $0 \le X \le L/2$

$$V_y = \omega X - \frac{3\omega L}{8}$$

$$M_z = \frac{\omega X^2}{2} - \frac{3\omega L X}{8}$$

Right Zone: $L/2 \le X \le L$

$$V_y = \frac{\omega L}{8}$$

$$M_z = \frac{\omega L X}{8} - \frac{\omega L^2}{8}$$

These expressions agree with those obtained in Example Problem 5.3 for the same beam by an equilibrium analysis.

We can see from the previous examples that the complexity of the algebra involved in developing expressions for V_y and M_z increases with the complexity of the loading on the beam. In some cases it might be easier to develop the desired expressions by performing an equilibrium analysis as in Example Problems 5.2 and 5.3, while in other cases it might be easier to develop the expressions by integration. It has been the experience of the author that an equilibrium analysis usually leads to fewer algebraic manipulations, thus reducing the chances of errors in the solution, particularly when more than two constants of integration must be evaluated.. Both procedures should give exactly the same results.

Equations for V_y and M_z for beams with various support conditions and loadings are available in a number of structural engineering design manuals (American Institute of Steel Construction, 1986, 1989) and reference books (Young, 1989). Any engineer using these published equations must be very careful to ensure that the sign convention for the equations is understood. Unfortunately, there is no universal sign convention which is used in all publications.

SHEAR FORCE AND BENDING MOMENT DIAGRAMS

Although the algebraic expressions for V_y and M_z can be used to compute the magnitudes of the shear force and bending moment at any point on a beam, it is not always obvious how these quantities vary over the length of the beam by merely looking at the equations. In particular, it is not obvious where the maximum values are located or where sign changes might occur. A simple process for eliminating these problems is to generate plots of V_y and M_z over the length of the beam. These plots are usually called *shear force diagrams* and *bending moment diagrams*. One way to obtain these diagrams is to plot the algebraic expressions for V_y and M_z along the beam. For example, Figure 5.9 shows the shear force and bending moment diagrams for several of the beams which were analyzed in the previous example problems. The diagrams are plotted for the specific case of a downward distributed force of 2 kN/m (i.e.; $\omega = -2$ kN/m) and a span length of 40 meters for each beam. A downward distributed force has been used since the majority of the beams which will be encountered in the day-to-day operations in most engineering design offices will be subjected to downward gravity loads. As we can see from these plots, the magnitudes and the variations of V_y and M_z are very different for each of the beams. The shape of the curves and the magnitudes of the ordinates are highly dependent on both the support conditions for the beam and the type of loading.

If algebraic expressions are available for V_y and M_z, it is a simple task to generate the shear force and bending moment diagrams for any beam. However, the derivation of these expressions, either by performing an equilibrium analysis or by direct integration, can be very tedious, particularly for cases where the loading cannot be expressed as one continuous algebraic expression over the entire length of the beam. Fortunately, it is possible to develop the shear force and bending mo-

Shear Force and Bending Moment Diagrams

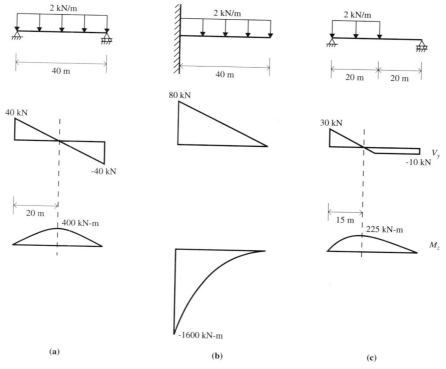

Figure 5.9 Shear force and bending moment diagrams.

ment diagrams without deriving any algebraic equations. The process consists of using a physical interpretation of the general relationships between q_y, V_y and M_z.

Before discussing this process, it is first necessary to develop two additional relationships between q_y, V_y and M_z from two equations which were developed previously. The first relationship will be developed by integrating both sides of Eq. (5.8) between two points defined by the coordinates X_1 and X_2 on a beam, where $X_2 > X_1$.

$$\Delta V_{(1/2), y} = V_{2y} - V_{1y} = \int_{X_1}^{X_2} q_y \, dX \tag{5.12}$$

while the second relationship can be obtained by performing the same operations on Eq. (5.10)

$$\Delta M_{(1/2), z} = M_{2z} - M_{1z} = \int_{X_1}^{X_2} V_y \, dX \tag{5.13}$$

The quantity $\Delta V_{(1/2), y}$ represents the change in the shear force V_y over the portion of the beam from X_1 to X_2, while the quantity $\Delta M_{(1/2), z}$ represents the change in the bending moment M_z. Using these relationships and the relationships shown previously in Eqs. (5.3) and (5.6), we can make the following observations concerning the physical properties of shear force and bending moment diagrams.

> Equation (5.3): The slope of the shear force diagram at any point is equal to the magnitude of the distributed force on the beam at that point. If the distributed force is up, the slope will be positive, while if the distributed force is down, the slope will be negative.
>
> Equation (5.12): The change in the shear force between any two points is equal to the area of the transverse distributed force between the points. This area will be equal to the total transverse force acting on the beam over that length. An upward force corresponds to a positive area while a downward force corresponds to a negative area.
>
> Equation (5.6): The slope of the bending moment diagram at any point is equal to the magnitude of the shear force at that point on the beam. If the shear force is positive, the slope will be positive, while if the shear force is negative, the slope will be negative.
>
> Equation (5.13): The change in the bending moment between any two points is equal to the area under the shear force diagram between the two points. Positive shear force corresponds to a positive area while negative shear force corresponds to a negative area.

We can now use these properties to develop the shear force and bending moment diagrams for any statically determinate beam without developing any algebraic equations for V_y and M_z.

Example Problem 5.7

> To demonstrate this process, we will develop the shear force and bending moment diagrams shown previously in Figure 5.9c, for the simply supported beam subjected to a partially uniform distributed force, by performing the following operations:
>
> 1. Compute the reactions for the beam and draw the load diagram. The load diagram is a free body diagram of the mathematical model of the beam which shows all of the active and reactive loads, as shown in Figure 5.10a. For this beam, these loads consist of a combination of concentrated and distributed forces.
> 2. The shear force diagram now can be constructed by starting at a point to the left of the left end of the beam, where the shear force is zero, and then computing the ordinates at each point along the beam by summing the area corresponding to the distributed forces in the load diagram up to that point. The specific operations are as follows:
>   (a) Add in the area corresponding to the concentrated reactive force at joint 1. This immediately leads to a problem since all of the previously developed relationships only consider the effect of a distributed force. However, the solution to this problem is very simple since concentrated forces cannot ex-

Shear Force and Bending Moment Diagrams

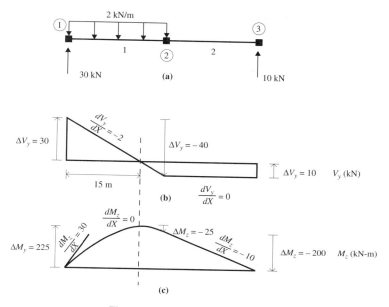

Figure 5.10 Example Problem 5.7.

ist on a real structure. They are merely mathematical quantities which are used during the analysis of the mathematical model of a structure. In reality, all forces acting on a structure are distributed over some finite length. Therefore, a concentrated force is actually a very high intensity distributed force acting over a very small distance. For our purposes here, we will consider a concentrated force to correspond to a distributed force of infinite intensity acting over a zero length, where the area under the load diagram at the point of application of the concentrated force will be equal to the magnitude of the force. Therefore, the change in the shear force at joint 1 will be equal to $+30$ kN and the slope of the shear force diagram at this point will be infinite. In other words, there will be an upward step of 30 kN in the shear force diagram at this point, as shown in Figure 5.10b.

(b) Add in the area corresponding to the distributed force from a point just to the right of joint 1 to joint 2. Since the only load on the beam over this portion is a uniform distributed force of magnitude -2 kN/m, the area corresponding to the distributed force, which is equal to the change in the shear force over this distance, will be -40 kN. Adding this ΔV_y to the previous value of V_y will result in a shear force of -10 kN at joint 2. The shear force will decrease at a uniform rate from joint 1 to joint 2 since q_y is constant (i.e.; the slope dV_y/dX will be equal to -2 kN/m over this distance). A simple calculation will show that the shear force is zero at a distance of 15 meters to the right of joint 1. The location of this point will be very useful when constructing the bending moment diagram.

(c) Add in the area corresponding to the distributed force from joint 2 to a point just to the left of joint 3. Since there is no distributed force over this portion of the beam, the area is zero and the shear force will have a constant value of −10 kN. Note that there is a kink in the shear diagram at joint 2 since the slope just to the left of this joint is −2 kN/m while the slope just to the right is zero. However, the ordinates are continuous since there is no concentrated force acting on the beam at this point.

  (d) Add in the area corresponding to the concentrated reactive force at joint 3. By using the same argument described previously for the concentrated force at joint 1, the area will be +10 kN. Adding this ΔV_y to the value of the shear force just to the left of the support results in a vertical step in the shear force diagram and a final shear force of zero just to the right of joint 3.

3. The bending moment diagram now can be constructed by starting at the left end of the beam, where the bending moment is zero, and then computing the ordinates at each point along the beam by summing the area under the shear force diagram. The specific operations are as follows:

  (a) Add in the area under the shear force diagram from joint 1 to the point where the shear force is zero. This point has been chosen since the ordinates of the shear force diagram are negative to the right of this point, which would result in a decreased value for ΔM_z. Using this point will allow us to compute the maximum bending moment in the beam. The area under the shear force diagram over this length is +225 kN m. Adding this ΔM_z to the previous value of zero at joint 1 results in a maximum positive bending moment of 225 kN m at the point 15 meters to the right of joint 1. Since the positive ordinates of the shear force diagram decrease with increasing X, the area under the bending moment diagram will increase at a decreasing rate. Therefore, the ordinates of the bending moment diagram vary, as shown in Figure 5.10c. The shape of this curve is further verified by Eq. (5.3), from which we can determine that the slope of the bending moment diagram is +30 kN m/m at joint 1 and then decreases continuously to a value of zero at the point of maximum moment.

  (b) Add in the area under the shear force diagram from the point of maximum bending moment to joint 2. The area under the shear force diagram over this length is −25 kN m. Adding this ΔM_z to the previous bending moment results in a bending moment of 200 kN m at joint 2. From Eq. (5.3), we can see that the slope of the bending moment diagram will be −10 kn m/m at this point. The bending moment will decrease at an increasing rate over this length.

  (c) Add in the area under the shear force diagram from joint 2 to joint 3. The area will be −200 kN m. Adding this ΔM_z to the bending moment at joint 2 results in a bending moment of zero at joint 3 as expected. The bending moment diagram will have a constant slope of −10 kN m/m over this length since the shear force is constant. The linear portion of the diagram to the right of joint 2 is tangent to the curved portion of the diagram to the left of joint 2. Therefore, both the ordinates and the slopes of the bending moment diagram are continuous at joint 2.

This same process can be used for any statically determinate beam. Although it took a lot of words to explain the process, the actual calculations can be performed very quickly. In some cases, the only limiting factor is how fast the analyst can draw. In fact, the whole process can be greatly simplified if you just do what the load diagram and the shear force diagram instruct you to do. For example, the shear force diagram can be developed for the previous example problem by merely following the action of the vertical forces shown in the load diagram.

1. Go up 30 kN at joint 1
2. Go down at a slope of −2 kN/m from joint 1 to joint 2. The change in the ordinate will be −40 kN.
3. Do nothing from joint 2 to joint 3
4. Go up 10 kN at joint 3

The bending moment diagram can be generated in a similar manner by following the instructions of the shear force diagram.

1. Start with a slope of 30 kN m/m at joint 1 and go up with a decreasing slope to a slope of zero at a point 15 meters to the right of joint 1. The change in the ordinate will be +225 kN m.
2. Go down with an increasing negative slope to a slope of −10 kN m/m at joint 2. The change in the ordinate will be −25 kN m.
3. Go down with a constant slope of −10 kN m/m to joint 3. The change in the ordinate will be −200 kN m.

By using this process, even very complex loadings can be handled with ease. Of course, as with anything new, practice makes perfect.

One of the nice properties of shear force and bending moment diagrams is that they are self checking. This is demonstrated by the three cases shown in Figure 5.11 for a simply supported beam subjected to a single concentrated force. The three cases are as follows:

Case I (Figure 5.11a). An error was made during the computation of the reactions so that ΣF_Y is not satisfied. When an attempt is made to generate the shear force diagram for this case, it will be found that the final value at the right end of the beam is −2 kips rather than zero. Since the shear force diagram does not close, there is a force equilibrium error in the reactions.

Case II (Figure 5.11b). An error was made during the computation of the reactions so that ΣM_Z is not satisfied. Force equilibrium is satisfied. It will be found for this case that the shear force diagram will close, but when an attempt is made to generate the bending moment diagram, the bending moment at the right end of the beam will be 40 kip feet rather than zero. Since the bending moment diagram does not close, there is a moment equilibrium error in the reactions.

Case III (Figure 5.11c). The reactions have been computed properly for this case. Both the shear force diagram and the bending moment diagrams will close.

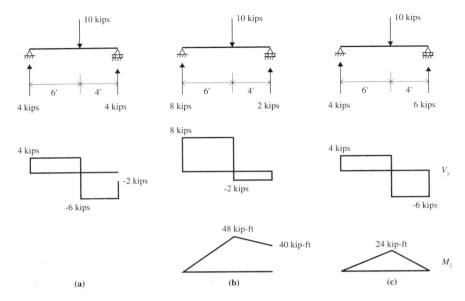

Figure 5.11 V_y and M_z check cases.

By using the fact that the shear force and bending moment diagrams must both close, many of the errors which might occur during an analysis can be found and eliminated. These errors might either be in the computation of the reactions, as shown by these three sample cases, or in the generation of the diagrams after the correct reactions have been found. Of course, there is always the possibility that two or more compensating errors will be made at some point in the analysis and the two diagrams will appear to close even though they are incorrect. However, the probability of this happening is small. An engineer usually can have confidence in the analysis if the shear force and bending moment diagrams close within acceptable roundoff error.

The following are some additional example problems which show the shear force and bending moment diagrams for some additional situations which might be encountered during the analysis of statically determinate beams.

Example Problem 5.8

Generate the shear force and bending moment diagrams for the beam whose mathematical model is shown in Figure 5.12a. This beam has four independent reactive quantities corresponding to vertical reactive forces at each of the supports and a horizontal reactive force at the left pinned support. The internal pin supplies an equation of condition which, when combined with the three overall equations of equilibrium, makes the beam statically determinate.

The load diagram is shown in Figure 5.12b while the shear force and bending moment diagrams are shown in Figures 5.12c and 5.12d, respectively. Note that the maximum positive and negative bending moments in the interior of the beam occur at points where the shear force diagram passes through zero. Also, the bending mo-

Shear Force and Bending Moment Diagrams

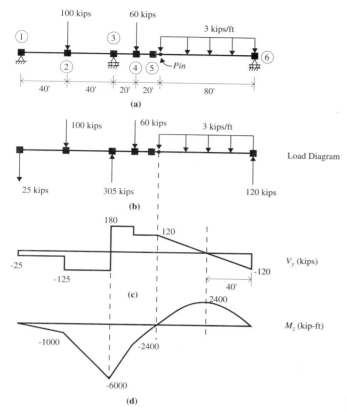

Figure 5.12 Example Problem 5.8.

ment diagram passes through zero at the location of the internal pin, while the shear force at this point is non zero. This is to be expected since an internal pin can resist transverse shear but it can not resist moment.

Example Problem 5.9

Generate the shear force and bending moment diagrams for the beam whose mathematical model is shown in Figure 5.13a. The first step in this analysis is to transfer the 3 kN horizontal force to the longitudinal center line of the beam as a force and a moment. The load diagram which shows this horizontal force and moment and the other active and reactive loads on the beam is shown in Figure 5.13b. The shear force and bending moment diagrams are shown in Figures 5.13c and 5.13d. The concentrated moment has no effect on the shear diagram at its point of application, but it does cause an upward step in the bending moment diagram. This upward step is present since the counterclockwise moment on the right end of the left segment of the beam would have to increase for a section cut just to the right of the point of application of the moment compared to a section cut just to the left of that point. According to the bending moment sign convention, a counterclockwise moment on the right end of a left segment is positive. If we had inserted a downward step in the bending moment diagram, the bending moment which would be computed at the right

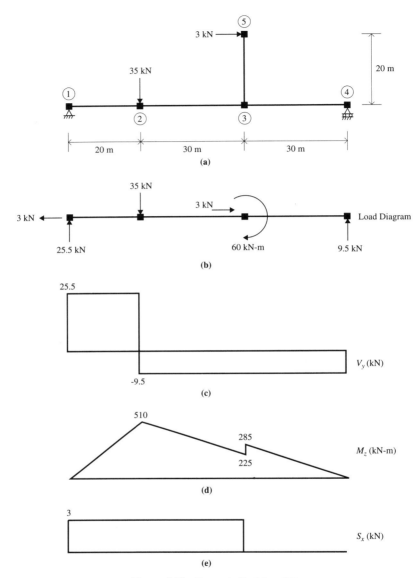

Figure 5.13 Example Problem 5.9.

end of the beam would be −120 kN m rather than the required value of zero. This would immediately indicate that something was wrong since the bending moment diagram did not close.

The horizontal active and reactive forces have no effect upon the shear force and bending moment diagrams. However, their effect on the beam can be demonstrated by an additional plot which shows the variation of the axial force in the beam

Shear Force and Bending Moment Diagrams

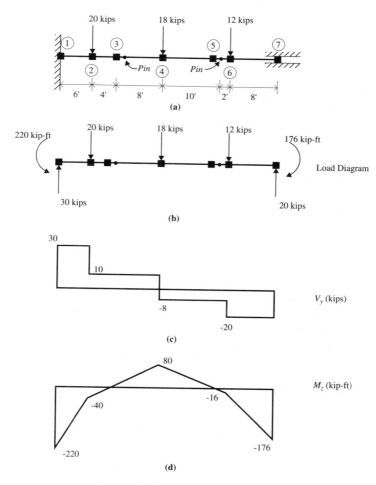

Figure 5.14 Example Problem 5.10.

at each point along its length, as shown in Figure 5.13e. This plot is called an *axial force diagram*. This diagram can be very useful during the design of beams which are subjected to large axial forces, particularly when the axial force varies over the length of the beam.

Example Problem 5.10

Generate the shear force and bending moment diagrams for the beam whose mathematical model is shown in Figure 5.14a. The support conditions for this beam consist of a fixed support at the left end and a horizontal slide at the right end, which results in five independent reactive quantities. The beam is statically determinate since two equations of condition can be generated at the internal pins.

The load diagram is shown in Figure 5.14b. In addition to the three downward active forces, the load diagram shows upward reactive forces and reactive moments at each end of the beam. The shear force and bending moment diagrams are shown in Figures 5.14c and 5.14d, respectively. The primary difference between this beam and the previous example beams is the presence of the moments at each end. The counterclockwise reactive moment at the left end results in a downward step in the bending moment diagram at that point, while the clockwise reactive moment at the right end results in an upward step. These end moments have no effect on the shear force diagram. It appears that the shear force and bending moment diagrams are correct since they both close. This is further verified since the bending moment diagram passes through zero at the locations of the two interior pins.

COMPUTER PROGRAM BEAMVM

The computer program BEAMVM, which is supplied for this book, will plot the shear force and bending moment diagrams for a statically determinate or statically indeterminate beam subjected to essentially any combination of concentrated vertical forces and concentrated moments at any locations on the beam and vertical distributed forces over any portions of the beam. The distributed forces may be uniform or they may vary linearly over the loaded length. The program can also plot the deformed shape of the beam, but we will only consider the shear force and bending moment plots at this time. The deformed shape plots will be considered in Chapter 8 when we discuss the computation of beam deflections. The analysis procedure which is used in the program will not be discussed here. The program is supplied so that the readers can verify their manual solutions for the suggested problems at the end of this chapter. The instructions for BEAMVM are contained in the ASCII disk file BEAMVM.DOC.

The input data file for BEAMVM is very similar to that required for the program BEAMIL, which was used in Chapter 2. The primary differences are: the modulus of elasticity of the materials must be specified; the moment of inertia of the member cross sections must be specified; and the loads acting on the beam must be specified. Since we will not be interested in the deformed shape for the statically determinate beams which will be analyzed in this chapter, the input for these problems can specify one material and one member cross section and any nonzero arbitrary values may be entered for the modulus of elasticity and the moment of inertia. A suggested value is 1000 for each quantity. Of course, if you are analyzing a beam for which you are interested in the deformed shape, then realistic values must be used for these quantities.

As in the program BEAMIL, the mathematical model for any beam in BEAMVM consists of a set of joints along a horizontal line which are connected by straight members. The program can plot the shear force and the bending moment diagrams for the whole beam, in which case the numerical values of the shear force and bending moment will only be listed at each joint, or it can plot the shear force and bending moment diagrams for any individual member in the mathematical model of the beam, in which case the numerical values will be listed for the ends of the member and for the tenth points in the interior of the member. If the

Computer Program BEAMVM

```
Example Problem 5.8 - Analysis by Program BEAMVM
6,1,1,1,3,2,1
Joint Coordinates
1,0.0
2,40.0
3,80.0
4,100.0
5,120.0
6,200.0
Material Data
1,1000.0
Cross Section Data
1,1000.0
Interior Pins
5
Support Restraints
1,1,1,0
3,0,1,0
6,0,1,0
Joint Loads
2,-100.0,0.0
4,-60.0,0.0
Distributed Member Loads
5,-3.0,-3.0
```

Figure 5.15 Example Problem 5.8, BEAMVM input.

shear force changes sign over the length of the member, the point of zero shear, which corresponds to the point of maximum moment, will also be specified.

Figure 5.15 shows the listing of the input data file for the analysis of the beam in Example Problem 5.8 by BEAMVM. The input data has units of feet and kips. The mathematical model of the beam has six joints and five members since joints are required at each of the support points, at the location of each of the concentrated forces and at the location of the internal pin. The joints at the internal pin and at the right support also satisfy the requirement that a joint is required at each end of the distributed force. Figure 5.16 shows the plots of the shear force and the bending moment diagrams on a VGA monitor for the whole beam, while Figure 5.17 shows a more detailed plot of the shear force and the bending moment diagrams for member 5, which extends between the interior pin at joint 5 and the right end of the beam at joint 6. This second plot shows that the point of zero shear force in this uniformly loaded member is $0.500L$ from the left end of the member and the maximum bending moment at that point is 2400 kip feet. The listed values for V_y and M_z in Figures 5.16 and 5.17 agree with those obtained in the manual solution.

The verification of the manual solution for the other example problems in this chapter is left as an exercise for the reader. Analyzing these problems, for which the solutions are known, will give the reader experience in the use of the program BEAMVM.

In some situations, it might be desirable to compute the values of the shear force and the bending moment at specific points in a beam. This can be easily accomplished by merely inserting a joint in the mathematical model of the beam at

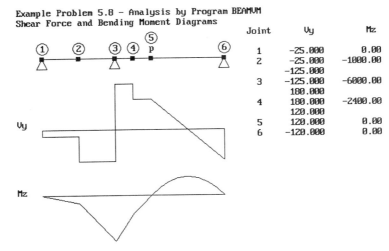

Figure 5.16 Example Problem 5.8, V_y and M_z diagrams from BEAMVM.

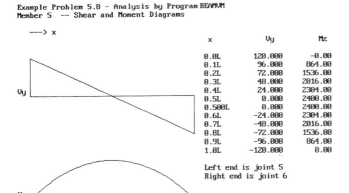

Figure 5.17 Example Problem 5.8, V_y and M_z diagrams from BEAMVM.

that point. The shear force and bending moment values will then be listed for that point in the whole beam plot.

INFLUENCE LINES FOR SHEAR FORCE AND BENDING MOMENT IN BEAMS

In Chapter 2, we discussed the development of influence lines for the vertical reactive forces for beams. We will now consider influence lines for shear force and bending moment.

Example Problem 5.11

To demonstrate the computation procedure, we will generate influence lines for the shear force and the bending moments at points a and b, which are 16 meters and 32 meters, respectively, from the left end of the 40 meter long simply supported beam whose mathematical model is shown in Figure 5.18a. To obtain the influence line ordinates, we must compute the magnitudes of the shear force and the bending moment at points a and b for a unit downward force at various positions on the beam. These quantities can be easily computed by cutting the beam at the two points and analyz-

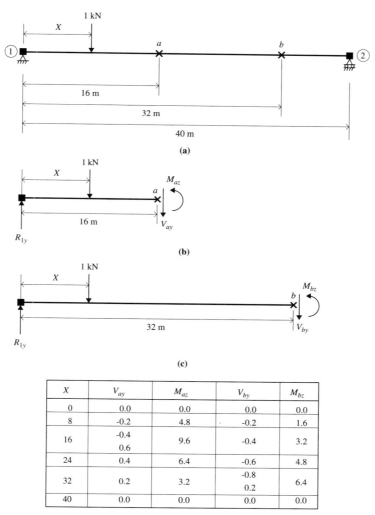

X	V_{ay}	M_{az}	V_{by}	M_{bz}
0	0.0	0.0	0.0	0.0
8	-0.2	4.8	-0.2	1.6
16	-0.4 0.6	9.6	-0.4	3.2
24	0.4	6.4	-0.6	4.8
32	0.2	3.2	-0.8 0.2	6.4
40	0.0	0.0	0.0	0.0

(d)

Figure 5.18 Example Problem 5.11.

ing the left beam segment for each cut. The free body diagrams for each of the left segments for cuts at points a and b are shown in Figures 5.18b and 5.18c respectively. Equations for computing V_y and M_z for each of the segments can be generated as follows:

V_{ay} and M_{az} for $0 \leq X \leq 16$

$$\Sigma F_Y \rightarrow V_{ay} = R_{1Y} - 1$$

$$\Sigma M_Z \rightarrow M_{az} = X - 16(1 - R_{1Y})$$

V_{ay} and M_{az} for $16 \leq X \leq 40$

$$\Sigma F_Y \rightarrow V_{ay} = R_{1Y}$$

$$\Sigma M_Z \rightarrow M_{az} = 16 R_{1Y}$$

V_{by} and M_{bz} for $0 \leq X \leq 32$

$$\Sigma F_Y \rightarrow V_{by} = R_{1Y} - 1$$

$$\Sigma M_Z \rightarrow M_{bz} = X - 32(1 - R_{1Y})$$

V_{by} and M_{bz} for $32 \leq X \leq 40$

$$\Sigma F_Y \rightarrow V_{by} = R_{1Y}$$

$$\Sigma M_Z \rightarrow M_{bz} = 32 R_{1Y}$$

As we see from these equations, the values of the shear forces and bending moments can be easily computed for any position of the unit force if the value of the reactive force at the left end of the beam can be determined. This reactive force can be determined for any position of the force by using the influence line for R_{1Y} shown in Figure 2.31c in Example Problem 2.8.

Figure 5.18d shows an influence table which contains the computed values for V_y and M_z at points a and b for positions of the unit force at the fifth points along the beam. Note that double values are listed for V_{ay} at point a and for V_{by} at point b, since there will be a discontinuity in the shear force diagram at each of these points when the unit force is at that point. The upper value corresponds to the shear force at the point for the unit force just to the left of the point, while the lower value corresponds to the shear force at the point for the unit force just to the right of the point. Only single values are listed for the bending moments for each force location since the bending moment diagrams are continuous for any position of the load on the beam.

The influence lines for V_{ay} and M_{az} are shown in Figure 5.19a while the influence lines for V_{by} and M_{bz} are shown in Figure 5.19b. These influence lines were obtained by plotting the values in the influence table for each position of the unit force. The points are connected by straight lines since we can see from the equations for the shear force and bending moment that both quantities vary linearly with X. The steps in the shear force influence lines at points a and b correspond to the instantaneous changes that occur in the value of the shear forces at these points as the unit force passes through these points as it moves along the beam.

The units of the ordinates of the shear force influence lines are force per unit force (e.g.; kN/kN or dimensionless) and the units of the ordinates of the bending

Influence Lines for Shear Force and Bending Moment in Beams

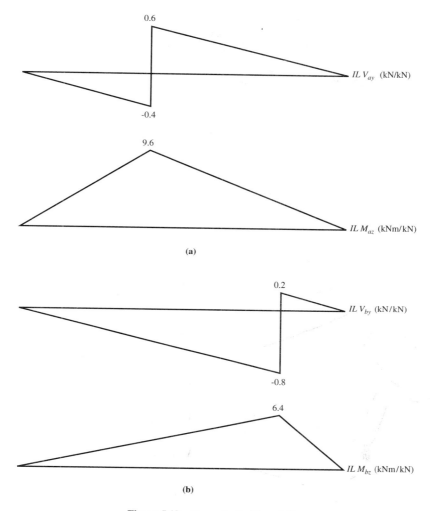

Figure 5.19 Example Problem 5.11.

moment influence lines are moment per unit force (e.g.; kN m/kN or m). Therefore, the units of the bending moment, which is computed from an influence line, will depend upon the units of the loads acting on the beam and the length units which were used for the beam when computing the influence line ordinates. The units of the shear force will only depend upon the units of the loads acting on the beam. Other than being careful with the units for the bending moments, these influence lines have the same properties and can be used in the same manner as described previously in Chapter 2 for reactive force influence lines. One of the common uses of influence lines of this type is to determine the maximum shear force or bending moment at a particular point on a beam as a set of loads, such as the wheel loads for a truck, moves along the beam.

Since the influence lines for a statically determinate beam will always be composed of straight lines, it is usually possible to draw the influence lines for most statically determinate beams with only a few computed points.

Example Problem 5.12

Generate the influence lines for the shear force and bending moment at point a which is 25 feet from the left end of the beam whose mathematical model is shown in Figure 5.20a. This is the same beam for which the influence lines for the vertical reactive forces were generated in Example Problem 2.7 in Chapter 2.

The values of V_{ay} and M_{az} can be computed, for any position of a downward unit force, in terms of the reactive force at the left support by using the same analysis procedure described in Example Problem 5.11. The free body diagram for the

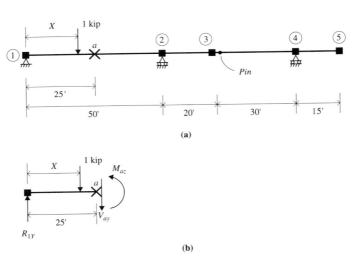

X	R_{1Y}	V_{ay}	M_{az}
0.0	1.0	0.0	0.0
25.0	0.5	−0.5 / 0.5	12.5
50.0	0.0	0.0	0.0
60.0	−0.2	−0.2	−5.0
70.0	−0.4	−0.4	−10.0
85.0	−0.2	−0.2	−5.0
100.0	0.0	0.0	0.0
107.5	0.1	0.1	2.5
115.0	0.2	0.2	5.0

(c)

Figure 5.20 Example Problem 5.12.

Influence Lines for Shear Force and Bending Moment in Beams

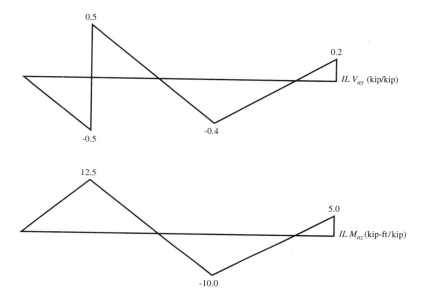

Figure 5.21 Example Problem 5.12.

left segment of the beam for a cut at point a is shown in Figure 5.20b. The equations for V_{ay} and M_{az} for any value of X are as follows.

For $0 \leq X \leq 25$:

$$V_{ay} = R_{1Y} - 1$$

$$M_{az} = X - 25(1 - R_{1Y})$$

For $25 \leq X \leq 115$:

$$V_{ay} = R_{1Y}$$

$$M_{az} = 25 R_{1Y}$$

The required values of the left reactive force R_{1Y} can be determined from the influence line shown in Figure 2.27c for Example Problem 2.7.

Figure 5.20c shows an influence table which contains the values of V_{ay} and M_{az} for various positions of the unit load along the beam, while the influence lines which were obtained by plotting these values are shown in Figure 5.21. As with the shear force influence lines in the previous example problem, there is a step in the influence line for V_{ay} at point a.

Equations for V_{ay} and M_{az} also could be derived by considering the equilibrium of the right segment of the beam for the cut at point a. These equations would express V_{ay} and M_{az} in terms of the reactive forces at joint 2 and joint 4. It is left as an exercise for the reader to derive these equations and to verify that they will give the same values for V_{ay} and M_{az} as listed in the influence table in Figure 5.20c. The required values for the reactive forces at joint 2 and joint 4 can be determined from the influence lines in Figure 2.27c.

164   Analysis of Statically Determinate Beams   Chap. 5

     A similar procedure to that used in these two example problems can be used to develop the influence lines for the shear force and bending moment at any point in any statically determinate beam. For most beams, it will be helpful to generate the influence lines for the reactions before the influence lines for the shear force and bending moment at any interior points are generated. The influence lines for the reactions usually must be computed anyway since they are needed to compute the maximum magnitudes of the reactions for the design of the supports for the structure.

COMPUTER PROGRAM BEAMIL

     The computer program BEAMIL, which was used in Chapter 2 to plot influence lines for the vertical reactive forces for beams, also can be used to plot influence lines for the shear force and the bending moment at any point in a beam. It is suggested that the reader review the instructions for this program in the disk file BEAMIL.DOC before attempting to use it again.

     During the analysis of any beam by BEAMIL, the program user must specify the number of evenly spaced interior calculation points to be assigned for each member in the mathematical model of the beam during the analysis. If N points are assigned, each member then will have $N + 2$ points along its length at which data will be available to plot the shear force and bending moment influence lines. These points consist of: the left end of the member; N evenly spaced interior points along the member: and the right end of the member. If an influence line for shear force or bending moment is desired at some point on the beam which cannot be conveniently defined by an interior member point, then the user should place a joint in the mathematical model at that point. That point can then be specified as either point $N + 2$ for the member to the left of the joint or point 1 for the member to the right of the joint.

     Figure 5.22 shows a listing of the input data file for the program BEAMIL for the beam which was analyzed in Example Problem 5.12. This is the same data file that was used for the analysis of this same beam in Chapter 2 by BEAMIL to obtain the influence lines for the vertical reactive forces. As explained previously in

```
Sample Beam for Program BEAMIL
5,1,1,1,3
Joint Coordinates
1,0.0
2,50.0
3,70.0
4,100.0
5,115.0
Interior Pins
3
Support Restraints
1,1,1,0
2,0,1,0
4,0,1,0
```

Figure 5.22 Example Problem 5.12, BEAMIL input.

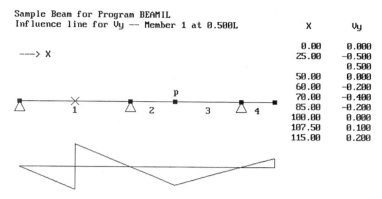

Figure 5.23 Example Problem 5.12, influence line from BEAMIL.

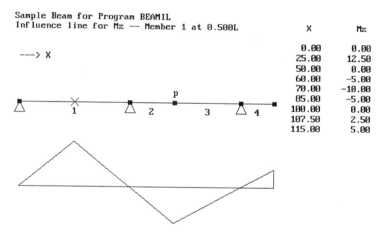

Figure 5.24 Example Problem 5.12, influence line from BEAMIL.

Chapter 2, the mathematical model of the beam for analysis by BEAMIL contains five joints and four members since joints are required at the support points, at the interior pin and at the free end. Figures 5.23 and 5.24 show plots of the influence lines for the shear force and the bending moment at a point $0.500L$ from the left end of member 1. This corresponds to point a in the beam in Example Problem 5.12. These influence lines were generated by specifying one interior calculation point for each member and then requesting the plots for Member 1, Point 2. The influence line ordinates which are listed on the right side of the plots correspond to the values at each joint and at each interior calculation point for each member. The shape of the influence lines and the listed ordinates agree exactly with the previous manual analysis.

Figures 5.25 and 5.26 show the influence lines for the shear force just to the left and just to the right of the support at joint 2. These two points correspond to

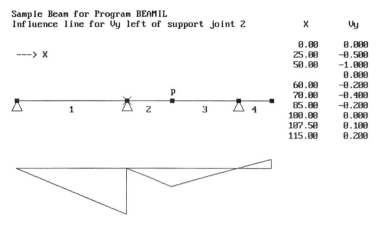

Figure 5.25 Example Problem 5.12, influence line from BEAMIL.

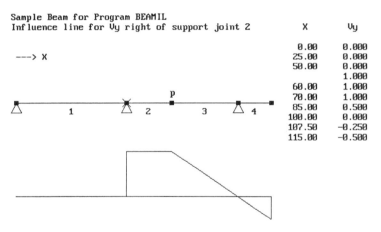

Figure 5.26 Example Problem 5.12, influence line from BEAMIL.

Member 1, Point 3, and Member 2, Point 1, respectively. The influence lines for the shear force on either side of this joint are different because of the step which occurs in the shear force diagram due to the vertical reactive force at this support for any position of a unit downward force on the beam. If the influence lines for the bending moments for these two points were plotted, they would be identical since the ordinates of a bending moment diagram will be continuous across the support. The manual verification of the ordinates of these two shear force influence lines is left as an exercise for the reader. (Hint: Cut the beam just to the left of joint 2 and sum the vertical forces for the left segment to find $V_{2y,\,\text{left}}$, and then cut the beam just to the right of joint 2 and sum the vertical forces for the right segment to find $V_{2y,\,\text{right}}$.)

Maximum Bending Moment in Beams   167

MAXIMUM BENDING MOMENT IN BEAMS

Although influence lines can be used to compute the maximum bending moment which will occur at a specific point in a beam due to a set of moving loads, they will not tell you where the maximum bending moment will occur in the beam. It is obvious that the maximum bending moment occurs at the center of a simply supported beam for a single concentrated force moving across the beam. However, this will not necessarily be the case for multiple loads, such as the wheel loads for a truck. If we are going to be able to design the beam to carry a set of moving concentrated forces, we must have a procedure for computing the maximum bending moment and its location. The first step in the analysis is to determine where the loads should be located on the beam to produce the maximum moment.

Consider a simply supported beam, which is subjected to a set of downward concentrated forces W_1 through W_n, with fixed spacings between the forces, at some position on the beam as shown in Figure 5.27a. The bending moment diagram for this loading will consist of a series of straight lines with kinks at the points of application of the concentrated forces, as shown in Figure 5.27b. The maximum bending moment in the beam will occur under one of the forces. Therefore, if we can determine under which force the maximum moment will occur and where the loads must be positioned on the beam to produce this maximum moment, it will be a simple task to compute the maximum moment as the loads move along the beam. To determine the position of the loads to produce the maximum bending moment under any specific load W_i we will use the situation shown in Figure 5.28, where: W is the resultant of all of the concentrated forces on the beam; W_L is the

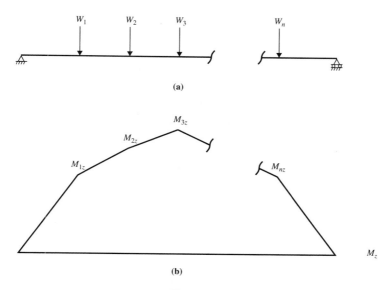

Figure 5.27

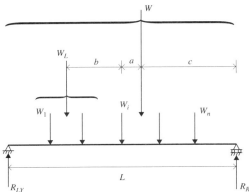

Figure 5.28

resultant of all of the forces on the beam to the left of the force W_i; a is the distance of the force W_i to the left of the resultant force W; b is the distance of the force W_L to the left of the force W_i; c is the distance of the resultant force W to the left of the right end of the beam; R_{LY} is the left reactive force; and R_{RY} is the right reactive force.

The left reactive force, which can be determined by summing moments about the right end of the beam, is equal to

$$R_{LY} = \frac{Wc}{L} \tag{5.14}$$

The moment directly under the force W_i can be determined by cutting the beam at the point of application of W_i and summing moments for the left segment:

$$M_{iz} = R_{LY}(L - a - c) - W_L b \tag{5.15}$$

which, upon substituting the expression for R_{LY} in Eq. (5.14), gives

$$M_{iz} = \frac{W}{L}(Lc - ac - c^2) - W_L b \tag{5.16}$$

To obtain the position of the set of forces to produce the maximum value of M_{iz}, we can set the derivative of M_{iz} with respect to c to zero

$$\frac{dM_{iz}}{dc} = \frac{W}{L}(L - a - 2c) = 0 \tag{5.17}$$

which, upon solving for c, gives

$$c = \frac{L}{2} - \frac{a}{2} \tag{5.18}$$

Therefore, the maximum moment will occur under any specific force W_i in the set when that force and the resultant of all of the forces on the beam are equal dis-

Maximum Bending Moment in Beams 169

tances from the center of the beam. The maximum bending moment which will occur in the beam, as the set of forces moves along the beam, can be found by positioning each force in turn according Eq. (5.18) and computing the bending moment under that force. The maximum bending moment will be the maximum of these bending moments.

The maximum shear force for a simply supported beam will always occur adjacent to one of the supports. Therefore, the maximum shear force can be found by computing the maximum values for R_{LY} and R_{RY} using the influence lines developed previously in Figure 2.31. The maximum shear force which will occur in the beam will be equal to the larger of these two values, since the shear force adjacent to either support is numerically equal to the vertical reactive force at that support.

Example Problem 5.13

Compute the maximum shear force and the maximum bending moment which will occur in a simply supported crane girder, with a span of 100 feet, due to the crane wheel loads shown in Figure 5.29a. The crane can move to any position on the girder, but it can only face in the direction shown.

It is obvious from the shapes of the influence lines for the reactive forces at the two ends of the beam, which were shown previously in Figure 2.31c, that the maximum shear force will occur at the right end of the beam when the 50 kip wheel load is just to the left of the right end. The magnitude of $V_{y,\max}$ can be computing by using the influence line ordinates for the right vertical reactive force at the positions of the two wheel loads:

$$V_{y,\max} = (50)(1.0) + (30)(0.8) = 74 \text{ kips}$$

The shear force adjacent to the right support is actually negative. However, for most design situations, the direction of the shear force on the cross section is not important and the absolute value usually can be used. If the sign of the shear force is important, the shear forces adjacent to both the left and right supports should be computed. The left shear force will be the maximum positive value and the right shear force will be the maximum negative value.

The position of the resultant of the two wheel loads can be found by summing moments of the vertical forces about either wheel and dividing by the total force. The resultant of 80 kips is located 12.5 feet from the left wheel load and 7.5 feet from the right wheel load, as shown in Figure 5.29b. Using this resultant location with Eq. (5.18) results in two possible positions for the wheel loads to produce the maximum bending moment in the girder. Figure 5.29c shows the load positions for the maximum bending moment under the right wheel load while Figure 5.29d shows the positions for the maximum bending moment under the left wheel load. The maximum moments for each of these cases are 1711.25 kip feet and 1531.25 kip feet, respectively. The verification of these values is left as an exercise for the reader. The maximum moment which will occur in the girder as the crane moves across the span is the larger of these two values. Therefore,

$$M_{z,\max} = 1711.25 \text{ kip feet}$$

170 Analysis of Statically Determinate Beams Chap. 5

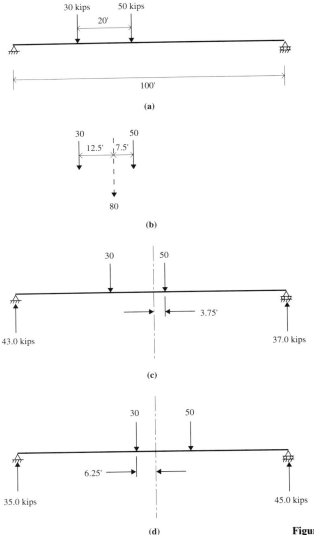

Figure 5.29 Example Problem 5.13.

This moment will occur at a point 46.25 feet from the right end of the girder when the 50 kip wheel load is directly over that point.

The maximum shear force and the maximum bending moment can be found for beams with other geometries and support conditions in a similar manner. The maximum shear force will always occur adjacent to one of the supports. The maximum value can be found by using the influence lines for the shear forces on either side of the supports, similar to those shown in Figures 5.25 and 5.26 from the analysis by the program BEAMIL. In order to find the position of the loads to produce the maximum bending moment, it will be necessary to derive an equation which relates the moment under any concentrated force to the position of the loads on the beam. Unfortunately, each different beam geometry is a special case and no universal equation can be developed. For any geometry other than a single span simply supported beam, the algebraic manipulations can be very tedious.

SHEAR FORCE AND BENDING MOMENT ENVELOPES

Although it is important during the design of a beam to know the magnitudes and the locations of the maximum shear force and bending moment, it would be even of more value if the maximum magnitudes which could occur for both the shear force and the bending moment were known at every point on the beam. It then would be possible to optimize the design of every cross section along the beam. To demonstrate the procedure which can be used to determine this information, we will consider the case of a single span simply supported beam subjected to a moving single downward concentrated force.

The influence lines for the shear force and the bending moment at any point on a single span simply supported beam of length L, at a distance X from the left end, are shown in Figure 5.30a. From the shear force influence line, we can determine that the magnitude of the maximum positive shear force at the point which is a distance X from the left end, for a moving downward concentrated force of magnitude W, will be equal to

$$V_{y, \max +} = \frac{WX}{L} \tag{5.19}$$

while the magnitude of the maximum negative shear force at that point will be

$$V_{y, \max -} = \frac{W(L - X)}{L} \tag{5.20}$$

From the bending moment influence line, we can see that the magnitude of the maximum bending moment at this point will be

$$M_{z, \max} = \frac{WX(L - X)}{L} \tag{5.21}$$

The bending moment will only have positive values over the entire span for any position of the concentrated force. If we now vary the value of X to represent a number of points along the beam, we can compute the maximum positive shear force, the maximum negative shear force and the maximum bending moment which will occur at each of these points as the force moves across the beam. The table in Figure 5.30b shows the results of an analysis of this type for a beam with a 100 foot span subjected to a moving concentrated force of 20 kips. If these maximum

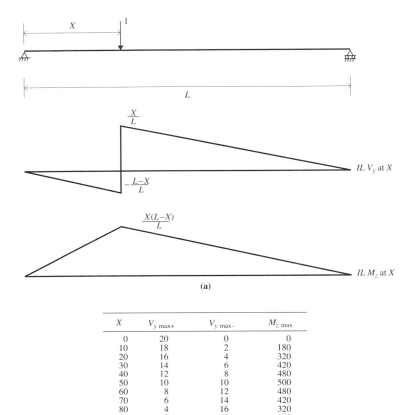

Figure 5.30

shear forces and bending moments are now plotted versus the distance X, the ordinates of the resulting curves will be equal to the maximum values for these quantities for every position on the beam due to the moving concentrated force. Figure 5.31a shows the plots of the maximum positive shear forces and the maximum negative shear forces while Figure 5.31b shows the plot of the maximum bending moments. These curves are known as the *shear force en elope* and the *bending moment en elope*.

As discussed previously in Example Problem 5.13, the sign of the shear force is usually not important for design purposes. Therefore, it is sometimes useful to generate a third curve in which the absolute value of the largest shear force for each position on the beam is plotted. This plot is shown in Figure 5.31c. This curve is known as the *absolute shear force en elope*. Since the sign of the bending moment is usually very important for design purposes, particularly for concrete

Shear Force and Bending Moment Envelopes

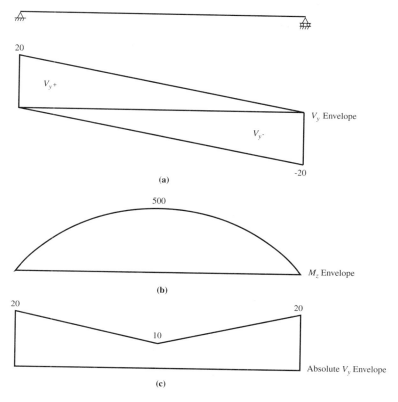

Figure 5.31

beams, both positive and negative bending moment envelopes should be plotted if the bending moment influence lines have both positive and negative ordinates along the beam.

A similar procedure can be used to generate shear force and bending moment envelopes for a set of concentrated loads, such as a H20-44 or HS20-44 truck loading, and for beams with other geometries. However, the calculations can become very tedious and time consuming since the analysis usually must be performed for a large number of points along the beam to obtain accurate plots and to ensure that the maximum values are obtained. Influence lines must be developed for each of the points and then the loads must be positioned on each of these influence lines to obtain the maximum value. If the influence lines have both positive and negative ordinates, then two maximum values must be computed for each influence line. Fortunately, computer programs are available which will generate shear force and bending moment envelopes for almost any type of beam subjected to a set of moving concentrated forces. Figures 5.32 to 5.34 show the signed shear force envelopes, the absolute shear force envelope and the signed bending moment envelopes for a statically indeterminate prismatic highway bridge with two 100 foot

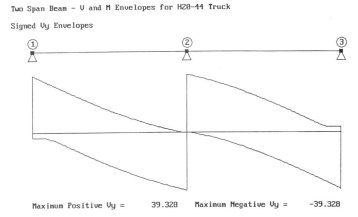

Figure 5.32 Two-span beam, signed shear-force envelope.

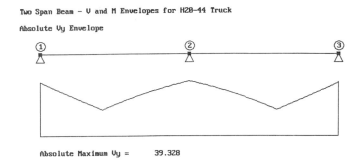

Figure 5.33 Two-span beam, absolute shear-force envelope.

spans, which are continuous over the center support, for a H20-44 truck which could face in either direction as it moved across the bridge. The computed shear forces are in kips and the bending moments are in kip feet. The program which was used to generate these diagrams was FATPAK V (Structural Software Systems, 1996), which is the Professional Edition of the program FATPAK II—Student Edition which is supplied for this book. The maximum shear forces and bending moments were computed and plotted for points at 1-foot intervals along the bridge to ensure that accurate plots were obtained. The program can use a much smaller interval if it is needed for accuracy. It can also superimpose the shear forces and the bending moments due to the dead weight of the structure on the values due to the moving forces. It would be impractical to even attempt to develop these curves for this structure manually. There are several other commercial structural analysis programs which can perform this type of analysis.

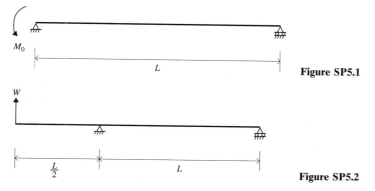

Figure 5.34 Two-span beam, signed bending-moment envelope.

REFERENCES

AMERICAN INSTITUTE OF STEEL CONSTRUCTION (1986). *Manual of Steel Construction—Load and resistance Factor Design.* Chicago: AISC.

AMERICAN INSTITUTE OF STEEL CONSTRUCTION (1989). *Manual of Steel Construction—Allowable Stress Design.* Chicago: AISC.

STRUCTURAL SOFTWARE SYSTEMS (1996). FATPAK V—Professional Edition. Monroeville, Pa.: Structural Software Systems.

YOUNG, WARREN C. (1989). *Roarke's Formulas for Stress and Strain.* New York: McGraw-Hill.

SUGGESTED PROBLEMS

SP5.1 to **SP5.7** Derive expressions for the variation of the shear force and bending moment along the beams shown in Figures SP5.1 to SP5.7. The origin of the global X axis is at the left end of each beam and extends toward the right. The loads are posi-

Figure SP5.1

Figure SP5.2

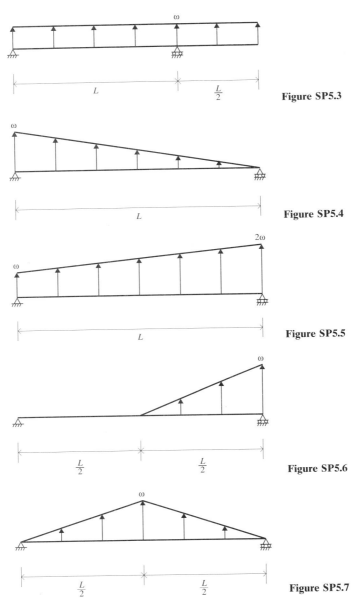

Figure SP5.3

Figure SP5.4

Figure SP5.5

Figure SP5.6

Figure SP5.7

tive as shown. Plot the shear force and bending moment diagrams from these expressions.

For the beams shown in Figures SP5.1 to SP5.3, derive the expressions using free body diagrams and equilibrium for various points along the beam.

For the beams shown in Figures SP5.4 to SP5.7, derive the expressions by direct integration using the equation

Suggested Problems

$$\frac{d^2 M_z}{dX^2} = q_y$$

SP5.8 to **SP5.18** Draw the shear force and bending moment diagrams for each of the beams shown in Figures SP5.8 to SP5.18. Verify the diagrams with the program BEAMVM. (Hint: Set E and I_z equal to unit values in the program input since you are not interested in the beam deflections.)

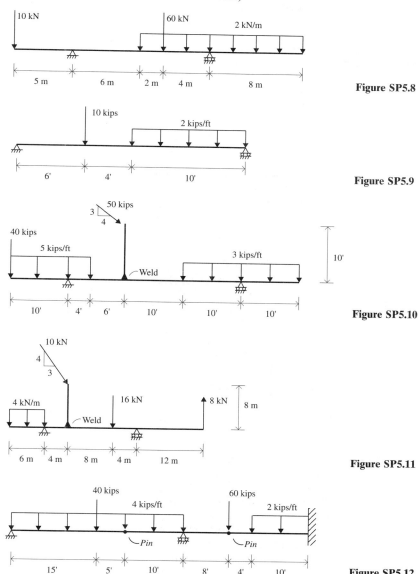

Figure SP5.8

Figure SP5.9

Figure SP5.10

Figure SP5.11

Figure SP5.12

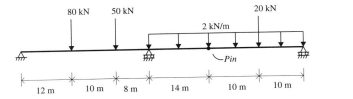

Figure SP5.13

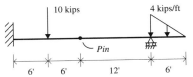

Figure SP5.14

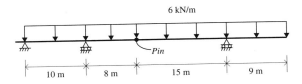

Figure SP5.15

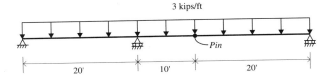

Figure SP5.16

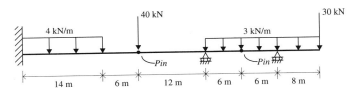

Figure SP5.17

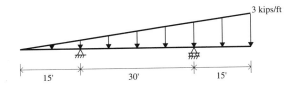

Figure SP5.18

SP5.19 to **SP5.22** Draw the influence lines for the indicated quantities for the beams shown in Figures SP5.19 to SP5.22. Verify your solution with the program BEAMIL.
For the beam in Figure SP5.19:
(a) Shear force and bending moment at point a
(b) Shear force to the left and right of the support at b
(c) Shear force and bending moment at point c
For the beam in Figure SP5.20:
(a) Shear force and bending moment at point a
(b) Shear force to the left and right of the support at b

Suggested Problems

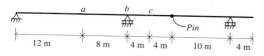

Figure SP5.19

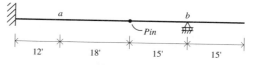

Figure SP5.20

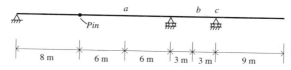

Figure SP5.21

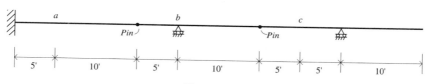

Figure SP5.22

For the beam in Figure SP5.21:
(a) Shear force and bending moment at point a
(b) Shear force and bending moment at point b
(c) Shear force to the left and right of the support at c

For the beam in Figure SP5.22:
(a) Shear force and bending moment at point a
(b) Shear force to the left and right of the support at b
(c) Shear force and bending moment at point c

SP5.23 Compute the maximum shear force and the maximum bending moment which will be produced in a simply supported highway bridge girder with a span of 150 feet due to a HS20-44 truck which can move to any position on the span. Do not consider the effect of the dead weight of the bridge.

6

Analysis of Statically Determinate Plane Frames

SHEAR FORCE AND BENDING MOMENT IN PLANE FRAMES

The shear force and bending moment diagrams for the individual members in a plane frame can be constructed in a similar manner to that shown in the previous chapter for a beam. The relationships between the loads acting on a plane frame member and the shear force and bending moment at any point in the member are the same as those developed previously for a beam, as expressed by Eqs. (5.3), (5.6), (5.12) and (5.13). The only difference is that a member in a plane frame may be at any orientation in the plane of the structure rather than lying on a horizontal line. However, by considering each individual member in the frame to be equivalent to a segment of a beam, the procedures developed in the previous chapter are still applicable.

Example Problem 6.1

Draw the shear force and bending moment diagrams for the four members in the plane frame whose mathematical model is shown in Figure 6.1a. This is the same plane frame which was analyzed previously in Example Problem 2.2 in Chapter 2 to demonstrate the analysis procedure for computing the reactions for a plane frame.

The first step in developing the shear force and bending moment diagrams for the frame, after the reactions have been computed, is to draw the free body diagrams for each joint and each member in the frame, as shown in the exploded view in Figure 6.1b. The easiest way to draw these free body diagrams is to start at one of the support joints by determining what loads the attached member must apply to the joint to balance the reactive loads and hold it in equilibrium. The free body diagrams for the members and the other joints then can be drawn in turn by proceeding in sequence

Shear Force and Bending Moment in Plane Frames

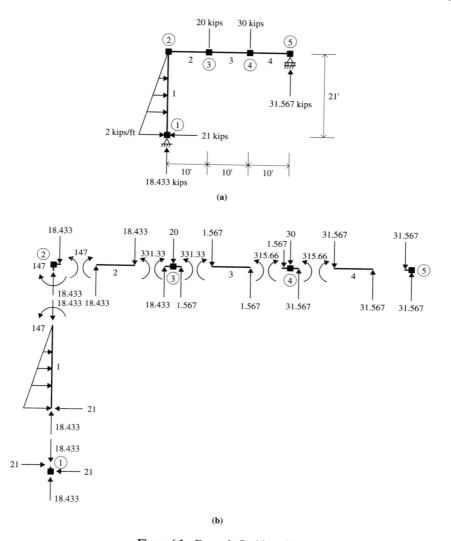

Figure 6.1 Example Problem 6.1.

through the structure. For example, if we start at joint 1, we have an upward vertical force of 18.433 kips and a horizontal force of 21 kips acting to the left which are applied to the joint by the pinned support. These reactive forces must be balanced by equal and opposite forces applied to the joint by the lower end of member 1. These forces will be a downward force of 18.433 kips and a horizontal force of 21 kips toward the right. In turn, joint 1 will apply an upward force of 18.433 kips and a horizontal force of 21 kips acting to the left to the lower end of member 1. These forces are equal and opposite to the forces that the lower end of member 1 applies to joint 1. If we now proceed to the upper end of member 1, we will find that, in order for the member to be in equilibrium, joint 2 must apply a downward force of 18.433 kips and a counterclockwise moment of 147 kip feet to the end of the member, while the member in turn will apply an upward force of 18.433 kips and a clockwise moment of 147 kip feet to the joint. Next, considering the equilibrium of joint 2, we can determine the loads that must be applied to the joint by the left end of member 2. By progressing through the frame in this manner, we finally arrive at joint 5, where the right end of member 4 applies a downward force of 31.567 kips and the support applies an upward force of the same magnitude. Since the loads on this joint balance, it appears that the free body diagrams which have been developed are correct. The analysis process is self checking, unless two or more compensating errors are made while analyzing the individual joints and members.

After the free body diagrams for the members have been developed, the shear force and bending moment diagrams for each member can be drawn, as shown in Figures 6.2 through 6.5 for members 1 through 4, respectively. The top diagram in each of these figures is the load diagram, which is merely a repeat of the free body diagram. Each member is considered to be equivalent to a segment of a beam with the local x axis extending along the member from the first joint toward the second joint and the local y axis lying in the plane of the structure. Therefore, the same procedures and sign conventions described in Chapter 5 can be used to develop the V_y and M_z diagrams.

Member 1 (Joint 1 to Joint 2; Figure 6.2). Since the distributed force varies linearly along the member, the shear force diagram will be a second order curve and the bending moment diagram will be third order. Due to the magnitude of the distributed force at each end of the member, the shear force diagram will have a slope of -2 kips per foot just to the right of the first end and a zero slope at the second end.

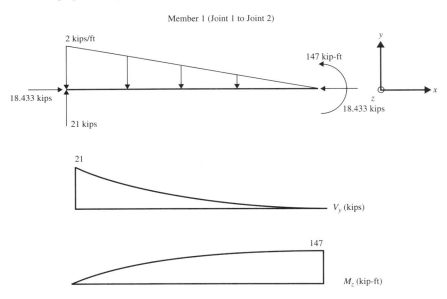

Figure 6.2 Example Problem 6.1.

Shear Force and Bending Moment in Plane Frames

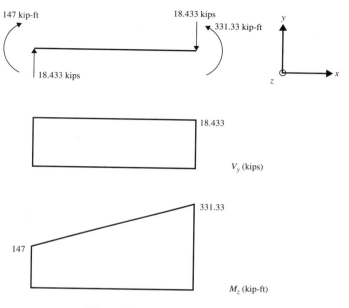

Figure 6.3 Example Problem 6.1.

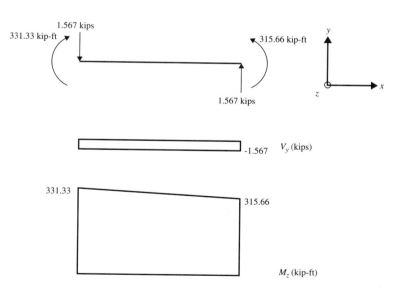

Figure 6.4 Example Problem 6.1.

184   Analysis of Statically Determinate Plane Frames   Chap. 6

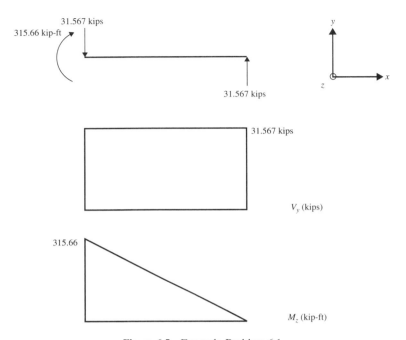

Figure 6.5 Example Problem 6.1.

Its ordinates will decrease in a decreasing manner, starting at an ordinate of −21 kips at the first end, due to the force applied to the end of the member by joint 1, and decreasing to zero since the area under the distributed force curve is exactly equal to 21 kips. The shear force diagram closes.

The bending moment diagram will have a zero ordinate at the first end, since no moment exists at the pinned support, and its ordinates will increase in a decreasing manner due to the shape of the shear force diagram. The slope of the bending moment diagram at the first end will be 21 kip feet per foot and the slope at the second end will be zero. Since the total area under the shear force diagram is 147 kip feet (i.e.; one third of the base times the height for a second order curve with a zero slope at one end), the ordinate of the bending moment diagram will be 147 kip feet at the second end of the member. This bending moment will be exactly balanced by the counterclockwise moment applied to the end of the member by joint 2. The bending moment diagram will close.

Member 2 (Joint 2 to Joint 3; Figure 6.3). The shear force diagram will have an upward step of 18.433 kips at the first end due to the upward force applied to the end of the member by joint 2. The ordinates of the diagram will remain constant over the entire length of the member, since there is no distributed force on the member, until a downward vertical step of 18.433 kips occurs at the second end

Shear Force and Bending Moment in Plane Frames 185

due to the downward force applied to the end of the member by joint 3. The shear force diagram closes.

The bending moment diagram will have an upward step of 147 kip feet at the first end due to the clockwise moment applied by joint 2. The ordinates will increase linearly to a value of 331.33 kip feet at the second end due to the constant shear force of 18.433 kips over the length of the member. The counterclockwise moment of 331.33 kip feet which is applied to the second end of the member by joint 3 will cause the bending moment diagram to close.

Member 3 (Joint 3 to Joint 4; Figure 6.4) and Member 4 (Joint 4 to Joint 5; Figure 6.5). The shear force and bending moment diagrams for these two members can be drawn in the same manner as the diagrams for member 2. The shear force diagram ordinates will be constant over the length of each member and the bending moment diagram ordinates will vary linearly over the length. The area under the shear force diagram for member 4 is -315.67 kip feet which will cause a closure error of -0.01 kip feet in the bending moment diagram for the member at the pinned support at joint 5. This is within acceptable roundoff error. To obtain a better closure would require expressing the reactive forces for the frame with more decimal places of accuracy. The accuracy of the calculations shown here should be acceptable for essentially all engineering design purposes.

Example Problem 6.2

Draw the shear force and bending moment diagrams for each member in the plane frame shown in Figure 6.6a. Although the frame has four independent reactive force components at the two pinned supports, it is statically determinate due to the equation of condition at the internal pin. The mathematical model for the frame consists of four joints and three members, as shown in Figure 6.6b. The internal pin is represented in the mathematical model by inserting a pin at the connection of member 3 to joint 3. For the determination of the reactions, the distributed force on member 2 can be considered to be equivalent to an inclined 60 kip concentrated force acting at the center of the member, which can be resolved into a 36 kip horizontal component acting toward the right and a downward 48 kip vertical component.

The reactions can be computed by using the equation of condition and the three independent global equilibrium equations for the frame as follows:

$$\Sigma_{3/right} M_Z \rightarrow R_{4X} = 0$$

$$\Sigma_1 M_Z \rightarrow R_{4Y} = 43.5 \text{ kips} \uparrow$$

$$\Sigma F_X \rightarrow R_{1X} = 6 \text{ kips} \leftarrow$$

$$\Sigma F_Y \rightarrow R_{1Y} = 4.5 \text{ kips} \uparrow$$

Using these reactions, the free body diagrams of the joints and the members can be drawn by starting at joint 1 and working clockwise through the frame for each succeeding member and joint. The process will be similar to that used in Example Problem 6.1, except that the components of the forces parallel and perpendicular to the longitudinal axis of member 2 must be used on the right side of joint 2 and the left side of joint 3 to correspond to the axial force and the shear force at the ends of member 2. The free body diagrams for each member and each joint are shown in Fig-

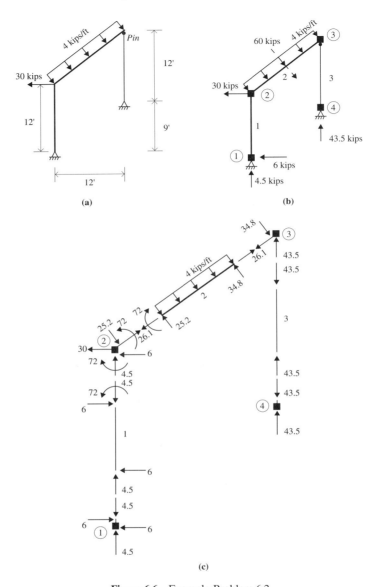

Figure 6.6 Example Problem 6.2.

ure 6.6c. Note that the final computed vertical force applied to joint 4 by member 3 exactly balances the reactive force at the joint. This indicates that the free body diagrams are correct.

The shear force and bending moment diagrams for members 1 and 2 are shown in Figures 6.7 and 6.8. The bending moment diagram for member 1 is a first order curve since there is no distributed force on the member. The vertical step in the

Shear Force and Bending Moment in Plane Frames

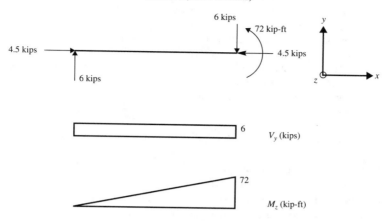

Figure 6.7 Example Problem 6.2.

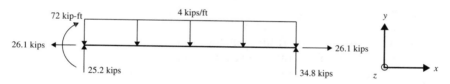

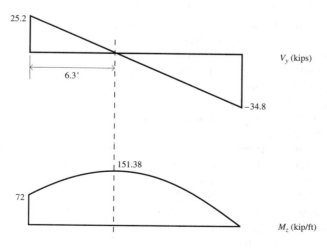

Figure 6.8 Example Problem 6.2.

bending moment diagram at the second end of the member is due to the moment applied to the end of the member by joint 2. The bending moment diagram for member 2 is a second order curve due to the uniform distributed force on the member. The step in the diagram at the first end is due to the moment applied to the end of the member by joint 2. Note that the maximum bending moment of 151.38 kip feet in the interior of member 2 occurs at the point of zero shear force, as expected. The shear force and bending moment diagrams are not shown for member 3 since the member is only subjected to an axial force at its ends. The shear force and the bending moment will be zero at all points on the member.

This same process can be used to draw the shear force and bending moment diagrams for any statically determinate and stable plane frame. The entire analysis procedure depends upon the ability of the analyst to compute the reactions and to draw the free body diagrams for the individual members. Extreme care must be exercised to ensure that the correct directions are assigned to the forces and moments when progressing through the structure from joint to member to joint to member to, and so on. Fortunately, there are several checks available since the free body diagram for the final joint must show that it is in equilibrium and the shear force and bending moment diagrams must close within acceptable roundoff error for each member.

It is often convenient for design purposes to summarize the bending moment diagrams for a plane frame by drawing them directly on the frame with the members oriented in the correct directions. However, in order to interpret the bending moments properly from drawings of this type, a convention must be established for the side of the member that the ordinates of the bending moment diagram will be plotted on; otherwise, it can become very confusing, particularly for frames which have members at a number of different orientations. If we look back at the sign convention which was used for the bending moment in a horizontal beam in Chapter 5, we will see that a positive moment on a cross section will produce compression bending stresses at the top of the cross section and tension bending stresses at the bottom. (It is assumed that the reader is familiar with the subject of bending stresses from a previous mechanics course.) Therefore, since we plotted positive moment up for a beam, we were actually plotting the bending moment diagram ordinates on the side of the beam which was in compression. To be consistent, we will use this same convention when plotting bending moment diagrams for the members in a plane frame. Figures 6.9a and 6.9b show the plots of the moment diagrams for the previous two example problems using this convention. For these particular plots, it was possible to plot the diagrams directly on the frame with the members connected since the diagrams did not overlap or interfere with each other. Figure 6.10 shows a plot of the bending moment diagrams for another plane frame in which the members of the frame are separated since the diagrams would overlap otherwise. Either type of plot will give the desired information in an understandable format. By looking at these plots, it is immediately obvious which portion of each side of a member is in tension or compression. This can be very useful during the design of the members, particularly for concrete frames.

Computer Program SDPFRAME

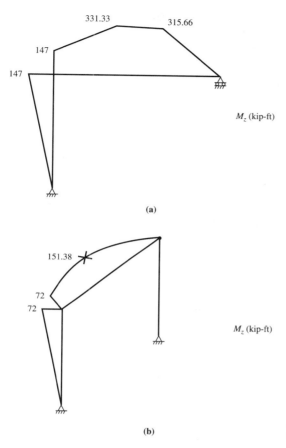

(a)

(b)

Figure 6.9 Total frame bending moment plots.

COMPUTER PROGRAM SDPFRAME

In Chapter 2, we used the program SDPFRAME to analyze plane frames. Our purpose for using the program at that time was to verify the manual solutions for the reactions for the frames. Therefore, we were only interested in the portion of the printed output that related to the reactions. However, if we look at this output closely, we will see that in addition to the reactions, it also lists the axial forces, the shear forces and the moments which are applied to the ends of each member by their end joints. In addition, as explained in the instructions for the program in the disk file SDPFRAME.DOC, a companion program named PLOTPFVM has been supplied which will read a data file created by SDPFRAME and plot the shear force and bending moment diagrams for any member in the mathematical model of the frame. Therefore, with the combination of these two programs we have the capability to perform a complete analysis of a statically determinate plane frame on the computer. The program SDPFRAME will list the axial force, the

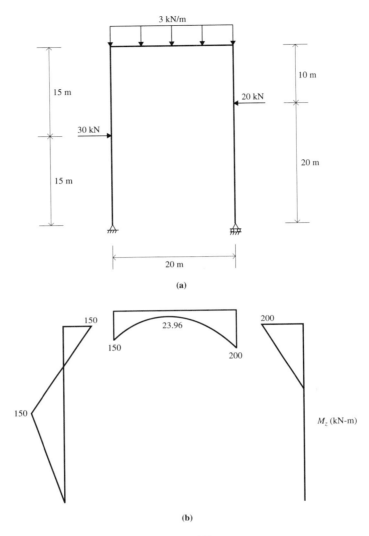

Figure 6.10

shear force and the bending moment at each end of each member and the reactions at the support joints. The program PLOTPFVM will show the variation of the shear force and the bending moment over the length of each member and list the values at the tenth points along the member. It will also show the location and magnitude of the maximum bending moment in the member if the shear force changes sign in the interior of the member.

Computer Program SDPFRAME

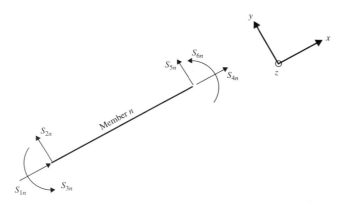

Figure 6.11 Plane frame member end loads.

Sign Convention for Program SDPFRAME

All active forces on the joints of a plane frame are expressed in the program input data as X and Y components in the global structure coordinate system, with the positive X axis extending toward the right and the positive Y axis extending up. All active moments on the joints are about the global Z axis, which extends outward from the plane of the structure, with positive moment acting in the counterclockwise direction. This sign convention corresponds to the right hand rule convention described in Chapter 2.

All loads applied to the ends of the members by the joints are expressed as components in the local member xyz right hand coordinate system, with the local x axis extending along the member from the first joint toward the second joint and the local z axis extending out of the plane of the structure in the same direction as the global Z axis. There are six end loads for any individual member n in the frame, as shown in Figure 6.11, where S_{1n} and S_{4n} are the axial forces along the local x axis, S_{2n} and S_{5n} are the shear forces along the local y axis and S_{3n} and S_{6n} are the moments about the local z axis. The directions for the end loads shown in this figure are considered to be the positive directions according to the right hand rule convention. All output values for the member end loads from SDPFRAME are expressed according to this sign convention. This was the most convenient sign convention to use during the development of the program. However, the companion program PLOTPFVM uses the designer sign convention for plotting the shear force and bending moment diagrams for any individual member, with the member shown in a horizontal orientation on the screen with the first joint on the left. This sign convention was shown in Figure 5.2 in Chapter 5.

All active distributed forces on any member are expressed as x and y components in the local member coordinate system, as shown in Figure 6.12 (this is a repeat of Figure 2.14 shown previously in Chapter 2), where the component q_x acts along the longitudinal axis of the member and the component q_y acts perpendicu-

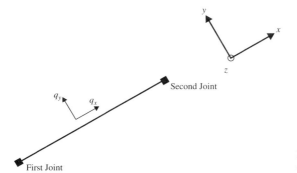

Figure 6.12 Plane frame member distributed force components.

lar to the member in the plane of the structure. It is extremely important that all force components be expressed with the correct signs in the program input data.

Stability and Static Determinacy of Plane Frames

Before attempting to perform an equilibrium analysis of a plane frame, it is necessary to establish that it is actually statically determinate. That is, the number of unknowns in the frame must be equal to the number of independent equilibrium equations.

The unknown quantities for any plane frame will consist of six end loads for each of the NM members in the frame and one reactive load component corresponding to each of the NR support restraints. Therefore, the total number of unknowns is

$$NUN = 6NM + NR \tag{6.1}$$

The independent equilibrium equations consist of three equations obtained by summing forces and moments in the global coordinate system for each of the NJ joints, three equations obtained by summing forces and moments in the local coordinate system for each of the NM members and one equation of condition for each of the NP pinned connections between the ends of the members and their end joints. Therefore, the number of independent equilibrium equations is

$$NEQ = 3NJ + 3NM + NP \tag{6.2}$$

Equating the number of equations to the number of unknowns and simplifying gives

$$3NJ = 3NM + NR - NP \tag{6.3}$$

which results in the following stability and static determinacy relationships for a plane frame:

if $3NJ > 3NM + NR - NP$, the plane frame is unstable

if $3NJ = 3NM + NR - NP$, the plane frame is statically determinate

Computer Program SDPFRAME

and

if $3NJ < 3NM + NR - NP$, the plane frame is statically indeterminate

The program SDPFRAME checks these conditions before attempting to analyze any frame. If either the first or the third condition is encountered, an appropriate message is printed and the analysis is terminated. If the count indicates that the frame is statically determinate, the program will attempt to generate a set of linear simultaneous equilibrium equations. If no solution can be found for the equilibrium equations, then the message "Frame appears to be geometrically unstable" will be printed and the analysis of the frame will be terminated. If the program can find a solution to the equations, it will print out the six end loads for each member and the reactive loads at the support joints. The member end loads for each individual member will be expressed as components in the local member coordinate system for that member, while the reactive loads will be expressed as components in the global coordinate system. The QBasic source code for the program is contained in the ASCII disk file SDPFRAME.BAS for those readers who are interested in seeing the details of the program operations. We will not discuss the details of the source code here.

Method of Analysis

The analysis procedure used in the program SDPFRAME is very similar to that used in the programs SDPTRUSS and SDSTRUSS which were discussed in Chapters 3 and 4. It will analyze a statically determinate plane frame by generating and solving a set of equilibrium equations with the member end loads S_{2n}, S_{3n}, S_{4n} and S_{6n}, for each individual member n in the frame, as the unknowns. After the equations have been solved for these member end loads, the additional end loads S_{1n} and S_{5n} are determined by summing forces in the local x and y directions for each member. By including only these four end loads as the unknowns in the equilibrium equations, only 4NM simultaneous equations must be solved rather than 6NM equations which would exist if all of the end loads for each member were included as unknowns. By solving 4NM equations, the program can analyze a frame with a maximum of 31 members, compared to a frame with only 21 members if 6NM equations had to be stored in the computer memory. The limit on the number of equations is due to the 64 kilobyte limit on the equation coefficient array in the QBasic language, as described previously in Chapter 3 in the description of the program SDPTRUSS. After the member end loads have been determined, the reactive forces and moments are computed by summing forces and moments in the restrained directions at the support joints.

The equilibrium equations are generated in the program by performing the following operations: summing forces in the unrestrained global X and Y directions and summing moments about the unrestrained global Z direction for each joint ($3NJ - NR$ equations); summing moments about local z at the first end of each member (NM equations); and setting the end moment to zero at any member end which is pinned to a joint (NP equations). The the total number of generated equations N is

$$N = 3NJ - NR + NM + NP \tag{6.4}$$

Comparing this expression to the requirement for a statically determinate plane frame in Eq. (6.3) shows that this process results in the required 4NM equations. These equations are in the form

$$[C]\{S\} = \{B\} \tag{6.5}$$

where the square matrix $[C]$ contains the equation coefficients, the column matrix $\{S\}$ contains the member end loads and the column matrix $\{B\}$ contains the right side constants for the equations. The equations are solved by the Gauss-Jordan Elimination Method using the same SUB Procedure which was used in the programs SDPTRUSS and SDSTRUSS.

Member Distributed Forces

The concentrated forces and moments acting on the joints can be used directly in the equations of equilibrium for the frame. However, in order to generate the joint equilibrium equations, any distributed forces acting on the members must be converted into equivalent concentrated joint loads. This can be accomplished by assuming that all joints in the mathematical model of the structure, at the ends of each member subjected to a distributed force over its length, are initially restrained against translation and rotation before the analysis is performed. For this situation, the loads applied to the joints at the end of any member by the distributed force on the member will be equal to the opposite of the reactive loads which would occur at the end of a beam with fixed supports at each end, as shown in Figure 6.13. Figure

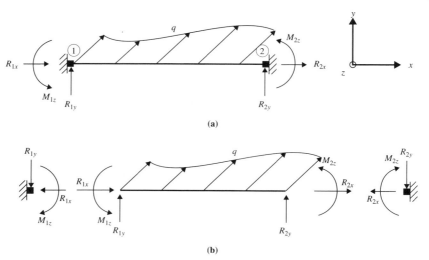

Figure 6.13 Plane frame fixed end member loads.

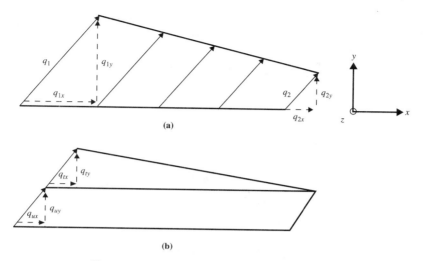

Figure 6.14 Plane frame member distributed load.

6.13a shows a beam with fixed ends, which is subjected to an inclined distributed force q, with horizontal and vertical reactive force components and a reactive moment at each end. Figure 6.13b shows an exploded view of the same beam, which shows the reactive loads applied to the end of the beam and the opposite loads applied to the end joints. These joint loads are the equivalent loads which can be used in the analysis to represent the effect of the distributed force on the member. These loads are usually called the *fixed end joint loads*. After the equilibrium analysis of the frame is performed using these equivalent joint loads, the total end loads for any member can be computed by superimposing the fixed end reactive end loads shown in Figure 6.13b and the end loads obtained from the equilibrium analysis.

SDPFRAME assumes that the distributed force on any member varies linearly over the length of the member, as shown in Figure 6.14a, where the local x and y components at the first end are q_{1x} and q_{1y} and the local components at the second end are q_{2x} and q_{2y}. For analysis purposes, the distributed force on the member can be assumed to be composed of a uniform portion and a triangular portion, as shown in Figure 6.14b, where the distributed force components at the two ends of the member, which correspond to the uniform portion, are

$$q_{ux} = q_{2x} \qquad (6.6)$$

$$q_{uy} = q_{2y} \qquad (6.7)$$

and the distributed force components at the first end of the member, which correspond to the triangular portion, are

$$q_{tx} = q_{1x} - q_{2x} \qquad (6.8)$$

$$q_{ty} = q_{1y} - q_{2y} \qquad (6.9)$$

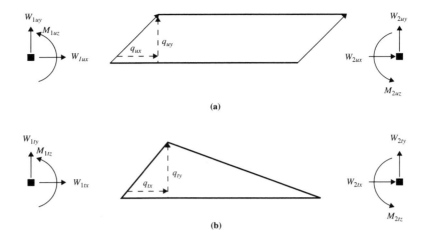

Figure 6.15 Plane frame member fixed-end joint loads.

The fixed end joint loads in the local member coordinate system, due to the uniform portion of the distributed force, as shown in Figure 6.15a, are

$$W_{1ux} = W_{2ux} = \frac{q_{ux}L}{2} \tag{6.10}$$

$$W_{1uy} = W_{2uy} = \frac{q_{uy}L}{2} \tag{6.11}$$

$$M_{1uz} = -M_{2uz} = \frac{q_{uy}L^2}{12} \tag{6.12}$$

while the fixed end joint loads in the local member coordinate system, due to the triangular portion of the distributed force, as shown in Figure 6.15b, are

$$W_{1tx} = \frac{q_{tx}L}{3} \tag{6.13}$$

$$W_{2tx} = \frac{q_{tx}L}{6} \tag{6.14}$$

$$W_{1ty} = \frac{7q_{ty}L}{20} \tag{6.15}$$

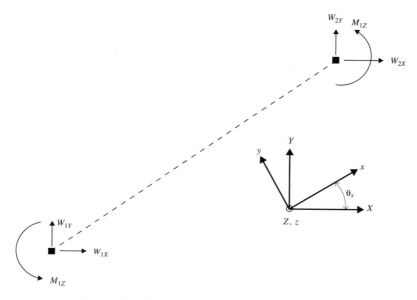

Figure 6.16 Plane frame global fixed-end joint loads.

$$W_{2ty} = \frac{3q_{ty}L}{20} \tag{6.16}$$

$$M_{1tz} = \frac{q_{ty}L^2}{20} \tag{6.17}$$

$$M_{2tz} = -\frac{q_{ty}L^2}{30} \tag{6.18}$$

The fixed end loads on the ends of the member have the same magnitudes but act in the opposite directions. We will not discuss the derivation of these expressions for the fixed end joint loads at this time since the analysis of a statically indeterminate fixed end beam is beyond our present capabilities. A procedure for computing the fixed end reactions for a beam subjected to a distributed force will be presented in Chapter 8.

The fixed end joint loads given in Eqs. (6.10) through (6.18) are expressed as components in the local member coordinate system. However, they must be converted to joint loads in the global coordinate system for use in the joint equilibrium equations. These global joint loads, as defined in Figure 6.16, are as follows:

$$W_{1X} = (W_{1ux} + W_{1tx})\cos\theta_X - (W_{1uy} + W_{1ty})\sin\theta_X \tag{6.19}$$

$$W_{2X} = (W_{2ux} + W_{2tx})\cos\theta_X - (W_{2uy} + W_{2ty})\sin\theta_X \tag{6.20}$$

$$W_{1Y} = (W_{1ux} + W_{1tx}) \sin \theta_X + (W_{1uy} + W_{1ty}) \cos \theta_X \quad (6.21)$$

$$W_{2Y} = (W_{2ux} + W_{2tx}) \sin \theta_X + (W_{2uy} + W_{2ty}) \cos \theta_X \quad (6.22)$$

$$M_{1Z} = M_{1uz} + M_{1tz} \quad (6.23)$$

$$M_{2Z} = M_{2uz} + M_{2tz} \quad (6.24)$$

where θ_X is the counterclockwise angle from the global X axis to the local member x axis. The signs of $\sin \theta_X$ and $\cos \theta_X$ will give the correct signs for the X and Y forces for any orientation of the member.

Computer Solution of Example Problem 6.1

Figure 6.17 shows the listing of the input data file for the analysis of the plane frame in Example Problem 6.1 by SDPFRAME. Except for the title, this is exactly the same as the data file listed for Example Problem 2.2 in Chapter 2. Figure 6.18 shows a plot of the mathematical model of the frame as described in the input file. Figure 6.19 shows the output from SDPFRAME for the member end loads and the reactions, while Figures 6.20 through 6.23 show the plots of the shear force and bending moments for members 1 through 4, respectively, from the program PLOTPFVM. The shear force is labeled V_y and the bending moment is labeled M_z to be consistent with their directions in the local member coordinate system. The computer results agree with those obtained in the manual solution.

Computer Solution of Example Problem 6.2

Figure 6.24 shows the listing of the input data file for Example Problem 6.2. A plot of the mathematical model of the frame by the program PLOTSDPF is

```
Example Problem 6.1 - Analysis by Program SDPFRAME
5,4,0,2,2,1
Joint Coordinates
1,0.0,0.0
2,0.0,21.0
3,10.0,21.0
4,20.0,21.0
5,30.0,21.0
Member Data
1,1,2
2,2,3
3,3,4
4,4,5
Support Restraints
1,1,1,0
5,0,1,0
Joint Loads
3,0.0,-20.0,0.0
4,0.0,-30.0,0.0
Distributed Member Loads
1,0.0,0.0,-2.0,0.0
```

Figure 6.17 Example Problem 6.1, SDPFRAME input.

Computer Program SDPFRAME

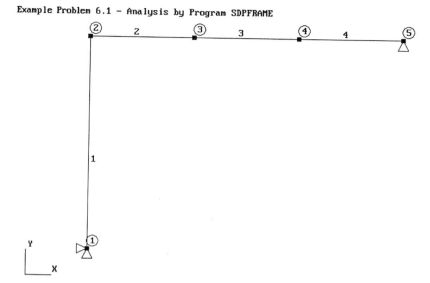

Figure 6.18 Example Problem 6.1, geometry plot from PLOTSDPF.

```
Example Problem 6.1 - Analysis by Program SDPFRAME
Member End Loads

Member    Joint          Sx              Vy              Mz

   1        1          18.433          21.000            0.00
            2         -18.433           0.000          147.00
   2        2          -0.000          18.433         -147.00
            3           0.000         -18.433          331.33
   3        3          -0.000          -1.567         -331.33
            4           0.000           1.567          315.67
   4        4           0.000         -31.567         -315.67
            5           0.000          31.567            0.00

Reactions

Joint                    RX              RY              MZ

   1                  -21.000          18.433            0.00
   5                    0.000          31.567            0.00
```

Figure 6.19 Example Problem 6.1, SDPFRAME output.

shown in Figure 6.25. The listing of the output from SDPFRAME is shown in Figure 6.26, and the plots of the shear force and bending moment diagrams for members 1 through 3 are shown in Figures 6.27 through 6.29. These results agree with the manual solution.

The program SDPFRAME can be used to verify the manual solutions for any of the suggested problems at the end of this chapter.

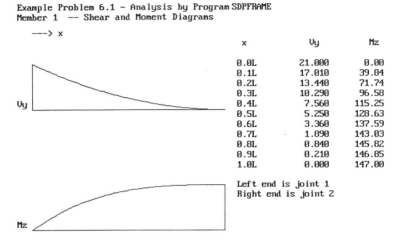

Figure 6.20 Example Problem 6.1, V_y and M_z diagrams from PLOTPFVM.

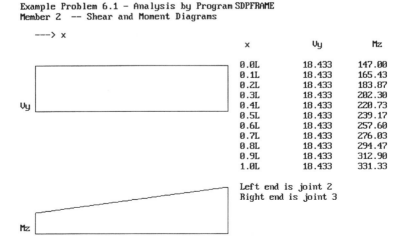

Figure 6.21 Example Problem 6.1, V_y and M_z diagrams from PLOTPFVM.

Example Problem 6.1 - Analysis by Program SDPFRAME
Member 3 -- Shear and Moment Diagrams

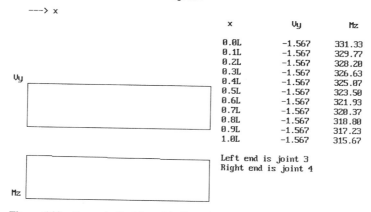

x	Vy	Mz
0.0L	-1.567	331.33
0.1L	-1.567	329.77
0.2L	-1.567	328.20
0.3L	-1.567	326.63
0.4L	-1.567	325.07
0.5L	-1.567	323.50
0.6L	-1.567	321.93
0.7L	-1.567	320.37
0.8L	-1.567	318.80
0.9L	-1.567	317.23
1.0L	-1.567	315.67

Left end is joint 3
Right end is joint 4

Figure 6.22 Example Problem 6.1, V_y and M_z diagrams from PLOTPFVM.

Example Problem 6.1 - Analysis by Program SDPFRAME
Member 4 -- Shear and Moment Diagrams

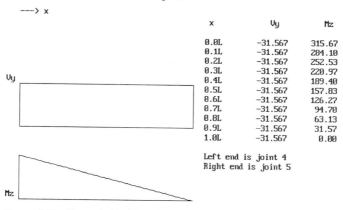

x	Vy	Mz
0.0L	-31.567	315.67
0.1L	-31.567	284.10
0.2L	-31.567	252.53
0.3L	-31.567	220.97
0.4L	-31.567	189.40
0.5L	-31.567	157.83
0.6L	-31.567	126.27
0.7L	-31.567	94.70
0.8L	-31.567	63.13
0.9L	-31.567	31.57
1.0L	-31.567	0.00

Left end is joint 4
Right end is joint 5

Figure 6.23 Example Problem 6.1, V_y and M_z diagrams from PLOTPFVM.

```
Example Problem 6.2 - Analysis by Program SDPFRAME
4,3,1,2,1,1
Joint Coordinates
1,0.0,0.0
2,0.0,12.0
3,12.0,21.0
4,12.0,9.0
Member Data
1,1,2
2,2,3
3,3,4
Pinned Members
3,0,1
Support Restraints
1,1,1,0
4,1,1,0
Joint Loads
2,-30.0,0.0,0.0
Distributed Member Loads
2,0.0,0.0,-4.0,-4.0
```

Figure 6.24 Example Problem 6.2, SDPFRAME input.

Example Problem 6.2 - Analysis by Program SDPFRAME

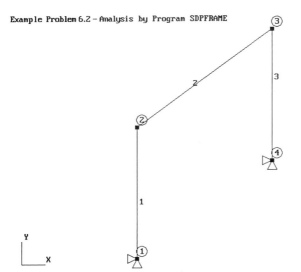

Figure 6.25 Example Problem 6.2, geometry plot from PLOTSDPF.

```
Example Problem 6.2 - Analysis by Program SDPFRAME
Member End Loads

Member    Joint         Sx              Vy              Mz
  1         1         4.500           6.000           0.00
            2        -4.500          -6.000          72.00
  2         2       -26.100          25.200         -72.00
            3        26.100          34.800           0.00
  3         3        43.500           0.000           0.00
            4       -43.500           0.000           0.00

Reactions

Joint              RX              RY              MZ
  1             -6.000           4.500           0.00
  4              0.000          43.500           0.00
```

Figure 6.26 Example Problem 6.2, SDPFRAME output.

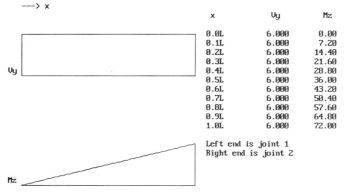

Figure 6.27 Example Problem 6.2, V_y and M_z diagrams from PLOTPFVM.

Suggested Problems

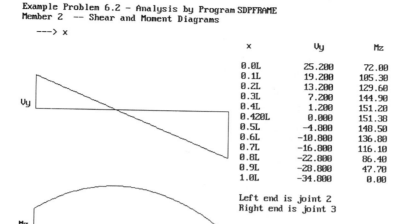

Figure 6.28 Example Problem 6.2, V_y and M_z diagrams from PLOTPFVM.

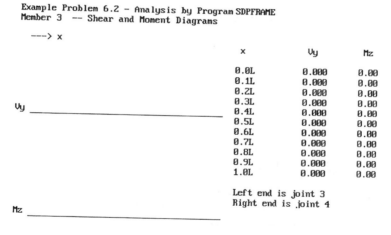

Figure 6.29 Example Problem 6.2, V_y and M_z diagrams from PLOTPFVM.

SUGGESTED PROBLEMS

SP6.1 to **SP6.8** Compute the reactions and draw the shear force and bending moment diagrams for each of the statically determinate plane frames shown in Figures SP6.1 to SP6.8. Verify the analysis results with the program SDPFRAME.

204 Analysis of Statically Determinate Plane Frames   Chap. 6

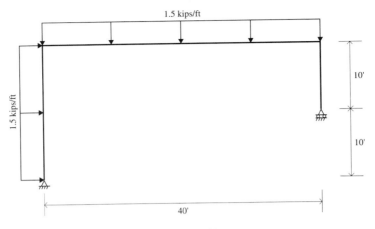

Figure SP6.1

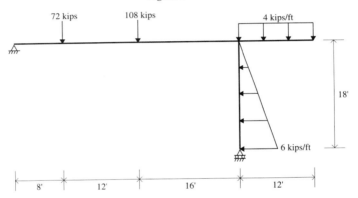

Figure SP6.2

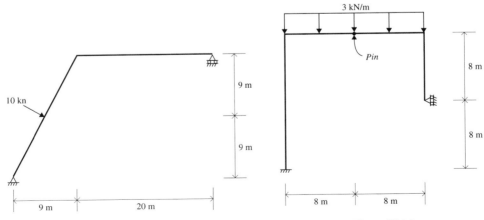

Figure SP6.3   **Figure SP6.4**

Suggested Problems

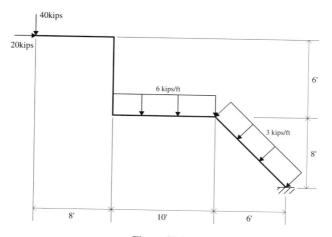

Figure SP6.5

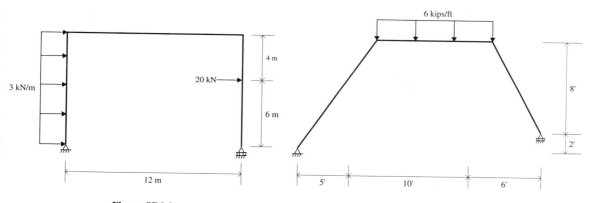

Figure SP6.6

Figure SP6.7

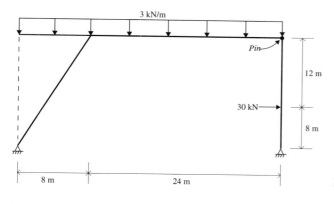

Figure SP6.8

7

Analysis of Statically Determinate Space Frames

SHEAR FORCE AND BENDING MOMENT IN SPACE FRAMES

The analysis of a space frame is much more complicated than the analysis of a plane frame since the joints may be at any location in three dimensional space and the members may be at any orientation. Each member in a space frame has six independent end loads at each end, consisting of three force components along the local x, y and z coordinate axes for the member and three moment components about these axes, as shown for a typical space frame member n in Figure 7.1. The end forces are shown in this figure as single headed vectors whose positive directions extend in the positive directions of the local axes, while the end moments are shown as double headed vectors whose positive directions are governed by the right hand rotation rule convention. All of the vectors in Figure 7.1 are shown acting in positive directions. Forces S_{1n} and S_{7n} correspond to the axial forces in the local x direction, forces S_{2n} and S_{8n} correspond to shear forces in the local y direction, forces S_{3n} and S_{9n} correspond to shear forces in the local z direction, moments S_{4n} and S_{10n} correspond to twisting moments about x, moments S_{5n} and S_{11n} correspond to bending moments about y and moments S_{6n} and S_{12n} correspond to bending moments about z.

Since there are two shear force components and two bending moment components at each end of a member, it is necessary to draw two sets of shear force and bending moment diagrams for each member in a space frame. One set will correspond to the shear force along the local y axis and the bending moment about the local z axis (V_y and M_z), while the other set will correspond to the shear force along the local z axis and the bending moment about the local y axis (V_z and M_y).

Shear Force and Bending Moment in Space Frames

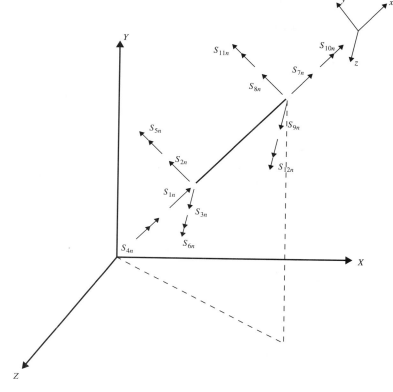

Figure 7.1 Space frame member end loads.

The shear force and bending moment diagrams can be developed for the members in a statically determinate space frame by using a procedure similar to that used to develop the shear force and bending moment diagrams for the members in a plane frame in Example Problems 6.1 and 6.2 in the previous chapter. That is, the free body diagrams for each member can be developed from which the shear forces and bending moments at any point on the member can be determined. The primary difference is that all calculations must be performed for a three dimensional load system and two sets of diagrams must be developed for each member.

Before we can perform the analysis of a space frame for the member shear force and bending moment diagrams, we must establish rules concerning the orientation of the global coordinate system for the overall structure and for the local coordinate systems for each member in the mathematical model of the frame. These rules are defined for consistency and convenience in the analysis procedure. The global coordinate system will be oriented such that the XZ plane is horizontal with the positive Y axis up. The local coordinate system for any member must be oriented so that the local x axis extends along the longitudinal axis of the member from the first joint toward the second joint, with the local z axis parallel to the global XZ plane. The local y axis must be oriented so that it projects a positive component on the global Y axis, except for vertical members, in which case, both the local y and z axes will be parallel to the global XZ plane. For this case, the local z axis must be oriented parallel to the global Z axis and extend in the positive Z direction. By following these rules there will be a unique orientation of the local coordinate system for each member in the frame so that there can be no confusion concerning which shear force and bending moment components are associated with which local member axes.

Example Problem 7.1

Draw the shear force and bending moment diagrams for the members in the space frame shown in Figure 7.2a. The mathematical model for the frame, which will be used in the analysis, will consist of four joints and three members, as shown in Figure 7.2b. The orientation of the global coordinate system, shown in the lower right corner of the figure, has been chosen for convenience so that the coordinate axes are parallel to the three members in the mathematical model. The reactive loads for this structure will consist of three force components along X, Y and Z and three moment components about these axes at joint 1. The frame is statically determinate since there are six unknown reactive quantities and six independent equations of equilibrium for a rigid body in space. The magnitudes and directions of the reactive forces and moments can be easily computed by summing forces in the global X, Y and Z directions and summing moments about the global X, Y and Z axes for the total structure:

$$R_{1X} = -5 \text{ kN}$$

$$R_{1Y} = 14 \text{ kN}$$

$$R_{1Z} = -5 \text{ kN}$$

$$M_{1X} = 10 \text{ kN-m}$$

$$M_{1Y} = 100 \text{ kN-m}$$

$$M_{1Z} = 160 \text{ kN-m}$$

The verification of these reactions is left as an exercise for the reader.

The free body diagrams for the members can be developed by starting at joint 1 with the known reactive loads and working sequentially along the frame to determine the equilibrium requirements for each joint and member in turn. Figures 7.3 to 7.5 show the free body diagrams for members 1, 2 and 3 and their end joints. All forces acting on the members and joints are represented by single headed vectors and all moments are represented by double headed vectors, as explained previously. The orientation of the local coordinate system, which is shown adjacent to each member, has been established according to the previously defined rules.

The two sets of shear force and bending moment diagrams for each member are shown in Figures 7.6 to 7.8. The top set in each figure corresponds to V_y and M_z

Shear Force and Bending Moment in Space Frames

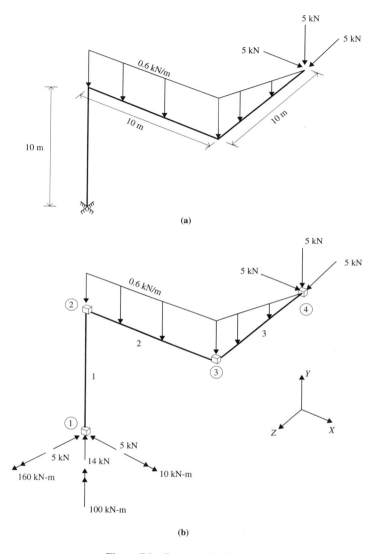

Figure 7.2 Example Problem 7.1.

while the bottom set corresponds to V_z and M_y. Note that the bending moment diagrams for M_z for member 2 and member 3 are second order and third order curves, respectively, due to the distributed forces on these members, while all other bending moment diagrams are first order curves. The sign convention for V_y and M_z in the diagrams is the same as established previously for a beam. That is, V_y is positive if it acts in the negative local y direction on the right end of a left segment and M_z is positive if it is counterclockwise on the right end of a left segment. This positive direc-

210   Analysis of Statically Determinate Space Frames   Chap. 7

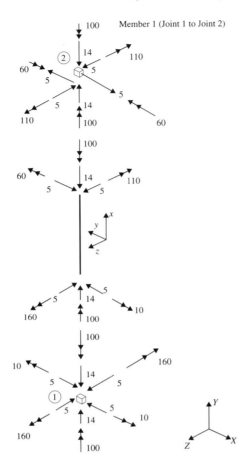

Figure 7.3 Example Problem 7.1.

tion for M_z corresponds to viewing the bending moment diagrams for M_z for each member from the positive local z directions, as shown by the orientation of the local axes to the right of the load diagrams corresponding to V_y and M_z. To be consistent for the bending moment diagrams for M_y for each member, we should view them from the positive local y directions. This will result in the relative orientations of the local x and z axes shown to the right of the load diagrams corresponding to V_z and M_y. Therefore, the sign convention which will be used for V_z and M_y is: V_z is positive if it acts in the positive local z direction on the right end of a left segment; and M_y is positive if it is counterclockwise on the right end of a left segment. By using this sign convention, the shear force diagram for V_z can be developed by following the instructions indicated by the local z load diagram and the bending moment diagram for M_y can be developed by following the instructions indicated by the shear force diagram for V_z, as demonstrated in Chapter 5 for the diagrams for V_y and M_z for a horizontal beam.

This same process can be used for any statically determinate space frame. However, the analysis procedure can become very tedious and errors can be easi-

Computer Program SDSFRAME

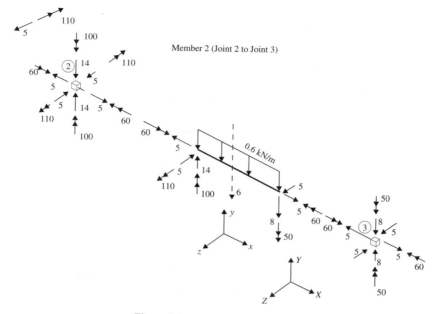

Figure 7.4 Example Problem 7.1.

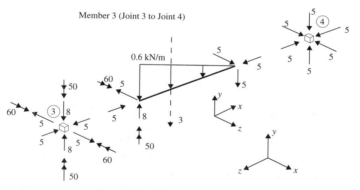

Figure 7.5 Example Problem 7.1.

ly made due to the difficulty of working with a three dimensional load system. Extreme care must be exercised in determining the correct magnitudes and directions for the member end forces and moments.

COMPUTER PROGRAM SDSFRAME

The computer program SDSFRAME will analyze a statically determinate space frame of arbitrary geometry and print out the 12 local end load components for each member and the 6 global reactive load components at each support joint. The companion program PLOTSDSF will plot the geometry of the mathematical model of the frame viewed from any desired view position and the program

212   Analysis of Statically Determinate Space Frames   Chap. 7

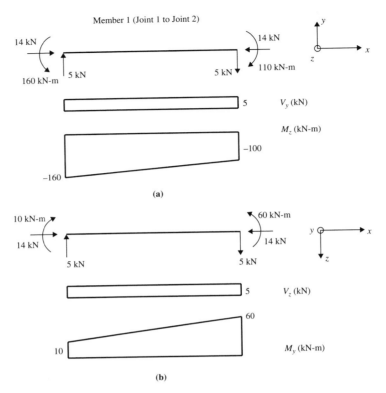

Figure 7.6 Example Problem 7.1, V and M diagrams.

PLOTSFVM will plot the shear force and bending moment diagrams for each member after the frame has been analyzed by SDSFRAME. The instructions for using SDSFRAME are contained in the ASCII disk file SDSFRAME.DOC and the QBasic source code is given in the ASCII disk file SDSFRAME.BAS. The program is limited to a frame with a maximum of 15 members due to the array size limitation in the QBasic language as described in previous chapters. This size limitation should be sufficient for most educational purposes.

Sign Convention for SDSFRAME

All active and reactive forces and moments on the joints of a space frame in SDSFRAME are expressed as components in the global coordinate system for the structure, with the global XZ plane horizontal and the global Y axis positive up. All loads applied to the ends of the members by the joints will be expressed as components in the local member coordinate system in the program output. The local coordinate system for each member will be oriented with respect to the global coordinate system, as explained previously for the manual analysis of Example

Computer Program SDSFRAME

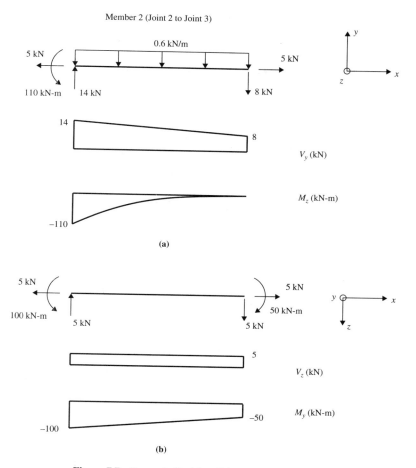

Figure 7.7 Example Problem 7.1, V and M diagrams.

Problem 7.1. The program user must be careful when using the computed results for the member end loads to ensure that the directions for the shear forces and the bending moments are interpreted properly, particularly for vertical members. The program PLOTSFVM will plot the shear force and bending moment diagrams using the sign convention described for the plots in Example Problem 7.1.

All active distributed forces on the members must be expressed as components in the local coordinate systems for the individual members. The forces may have three components, where q_x acts along the member and q_y and q_z act perpendicular to the member. Each force component may have a linear variation over the length of the member.

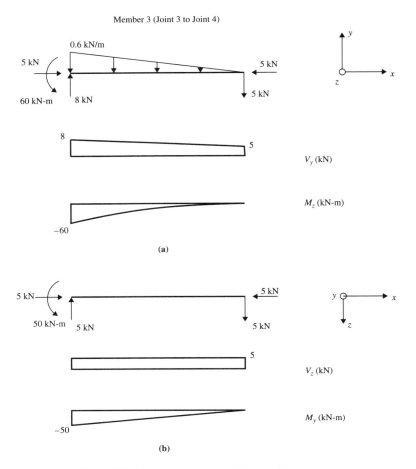

Figure 7.8 Example Problem 7.1, V and M diagrams.

Stability and Static Determinacy of Space Frames

The unknown quantities for a space frame consist of the 12 end load components for each of the NM members in the frame and one reactive load component for each of the NR support restraints. Therefore, the total number of unknowns is

$$\text{NUN} = 12\text{NM} + \text{NR} \tag{7.1}$$

The independent equilibrium equations consist of six equations obtained by summing forces and moments in the global directions for each of the NJ joints, six equations obtained by summing forces and moments in the local directions for each of the NM members and one equation of condition for each of the NP pinned con-

Computer Program SDSFRAME

nections for either the local y or z axis at the ends of the members. Therefore, the total number of independent equilibrium equations is

$$\text{NEQ} = 6\text{NJ} + 6\text{NM} + \text{NP} \tag{7.2}$$

Equating the number of equations to the number of unknowns and simplifying gives

$$6\text{NJ} = 6\text{NM} + \text{NR} - \text{NP} \tag{7.3}$$

which results in the following stability and static determinacy relationships for a space frame:

if $6\text{NJ} > 6\text{NM} + \text{NR} - \text{NP}$, the space frame is unstable;

if $6\text{NJ} = 6\text{NM} + \text{NR} - \text{NP}$, the space frame is statically determinate;

and

if $6\text{NJ} < 6\text{NM} + \text{NR} - \text{NP}$, the space frame is statically indeterminate.

The program SDSFRAME will only attempt to analyze a space frame if the second condition is satisfied. If either the first or the third condition is encountered, the analysis is terminated. If the frame appears to be statically determinate and the program cannot solve the set of equilibrium equations, the message "Frame appears to be geometrically unstable" will be printed and the analysis will be terminated. If the program can solve the equations, it will print out the 12 local end load components for each member and the 6 global reactive load components at each support joint.

Method of Analysis

The program SDSFRAME is almost identical to the plane frame analysis program SDPFRAME, which was described in Chapter 6 except for the differences which are required to handle the third dimension. The program generates a set of 8NM equilibrium equations with the member end loads S_{2n}, S_{3n}, S_{5n}, S_{6n}, S_{7n}, S_{10n}, S_{11n} and S_{12n} for each individual member n in the frame as the unknowns. After the set of equilibrium equations has been solved, the end forces S_{1n}, S_{8n} and S_{9n} are determined by summing forces in the local x, y and z directions, respectively, for each member, and the end moment S_{4n} is determined by summing moments about the local x axis for each member.

The equilibrium equations are generated in the program by performing the following operations: summing forces and moments in the unrestrained global X,

Y and Z directions for each joint (6NJ − NR equations); summing moments about the local y and z axes at the first end of each member (2NM equations); and setting the end moments to zero for any member end which is pinned to a joint about either the local y or z axis (NP equations). The total number of equations N is

$$N = 6NJ - NR + 2NM + NP \tag{7.4}$$

Comparing this expression to the requirement for a statically determinate space frame in Eq. (7.3) shows that this process results in the required 8NM equations. The equations are solved by the Gauss-Jordan Elimination Method using the same SUB Procedure used in programs SDPTRUSS, SDSTRUSS and SDPFRAME.

When setting up the equilibrium equations, the concentrated joint loads are used directly. However, the distributed member forces must be converted into equivalent concentrated joint loads before they can be used in the joint equilibrium equations. The same procedure described in Chapter 6 for the program SDPFRAME is used, except three distributed force components along the local x, y and z axes must be considered rather than the two components for a plane frame member.

Computer Analysis of Example Problem 7.1

Figure 7.9 shows a listing of the input data file for the analysis of the space frame in Example Problem 7.1. A plot of the geometry of the mathematical model of the frame by the program PLOTSDSF is shown in Figure 7.10. This is an isometric view with a horizontal view angle of 45° and a vertical view angle of 30°. The translation restraints at the support joints are indicated by small triangular wedges in the plot. The rotation restraints are not shown due to the difficulty of interpreting them in three dimensions. The listing of the output from SDSFRAME is shown in Figure 7.11 and the plots of the shear force and bending moment dia-

```
Example Problem 7.1 - Analysis by Program SDSFRAME
4,3,0,1,1,2
Joint Coordinates
1,0.0,0.0,0.0
2,0.0,10.0,0.0
3,10.0,10.0,0.0
4,10.0,10.0,-10.0
Member Data
1,1,2
2,2,3
3,3,4
Support Restraints
1,1,1,1,1,1,1
Joint Loads
4,5.0,-5.0,5.0,0.0,0.0,0.0
Distributed Member Loads
2,0.0,0.0,-0.6,-0.6,0.0,0.0
3,0.0,0.0,-0.6,0.0,0.0,0.0
```

Figure 7.9 Example Problem 7.1, SDSFRAME input.

Computer Program SDSFRAME

Example Problem 7.1 - Analysis by Program SDSFRAME

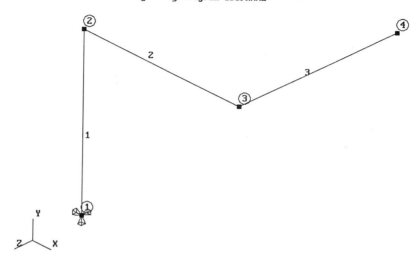

Figure 7.10 Example Problem 7.1, geometry plot from PLOTSDSF.

```
Example Problem 7.1 - Analysis by Program SDSFRAME
Member End Loads
```

Member	Joint	Sx	Vy	Vz	Mx	My	Mz
1	1	14.000	5.000	-5.000	100.00	-10.00	160.00
	2	-14.000	-5.000	5.000	-100.00	60.00	-110.00
2	2	-5.000	14.000	-5.000	60.00	100.00	110.00
	3	5.000	-8.000	5.000	-60.00	-50.00	0.00
3	3	5.000	8.000	-5.000	0.00	50.00	60.00
	4	-5.000	-5.000	5.000	0.00	0.00	0.00

Reactions

Joint	RX	RY	RZ	MX	MY	MZ
1	-5.000	14.000	-5.000	10.00	100.00	160.00

Figure 7.11 Example Problem 7.1, SDSFRAME output.

grams for each member by the program PLOTSFVM are shown in Figures 7.12 through 7.17. These diagrams agree with those which were generated in the manual analysis. The program first generates the plot of the diagrams for V_y and M_z. The diagrams for V_z and M_y are plotted after the Enter key is pressed.

The program SDSFRAME can be used to verify the manual solutions for any of the suggested problems at the end of this chapter.

218   Analysis of Statically Determinate Space Frames   Chap. 7

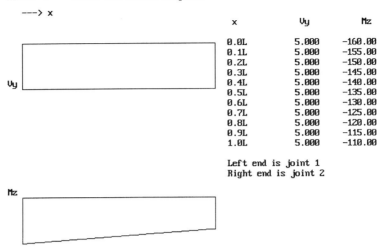

Figure 7.12 Example Problem 7.1, V_y and M_z diagram, from PLOTSFVM.

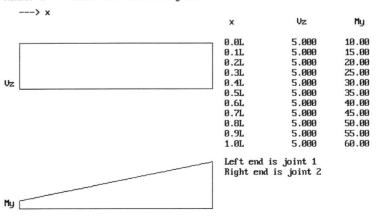

Figure 7.13 Example Problem 7.1, V_z and M_y diagrams from PLOTSFVM.

Computer Program SDSFRAME

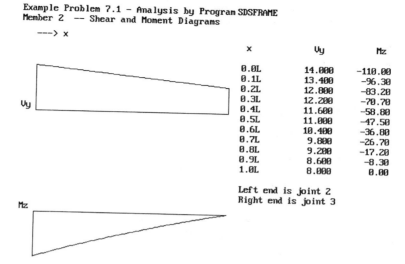

Figure 7.14 Example Problem 7.1, V_y and M_z diagrams from PLOTSFVM.

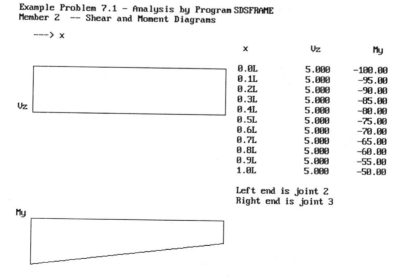

Figure 7.15 Example Problem 7.1, V_z and M_y diagrams from PLOTSFVM.

220 Analysis of Statically Determinate Space Frames   Chap. 7

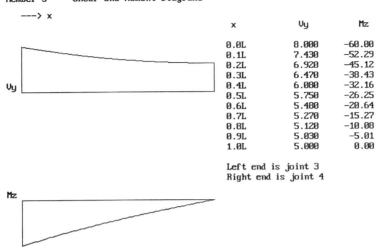

Figure 7.16 Example Problem 7.1, V_y and M_z diagrams from PLOTSFVM.

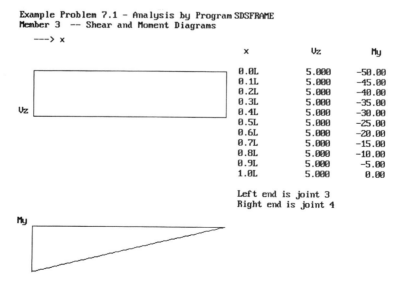

Figure 7.17 Example Problem 7.1, V_z and M_y diagrams from PLOTSFVM.

SUGGESTED PROBLEMS

SP7.1 and **SP7.2** Compute the reactions and draw the shear force and bending moment diagrams for each of the statically determinate space frames shown in Figures SP7.1 and SP7.2. Verify the analysis results with the program SDSFRAME.

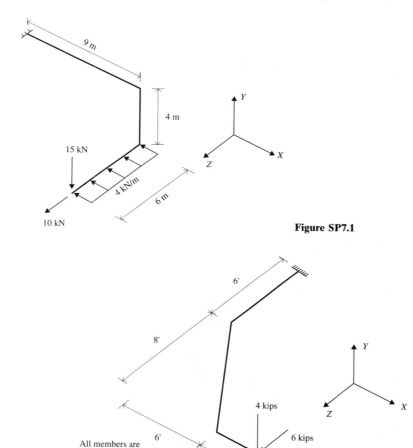

Figure SP7.1

Figure SP7.2

8

Deflection of Beams

DEFLECTIONS OF STRUCTURAL SYSTEMS

In the analyses in the previous chapters, we saw that when a set of loads is applied to the joints in the mathematical model of a structure, the joints in turn will apply loads to the ends of the individual members. These end loads will produce stresses in the material, which will cause the material to deform. The type and magnitude of the material deformation will depend upon both the type and magnitude of the stresses and the properties of the material. It is assumed that the reader has some background in the computation of stresses in various types of structural members and in the stress-strain relationships for common structural materials. This information is usually covered in one or more sophomore level mechanics courses in most civil engineering and mechanical engineering undergraduate programs.

If we are to fully understand how a structure behaves under a set of loads, we must not only be able to determine the end loads for the individual members, but we must also be able to determine what types of deformations are produced in the members and how these member deformations affect the final deflected shape of the structure. In this chapter, we will restrict our discussion to the deflection of horizontal beams. In Chapter 9, we will consider deflections in trusses and frames.

DIFFERENTIAL EQUATION OF BENDING

Before we can determine the deflections which occur in a beam under any set of applied loads, we must derive a relationship between the loads acting on the beam

Differential Equation of Bending

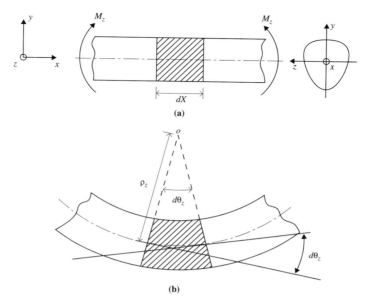

Figure 8.1 Deformed beam in bending.

and the deformed shape of the beam. To develop the required relationship, we will consider a segment of a beam of infinitesimal length dX which is subjected to pure bending (i.e.; the vertical shear force is zero), as shown in Figure 8.1a. The positive sense of the moments on each end of the segment is defined by the designer sign convention described previously in Chapter 5. The global X axis extends along the length of the beam toward the right and the positive global Y axis is up. The origin of the global coordinate system will be at the level of the centroid of the cross section, so that as the beam deforms under the applied loads, the vertical distance Y from the global origin will be equal to the deflection of the centroidal axis at any point along the beam. The local x, y and z axes, for each individual member in the mathematical model of the beam, are in the same directions as the global X, Y and Z axes, respectively, with the local x axis passing through the centroid of the cross section of the member. The following restrictions are placed on the beam and the material:

1. The local y and z axes for each individual member cross section correspond to the principal axes, so that the product of inertia I_{yz} of the cross section is zero.
2. The beam is initially straight.
3. The beam material is linear elastic.

The deformed shape of the segment of the beam is shown in Figure 8.1b. Although it will not be proved analytically, it can be shown by experimental observation that the sides of the segment, when subjected to pure bending, remain straight and rotate so that they remain perpendicular to the deformed line through the centroid of the cross section of the beam (i.e.; plane sections before bending are plane after bending). Also, from our previous sophomore level mechanics course, we know that the bending stress σ_x perpendicular to the cross section at any level in a beam, which meets the previous restrictions is

$$\sigma_x = -\frac{M_z y}{I_z} \tag{8.1}$$

where the negative sign is present due to the positive directions for the local y axis and the bending moment M_z. A positive sign for the stress indicates tension while a negative sign indicates compression. Therefore, compression bending stresses occur in the upper portion of a cross section and tension bending stresses occur in the lower portion when the cross section is subjected to a positive bending moment. The line of zero stress which separates the two stress zones, at the level of the centroid of the cross section, is known as the *neutral axis*. The deformed position of the neutral axis along the length of the beam is usually called the *elastic cur e* for the beam.

If the two sides of the deformed segment are extended, as shown by the dashed lines in Figure 8.1b, the lines will intersect at point o. The angle between these two lines is equal to $d\theta_z$. This angle is equal to the change in the slope angle of the deformed beam over the length dX since the two tangent lines which define the slope at either end of the segment are perpendicular to the sides of the deformed segment. The distance from point o to the neutral axis of the beam is equal to the *radius of cur ature* ρ_z. The length of the arc at the neutral axis can be expressed in terms of the angle $d\theta_z$ and the radius of curvature as

$$ds = \rho_z d\theta_z \tag{8.2}$$

or, since the length of the neutral axis does not change during the deformation of the segment, the length of this arc is equal to the undeformed length of the segment. Therefore,

$$dX = \rho_z d\theta_z \tag{8.3}$$

from which we can conclude that

$$\frac{d\theta_z}{dX} = \frac{1}{\rho_z} \tag{8.4}$$

The physical interpretation of this relationship is that the rate of change of the slope angle at any point along the beam is equal to the inverse of the radius of curvature at that point.

The change in the length of any fiber of the segment as it deforms under the action of the moment M_z, at a distance y from the neutral axis of the beam, is

Differential Equation of Bending

$$dL = -y\, d\theta_z \tag{8.5}$$

which, upon dividing by the original length dX, and recognizing that the strain ε_x in the fiber is equal to the change in length divided by the original length, gives

$$\varepsilon_x = \frac{dL}{dX} = -y\frac{d\theta_z}{dX} \tag{8.6}$$

For a linear elastic material, the strain can be expressed in terms of the stress as

$$\varepsilon_x = \frac{\sigma_x}{E} \tag{8.7}$$

where E is the modulus of elasticity of the material. Substituting Eq. (8.7) into Eq. (8.6) and rearranging gives

$$\frac{d\theta_z}{dX} = -\frac{\sigma_x}{Ey} \tag{8.8}$$

which, upon substituting the relationship from Eq. (8.1) for σ_x, results in

$$\frac{d\theta_z}{dX} = \frac{M_z}{EI_z} \tag{8.9}$$

If we now compare Eq. (8.4) and Eq. (8.9), we can write a simple relationship which expresses the deformed shape of the segment in terms of the applied moment:

$$\frac{1}{\rho_z} = \frac{M_z}{EI_z} \tag{8.10}$$

This load deformation relationship is valid at any cross section along a beam as long as the restrictions stated previously are satisfied. However, if the cross section is also subjected to a vertical shear force V_y, this expression is not correct since there will be an additional deformation in the cross section due to the effect of the shear strains. The primary effect of the shear strains is that the cross section will deform out of the plane (i.e.; plane sections before bending will not be plane after bending). However, for most structures, the effects of the shear deformations are small and Eq. (8.10) will result in a solution which is within acceptable accuracy. Therefore, we will ignore the effects of shear deformations for the present. A procedure for considering the effect of the vertical shear forces on the deflections of a beam will be presented in Chapter 9. It will be shown at that time that the assumption of ignoring shear deformations is reasonable for most situations.

One of the basic topics which is usually discussed in any introductory calculus course is the relationship between the radius of a curve at any point and its derivatives:

$$\frac{1}{\rho_z} = \frac{\dfrac{d^2Y}{dX^2}}{\left[1 + \left(\dfrac{dY}{dX}\right)^2\right]^{\frac{3}{2}}} \qquad (8.11)$$

If we now equate Eq. (8.10) and Eq. (8.11), we will obtain the following relationship between the moment at any point on a beam and the derivatives of the elastic curve at that point:

$$\frac{\dfrac{d^2Y}{dX^2}}{\left[1 + \left(\dfrac{dY}{dX}\right)^2\right]^{\frac{3}{2}}} = \frac{M_z}{EI_z} \qquad (8.12)$$

Theoretically, all we have to do now to determine an equation for the elastic curve for any beam is to substitute an expression for M_z as a function of X into the preceding equation and solve for Y. However, this is easier said than done due to the complexity of the equation. It is essentially impossible to solve this equation algebraically for a beam with even the simplest loading and support conditions. The only practical approach is to use some type of numerical analysis technique to obtain an approximate numerical solution. This is beyond the scope of this book.

Fortunately, it is possible to obtain an approximate solution to this equation by recognizing that the deflections of the types of beams which are encountered in engineering practice are usually very small compared to their length. For this situation, the slope of the beam, which is represented by the quantity dY/dX, will be a number much less than 1, which results in the denominator on the left side of Eq. (8.12) being essentially equal to 1. Therefore, for the beams which normally occur in structural systems such as bridges and buildings, an approximation for Eq. (8.12) is

$$\frac{d^2Y}{dX^2} = \frac{M_z}{EI_z} \qquad (8.13)$$

which can be easily solved by integrating each side twice and then solving for the two constants of integration by using the known displacement conditions for the beam at the supports. For example, at a pinned support the deflection Y will be zero, while at a fixed support both the deflection Y and the slope dY/dX will be zero. The solution of this equation falls into the general classification known as a *boundary alue problem* since the information which is used to evaluate the constants of integration is known at the boundaries of the zone for which the equation is valid.

An additional relationship between the deflection Y and the loads acting on the beam can be developed from Eq. (8.13) by taking the derivatives of each side of the equation with respect to X:

Differential Equation of Bending

$$\frac{d^3Y}{dX^3} = \frac{1}{EI_z}\left(\frac{dM_z}{dX}\right) + M_z \frac{d\left(\frac{1}{EI_z}\right)}{dX} \qquad (8.14)$$

Unfortunately, this equation is very difficult to use in its present form. However, if we restrict the analysis to the case of a *prismatic beam* (i.e.; EI_z is constant over the length of the beam) and recognize from Eq. (5.6) in Chapter 5 that

$$\frac{dM_z}{dX} = V_y \qquad (8.15)$$

the relationship can be reduced to

$$\frac{d^3Y}{dX^3} = \frac{V_y}{EI_z} \qquad (8.16)$$

which can be solved for Y by substituting V_y as a function of X and integrating three times. If we now take an additional derivative and recognize from Eq. (5.3) in Chapter 5 that

$$\frac{dV_y}{dX} = q_y \qquad (8.17)$$

we obtain

$$\frac{d^4Y}{dX^4} = \frac{q_y}{EI_z} \qquad (8.18)$$

from which Y can be determined by integrating four times. This equation is commonly known as the *differential equation of bending*. This equation is valid as long as EI_z is constant over the portion of the beam for which it is being applied.

Analysis of Statically Determinate Beams

The following four example problems demonstrate the application of the differential equation of bending to the analysis of statically determinate beams.

Example Problem 8.1

The integration of Eq. (8.18) will be demonstrated by deriving the expression for the elastic curve for the prismatic beam subjected to a positive triangular distributed force shown in Figure 8.2a. The mathematical model of the cantilever beam and the location of the origin of the global coordinate system are shown in Figure 8.2b. The origin of the global coordinate system could have been located at any point. The location used here was chosen to simplify the algebraic manipulations during the analysis.

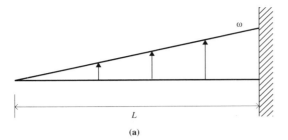

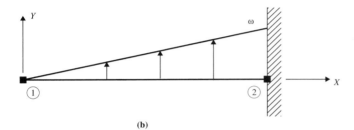

Figure 8.2 Example Problem 8.1.

The triangular distributed force, expressed as a function of X, is

$$q_y = \frac{\omega X}{L}$$

Substituting into Eq.(8.18) and integrating four times gives

$$\frac{d^4Y}{dX^4} = \frac{1}{EI_z}\left(\frac{\omega X}{L}\right)$$

$$\frac{d^3Y}{dX^3} = \frac{1}{EI_z}\left(\frac{\omega X^2}{2L} + C_1\right)$$

$$\frac{d^2Y}{dX^2} = \frac{1}{EI_z}\left(\frac{\omega X^3}{6L} + C_1 X + C_2\right)$$

$$\frac{dY}{dX} = \frac{1}{EI_z}\left(\frac{\omega X^4}{24L} + C_1 \frac{X^2}{2} + C_2 X + C_3\right)$$

$$Y = \frac{1}{EI_z}\left(\frac{\omega X^5}{120L} + C_1 \frac{X^3}{6} + C_2 \frac{X^2}{2} + C_3 X + C_4\right)$$

where C_1, C_2, C_3 and C_4 are the constants which are introduced at each step in the integration process. These constants can be determined by using the boundary conditions at the two ends of the beam as follows:

$$V_y = 0 \text{ at } X = 0 \rightarrow \frac{d^3Y}{dX^3} = 0 \text{ at } X = 0 \rightarrow C_1 = 0$$

Differential Equation of Bending

$$M_z = 0 \text{ at } X = 0 \to \frac{d^2Y}{dX^2} = 0 \text{ at } X = 0 \to C_2 = 0$$

$$\frac{dY}{dX} = 0 \text{ at } X = L \to C_3 = -\frac{\omega L^3}{24}$$

$$Y = 0 \text{ at } X = L \to C_4 = \frac{\omega L^4}{30}$$

Note that due to the specific form of the boundary conditions for this beam and the location of the origin of the global coordinate system, it was a simple task to determine each constant of integration independently. Unfortunately, it will not always be this easy. For some beams, it will be necessary to solve a set of simultaneous algebraic equations to determine these constants.

Substituting these values back into the expressions for the deflection and the slope gives

$$Y = \frac{\omega X^5}{120 L E I_z} - \frac{\omega L^3 X}{24 E I_z} + \frac{\omega L^4}{30 E I_z}$$

$$\frac{dY}{dX} = \frac{\omega X^4}{24 L E I_z} - \frac{\omega L^3}{24 E I_z}$$

We can now use these expressions to compute the deflection and the slope at any point along the beam. For example,

$$Y = \frac{\omega L^4}{30 E I_z} \text{ at } X = 0$$

$$\frac{dY}{dX} = -\frac{\omega L^3}{24 E I_z} \text{ at } X = 0$$

$$Y = \frac{49 \omega L^4}{3840 E I_z} \text{ at } X = \frac{L}{2}$$

$$\frac{dY}{dX} = -\frac{15 \omega L^3}{384 E I_z} \text{ at } X = \frac{L}{2}$$

Note that the deflection is positive and the slope is negative at any point along the beam due to the sign convention being used for q_y and Y.

The shear force and bending moment at any point in the beam can be determined from the final expression for Y by using Eqs. (8.13) and (8.16):

$$M_z = E I_z \left(\frac{d^2Y}{dX^2} \right) = \frac{\omega X^3}{6L}$$

$$V_y = E I_z \left(\frac{d^3Y}{dX^3} \right) = \frac{\omega X^2}{2L}$$

These expressions can be easily verified by cutting the beam at a distance X from the left end and summing forces and moments for the left segment of the beam. This is left as an exercise for the reader.

Example Problem 8.2

Derive an equation for the elastic curve for a simply supported prismatic beam which is subjected to a positive uniform distributed force, as shown in Figure 8.3a. The mathematical model of the beam and the location of the origin of the global coordinate system are shown in Figure 8.3b.

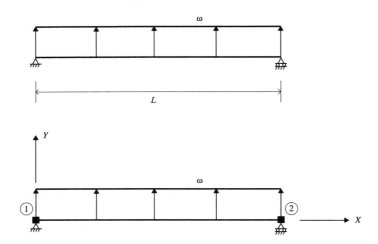

Figure 8.3 Example Problem 8.2.

The uniform distributed force on the beam can be expressed as

$$q_y = \omega$$

which, upon substituting into Eq. (8.18) and integrating, gives

$$\frac{d^4Y}{dX^4} = \frac{1}{EI_z}(\omega)$$

$$\frac{d^3Y}{dX^3} = \frac{1}{EI_z}(\omega X + C_1)$$

$$\frac{d^2Y}{dX^2} = \frac{1}{EI_z}\left(\frac{\omega X^2}{2} + C_1 X + C_2\right)$$

$$\frac{dY}{dX} = \frac{1}{EI_z}\left(\frac{\omega X^3}{6} + C_1 \frac{X^2}{2} + C_2 X + C_3\right)$$

$$Y = \frac{1}{EI_z}\left(\frac{\omega X^4}{24} + C_1 \frac{X^3}{6} + C_2 \frac{X^2}{2} + C_3 X + C_4\right)$$

Differential Equation of Bending

The constants of integration can be determined by using the boundary conditions at the two pinned supports at the ends of the beam.

$$M_z = 0 \text{ at } X = 0 \rightarrow C_2 = 0$$

$$M_z = 0 \text{ at } X = L \rightarrow \frac{\omega L^2}{2} + C_1 L + C_2 = 0$$

$$Y = 0 \text{ at } X = 0 \rightarrow C_4 = 0$$

$$Y = 0 \text{ at } X = L \rightarrow \frac{\omega L^4}{24} + C_1 \frac{L^3}{6} + C_2 \frac{L^2}{2} + C_3 L + C_4 = 0$$

Solving these equations for C_1 through C_4 gives

$$C_1 = -\frac{\omega L}{2}$$

$$C_2 = 0$$

$$C_3 = \frac{\omega L^3}{24}$$

$$C_4 = 0$$

from which the final expressions for the deflection and slope are

$$Y = \frac{\omega X^4}{24EI_z} - \frac{\omega L X^3}{12EI_z} + \frac{\omega L^3 X}{24EI_z}$$

$$\frac{dY}{dX} = \frac{\omega X^3}{6EI_z} - \frac{\omega L X^2}{4EI_z} + \frac{\omega L^3}{24EI_z}$$

An important consideration during the design of many beams is the magnitude of the maximum deflection, since in some cases, a deflection limitation will determine the allowable load on a structure. The location of the maximum deflection can be determined for this beam by setting the slope to zero and solving the resulting expression for X:

$$\frac{dY}{dX} = 0 \rightarrow \frac{\omega X^3}{6EI_z} - \frac{\omega L X^2}{4EI_z} + \frac{\omega L^3}{24EI_z} = 0 \rightarrow X = \frac{L}{2}$$

Therefore, the maximum deflection occurs at the center of the beam, as expected due to the symmetry of the vertical supports and the loading. Substituting this value for X into the expression for Y gives

$$Y_{max} = \frac{5\omega L^4}{384EI_z}$$

This is an expression which is worth remembering since a simply supported beam with a uniform distributed force is a very common configuration in engineering practice.

For most beams which are encountered in actual practice, the uniform force will be down due to gravity. This will result in a negative (i.e.; downward) deflection when a negative value for ω is substituted into the expression for Y_{max}.

The bending moment and the shear force at any point in the beam are

$$M_z = EI_z \left(\frac{d^2Y}{dX^2} \right) = \frac{\omega X^2}{2} - \frac{\omega LX}{2}$$

$$V_y = EI_z \left(\frac{d^3Y}{dX^3} \right) = \omega X - \frac{\omega L}{2}$$

These expressions agree with the expressions which were derived previously using an equilibrium analysis of this same beam in Example Problem 5.2 in Chapter 5.

Example Problem 8.3

Derive an algebraic expression for the elastic curve for a prismatic cantilever beam which is subjected to a positive concentrated force at its free end, as shown in Figure 8.4a. The mathematical model of the beam and the location of the origin of the global coordinate system are shown in Figure 8.4b. The primary difference between this beam and the beam which was analyzed in Example Problems 8.1 and 8.2 is that this beam is subjected to a concentrated force rather than a distributed force. Therefore, it is not convenient to use Eq. (8.18) in the analysis since the concentrated force actually corresponds to a distributed force of infinite intensity acting over a

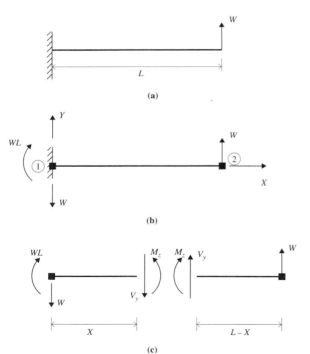

Figure 8.4 Example Problem 8.3.

Differential Equation of Bending

zero contact length, as discussed previously in Chapter 5. However, this does not present any significant difficulty, since we can easily determine an algebraic expression for either the shear force V_y or the bending moment M_z for this beam and then start with either Eq. (8.16) or Eq. (8.13) to perform the integration. In order to reduce the algebra involved in the analysis, it is usually more convenient to start with Eq. (8.13) since it will be one step further along in the solution.

The expression for M_z can be determined by cutting the beam at a point a distance X from the left end, as shown in Figure 8.4c, and summing moments for either the left or the right segments of the beam:

$$M_z = -WX + WL$$

Substituting this expression into Eq. (8.13) and integrating twice gives

$$\frac{d^2Y}{dX^2} = \frac{1}{EI_z}(-WX + WL)$$

$$\frac{dY}{dX} = \frac{1}{EI_z}\left(-\frac{WX^2}{2} + WLX + C_1\right)$$

$$Y = \frac{1}{EI_z}\left(-\frac{WX^3}{6} + \frac{WLX^2}{2} + C_1 X + C_2\right)$$

The constants of integration can be determined from the known slope and deflection at the fixed end:

$$\frac{dY}{dX} = 0 \text{ at } X = 0 \rightarrow C_1 = 0$$

$$Y = 0 \text{ at } X = 0 \rightarrow C_2 = 0$$

Therefore, the expressions for the deflection and slope are

$$Y = -\frac{WX^3}{6EI_z} + \frac{WLX^2}{2EI_z}$$

$$\frac{dY}{dX} = -\frac{WX^2}{6EI_z} + \frac{WLX}{EI_z}$$

It is obvious that the maximum deflection and slope will occur at the free end for this beam:

$$Y_{max} = \frac{WL^3}{3EI_z}$$

$$\frac{dY}{dX}_{max} = \frac{WL^2}{2EI_z}$$

As in the previous example problem, these are also expressions which are worth remembering.

234   Deflection of Beams   Chap. 8

Note that if the support and the downward concentrated force were switched end for end for this beam so that the free end was at the left, the algebraic expressions for the deflection and the slope at any point along the beam would be completely different. However, the expressions for the maximum deflection and the maximum slope at the free end would be the same except that the slope would be negative. The verification of this statement is left as an exercise for the reader.

Example Problem 8.4

Determine the deflection at any point in the simply supported prismatic beam shown in Figure 8.5a. This is the same beam which was analyzed in Example Problems 5.3 and 5.5 in Chapter 5 to determine expressions for V_y and M_z. The mathematical model of the beam and the location of the origin of the global coordinate system are shown in Figure 8.5b.

The analysis of this beam is more complicated than those in the previous example problems since it is not possible to develop one expression for Y which is valid over the entire length of the beam. One expression must be developed for the left zone for $0 \leq X \leq L/2$ and another expression must be developed for the right zone for $L/2 \leq X \leq L$. Although it is possible to start with Eq. (8.18) and integrate four times for each zone, the analysis of this particular beam will be simplified if we first develop expressions for M_z in the two zones and then use Eq. (8.13) for each zone. The expressions for M_z can be easily developed by an equilibrium analysis, as shown previously in Example Problem 5.3.

Left zone $(0 \leq X \leq L/2)$:

$$M_z = \frac{\omega X^2}{2} - \frac{3\omega L X}{8}$$

Right zone $(L/2 \leq X \leq L)$:

$$M_z = \frac{\omega L X}{8} - \frac{\omega L^2}{8}$$

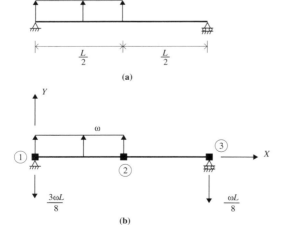

Figure 8.5 Example Problem 8.4.

Differential Equation of Bending

We can now substitute these two expressions into Eq. (8.13) and integrate twice for each zone. To eliminate confusion, the displacement in the left zone will be designated Y_L and the displacement in the right zone will be designated Y_R.

Left zone ($0 \leq X \leq L/2$):

$$\frac{d^2 Y_L}{dX^2} = \frac{1}{EI_z}\left(\frac{\omega X^2}{2} - \frac{3\omega L X}{8}\right)$$

$$\frac{dY_L}{dX} = \frac{1}{EI_z}\left(\frac{\omega X^3}{6} - \frac{3\omega L X^2}{16} + C_1\right)$$

$$Y_L = \frac{1}{EI_z}\left(\frac{\omega X^4}{24} - \frac{\omega L X^3}{16} + C_1 X + C_2\right)$$

Right zone ($L/2 \leq X \leq L$):

$$\frac{d^2 Y_R}{dX^2} = \frac{1}{EI_z}\left(\frac{\omega L X}{8} - \frac{\omega L^2}{8}\right)$$

$$\frac{dY_R}{dX} = \frac{1}{EI_z}\left(\frac{\omega L X^2}{16} - \frac{\omega L^2 X}{8} + C_3\right)$$

$$Y_R = \frac{1}{EI_z}\left(\frac{\omega L X^3}{48} - \frac{\omega L^2 X^2}{16} + C_3 X + C_4\right)$$

The constants of integration can be determined from the deflection conditions at the two ends of the beam:

$$Y_L = 0 \text{ at } X = 0 \rightarrow C_2 = 0$$

$$Y_R = 0 \text{ at } X = L \rightarrow -\frac{\omega L^4}{24} + C_3 L + C_4 = 0$$

and from the continuity of the slopes and the deflections on either side of the center of the beam:

$$\frac{dY_L}{dX} = \frac{dY_R}{dX} \text{ at } X = \frac{L}{2} \rightarrow -\frac{\omega L^4}{48} - C_1 + C_3 = 0$$

$$Y_L = Y_R \text{ at } X = \frac{L}{2} \rightarrow -\frac{\omega L^4}{64} - C_1 L - 2C_2 + C_3 L + 2C_4 = 0$$

Solving these equations for C_1 through C_4 gives

$$C_1 = \frac{9\omega L^3}{384}$$

$$C_2 = 0$$

$$C_3 = \frac{17\omega L^3}{384}$$

$$C_4 = -\frac{\omega L^4}{384}$$

from which the final expressions for the deflection and the slope in each zone of the beam are as follows.

Left zone ($0 \leq X \leq L/2$):

$$Y_L = \frac{\omega X^4}{24EI_z} - \frac{\omega L X^3}{16EI_z} + \frac{3\omega L^3 X}{128EI_z}$$

$$\frac{dY_L}{dX} = \frac{\omega X^3}{6EI_z} - \frac{3\omega L X^2}{16EI_z} + \frac{3\omega L^3}{128EI_z}$$

Right zone ($L/2 \leq X \leq L$):

$$Y_R = \frac{\omega L X^3}{48EI_z} - \frac{\omega L^2 X^2}{16EI_z} + \frac{17\omega L^3 X}{384EI_z} - \frac{\omega L^4}{384EI_z}$$

$$\frac{dY_R}{dX} = \frac{\omega L X^2}{16EI_z} - \frac{\omega L^2 X}{8EI_z} + \frac{17\omega L^3}{384EI_z}$$

If we had started with Eq. (8.18) in Example Problem 8.4 instead of Eq. (8.13), it would have been necessary to solve for eight constants of integration. Although we would have ended up with the same final expressions for Y and dY/dX, it would have been a much more difficult task. It will usually be found that the analysis of a statically determinate beam can be considerably simplified by starting with Eq. (8.13). The task of developing algebraic expressions for the bending moment in the beam by an equilibrium analysis is usually much easier than solving the additional equations to determine the increased number of constants of integration.

Analysis of Statically Indeterminate Beams

If we look back at the derivation of the differential equation of bending, as expressed by Eq. (8.18), we will see that at no point in the development of the equation was the static determinacy or static indeterminacy of the beam considered. The equation merely relates the deformed shape at any point in a linear elastic straight beam to the distributed load at that point. Therefore, the equation should be equally applicable to any stable beam. As we will see in the next example problem, it will be possible to determine the constants of integration from the known boundary conditions even though the beam is statically indeterminate.

Example Problem 8.5

Derive an expression for the elastic curve of a fixed-pinned statically indeterminate prismatic beam which is subjected to a positive uniform distributed force, as shown in Figure 8.6a. The mathematical model of the beam and the location of the origin of

Differential Equation of Bending

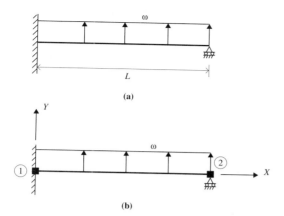

Figure 8.6 Example Problem 8.5.

the global coordinate system are shown in Figure 8.6b. Since we cannot develop expressions for either V_y or M_z for this beam using only the equilibrium equations for the beam, it is necessary to start the analysis with Eq. (8.18)

The distributed force at any point on the beam is

$$q_y = \omega$$

Substituting into Eq. (8.18) and integrating four times will give exactly the same expressions for the derivatives, before the constants of integration are evaluated, which were obtained in Example Problem 8.2.

$$\frac{d^4Y}{dX^4} = \frac{1}{EI_z}(\omega)$$

$$\frac{d^3Y}{dX^3} = \frac{1}{EI_z}(\omega X + C_1)$$

$$\frac{d^2Y}{dX^2} = \frac{1}{EI_z}\left(\frac{\omega X^2}{2} + C_1 X + C_2\right)$$

$$\frac{dY}{dX} = \frac{1}{EI_z}\left(\frac{\omega X^3}{6} + C_1 \frac{X^2}{2} + C_2 X + C_3\right)$$

$$Y = \frac{1}{EI_z}\left(\frac{\omega X^4}{24} + C_1 \frac{X^3}{6} + C_2 \frac{X^2}{2} + C_3 X + C_4\right)$$

The differences in this beam and the statically determinate beam from Example Problem 8.2 are the boundary conditions at the supports.

$$M_z = 0 \text{ at } X = L \rightarrow \frac{d^2Y}{dX^2} = 0 \text{ at } X = L \rightarrow \frac{\omega L^2}{2} + C_1 L + C_2 = 0$$

$$\frac{dY}{dX} = 0 \text{ at } X = 0 \rightarrow C_3 = 0$$

$Y = 0$ at $X = 0 \rightarrow C_4 = 0$

$Y = 0$ at $X = L \rightarrow \dfrac{\omega L^4}{24} + C_1 \dfrac{L^3}{6} + C_2 \dfrac{L^2}{2} + C_3 L + C_4 = 0$

Solving these equations for C_1 through C_4 gives

$$C_1 = -\dfrac{5\omega L}{8}$$

$$C_2 = \dfrac{\omega L^2}{8}$$

$$C_3 = 0$$

$$C_4 = 0$$

from which the final expressions for the deflection and slope are

$$Y = \dfrac{\omega X^4}{24 EI_z} - \dfrac{5\omega L X^3}{48 EI_z} + \dfrac{\omega L^2 X^2}{16 EI_z}$$

$$\dfrac{dY}{dX} = \dfrac{\omega X^3}{6 EI_z} - \dfrac{5\omega L X^2}{16 EI_z} + \dfrac{\omega L^2 X}{8 EI_z}$$

In addition, expressions for M_z and V_y can be determined from Eq. (8.13) and Eq. (8.16):

$$M_z = EI_z \left(\dfrac{d^2 Y}{dX^2} \right) = \dfrac{\omega X^2}{2} - \dfrac{5\omega L X}{8} + \dfrac{\omega L^2}{8}$$

$$V_y = EI_z \left(\dfrac{d^3 Y}{dX^3} \right) = \omega X - \dfrac{5\omega L}{8}$$

These expressions can now be used to plot the shear force and bending moment diagrams for the beam. Figure 8.7 shows the shear force and bending moment diagrams for a downward uniform force of 2kN/m (i.e., $\omega = -2$ kN/m) and a span length of 40 meters. The shear force diagram is a first order curve and the bending moment diagram is a second order curve. The bending moment changes sign along the beam with the maximum negative bending moment at the left support and the maximum positive bending moment at a point 25 meters ($5L/8$) from the left end. Note that the maximum interior bending moment occurs at the point of zero shear force, as expected. The sign change in the bending moment occurs at a point 10 meters ($L/4$) from the left end. This point is known as an *inflection point* since the curvature of the beam is zero at this point.

The reactive forces and moments at the supports can be determined from the ordinates of the shear force and the bending moment diagrams at the end points of the beam as

Differential Equation of Bending

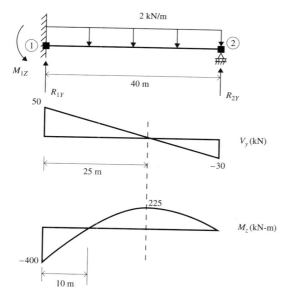

Figure 8.7 Example Problem 8.5, V_y and M_z diagrams.

$$R_{1Y} = 50 \text{ kN up}$$
$$M_{1Z} = 400 \text{ kN-m counterclockwise}$$
$$R_{2Y} = 30 \text{ kN up}$$
$$M_{2Z} = 0$$

A quick check will show that these reactions satisfy the equilibrium requirements for the beam.

Example Problem 8.6

Derive an expression for the elastic curve of a fixed-fixed statically indeterminate prismatic beam which is subjected to a uniform distributed force, as shown in Figure 8.8a.

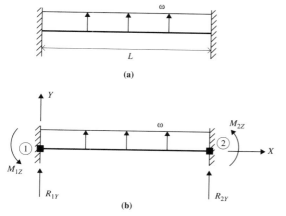

Figure 8.8 Example Problem 8.6.

The mathematical model for the beam and the location of the origin of the global coordinate system are shown in Figure 8.8b. The analysis procedure for this beam will be exactly the same as for the beam in Example Problem 8.5, except that one boundary condition will be different for determining the constants of integration. In the previous beam, the moment M_z was equal to zero at the right end of the beam, while in this beam, the slope is zero at that point. The other three boundary conditions are the same.

The final expressions for the deflection, slope, bending moment and shear force, after solving for the constants of integration, are

$$Y = \frac{\omega X^4}{24EI_z} - \frac{\omega L X^3}{12EI_z} + \frac{\omega L^2 X^2}{24EI_z}$$

$$\frac{dY}{dX} = \frac{\omega X^3}{6EI_z} - \frac{\omega L X^2}{4EI_z} + \frac{\omega L^2 X}{12EI_z}$$

$$M_z = \frac{\omega X^2}{2} - \frac{\omega L X}{2} + \frac{\omega L^2}{12}$$

$$V_y = \omega X - \frac{\omega L}{2}$$

The values for the shear force and the bending moment at the left end of the beam are

$$V_y = -\frac{\omega L}{2} \text{ at } X = 0$$

$$M_z = \frac{\omega L^2}{12} \text{ at } X = 0$$

while, and at the right end, they are

$$V_y = \frac{\omega L}{2} \text{ at } X = L$$

$$M_z = \frac{\omega L^2}{12} \text{ at } X = L$$

from which the reactive forces and moments at the two supports are

$$R_{1Y} = -\frac{\omega L}{2}$$

$$M_{1Z} = -\frac{\omega L^2}{12}$$

$$R_{2Y} = -\frac{\omega L}{2}$$

$$M_{2Z} = \frac{\omega L^2}{12}$$

Differential Equation of Bending

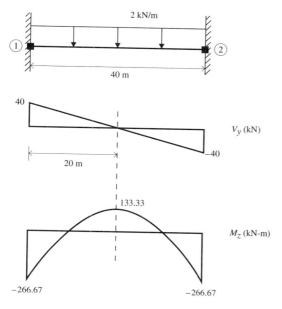

Figure 8.9 Example Problem 8.6, V_y and M_z diagrams.

Note that these reactions agree with the fixed end joint loads, with opposite signs, given previously in Eqs. (6.11) and (6.12) in Chapter 6 for a uniformly loaded member in a plane frame. The difference in the signs is because the loads applied to the end joints by the member, which were used in the frame analysis in SDPFRAME, are always in the opposite directions of the loads applied to the ends of the member by the joints. The reactions shown here correspond to the loads applied to the ends of the beam by the support joints. The same procedure can be used to verify the fixed end joint loads given in Eqs. (6.15) through (6.18) for a triangular distributed load. This verification will be left for one of the suggested problems at the end of this chapter.

Figure 8.9 shows the shear force and bending moment diagrams for a fixed-fixed beam for the specific case of a 2 kN/m downward distributed force and a span length of 40 meters. The shear force is identical to the shear force for the simply supported beam shown earlier in Figure 5.9a. The bending moment diagram is also the same shape as for the simply supported beam except that the zero axis has been shifted up 266.67 kN m. Therefore, the fixed-fixed beam subjected to a 2 kN/m downward uniform force is equivalent to a simply supported beam with the same uniform force and counterclockwise and clockwise moments of 266.67 kN-m at the left and right ends respectively.

Analysis of Nonprismatic Beams

Although Eqs. (8.16) and (8.18) are restricted to the analysis of prismatic beams, a nonprismatic beam can be analyzed by Eq. (8.13) as long as the variation in EI_z is considered during the integration.

Example Problem 8.7

The application of Eq. (8.13) to the analysis of a nonprismatic beam will be demonstrated by developing the expressions for the slope and deflection at any point in the simply supported beam whose mathematical model is shown in Figure 8.10a. The moment of inertia of the cross section for the portion of the beam to the left of the center is I_z, while it is equal to $2I_z$ for the portion of the beam to the right of the center. The material modulus of elasticity is constant over the length of the beam.

For this beam, we will work with the actual numbers corresponding to the concentrated force and the span length rather than with algebraic symbols as in the previous example problems. The bending moment diagram for the beam is shown in Figure 8.10b and the expressions for the bending moment in the left and right zones of the beam, expressed in kn-m units, are as follows:

Left zone ($0 \le X \le 30$):

$$M_z = 7.5X$$

Right zone ($30 \le X \le 60$):

$$M_z = -7.5X + 450$$

which, upon substituting into Eq. (8.13) for the left and right zones and integrating, gives

Left zone ($0 \le X \le 30$):

$$\frac{d^2 Y_L}{dX^2} = \frac{1}{EI_z}(7.5X)$$

$$\frac{dY_L}{dX} = \frac{1}{EI_z}(3.75X^2 + C_1)$$

$$Y_L = \frac{1}{EI_z}(1.25X^3 + C_1 X + C_2)$$

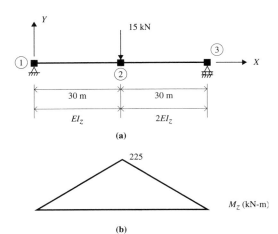

Figure 8.10 Example Problem 8.7.

Differential Equation of Bending

Right zone ($30 \leq X \leq 60$):

$$\frac{d^2Y_R}{dX^2} = \frac{1}{2EI_z}(-7.5X + 450) = \frac{1}{EI_z}(-3.75X + 225)$$

$$\frac{dY_R}{dX} = \frac{1}{EI_z}(-1.875X^2 + 225X + C_3)$$

$$Y_R = \frac{1}{EI_z}(-0.625X^3 + 112.5X^2 + C_3X + C_4)$$

The constants of integration can be determined by using the deflection conditions at the two ends of the beam and the continuity of the slope and the deflection at the center, as shown previously in Example Problem 8.4.

$Y_L = 0$ at $X = 0 \rightarrow C_2 = 0$

$Y_R = 0$ at $X = 60 \rightarrow 270{,}000 + 60C_3 + C_4 = 0$

$\dfrac{dY_L}{dX} = \dfrac{dY_R}{dX}$ at $X = 30 \rightarrow -1687.5 + C_1 - C_3 = 0$

$Y_L = Y_R$ at $X = 30 \rightarrow -50{,}625 + 30C_1 + C_2 - 30C_3 - C_4 = 0$

Solving these equations for C_1 through C_4 gives

$$C_1 = -2812.5$$
$$C_2 = 0$$
$$C_3 = -4500$$
$$C_4 = 0$$

from which the final expressions for the slope and deflection in each zone are

Left zone ($0 \leq X \leq 30$):

$$\frac{dY_L}{dX} = \frac{1}{EI_z}(3.75X^2 - 2812.5)$$

$$Y_L = \frac{1}{EI_z}(1.25X^3 - 2812.5X)$$

Right zone ($30 \leq X \leq 60$):

$$\frac{dY_R}{dX} = \frac{1}{EI_z}(-1.875X^2 + 225X - 4500)$$

$$Y_R = \frac{1}{EI_z}(-0.625X^3 + 112.5X^2 - 4500X)$$

The previous procedure can be used to compute the deflections in any linear elastic straight beam which is bending about a principal axis. However, the algebra involved in determining the expressions for the elastic curve can become very tedious and error prone, particularly for beams in which different expressions must be determined for the deflection for a number of different zones along the beam. The solution of the simultaneous equations to determine the constants of integration can be extremely difficult for many beams. The next section in this chapter describes a semi-graphical procedure which can be used to compute the slopes and deflections at specific points in a beam while eliminating much of the tedious algebra which is involved in integrating and solving for the constants of integration.

THE MOMENT AREA METHOD

In many situations during the analysis of a beam, it is necessary to compute the slopes and/or deflections at specific points along the beam, but a complete set of equations which describe the deflection at all points on the beam are not needed. For these situations, the computation of the slopes and deflections often can be simplified, compared to the algebraic manipulations which are required for a solution of the differential equation of bending, by making use of two simple geometric relationships which express the deformed shape of a segment of the beam in terms of the variation of the bending moment M_z over the segment. These relationships are known as the *Moment Area Theorems* and the general analysis process is known as the *Moment Area Method*.

Moment Area Theorems

Theorem I. The *First Moment Area Theorem* relates the change in the slope of the elastic curve over a segment of a beam to the moment acting on the segment. To develop this theorem, we will consider the infinitesimal segment of a beam of length dX, shown previously in Figure 8.1b, for which the change in the slope angle $d\theta_z$ can be expressed by rearranging Eq. (8.9) as

$$d\theta_z = \frac{M_z}{EI_z} dX \tag{8.19}$$

This expression now can be used to determine the total change in the slope angle $\Delta\theta_{(a/b)z}$ between two points a and b on a beam, as defined in Figure 8.11, by summing the contributions to the change in slope for each infinitesimal length of the beam between the two points. This summation corresponds to the integration of Eq. (8.19) between the limits of X_a and X_b:

$$\Delta\theta_{(a/b)z} = \theta_{bz} - \theta_{az} = \int_{X_a}^{X_b} \frac{M_z}{EI_z} dX \tag{8.20}$$

As stated previously during the derivation of the differential equation of bending, the displacements for most beams will be very small compared to their length,

The Moment Area Method

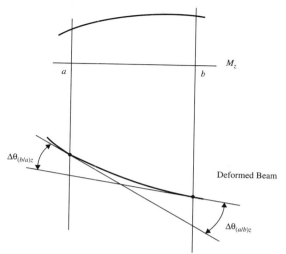

Figure 8.11 First Moment Area Theorem.

which results in the slope angle at any point on the beam being very small. From basic trigonometry, we know that for very small angles, the magnitude of the angle, expressed in radians, is essentially equal to the tangent of the angle. Therefore, an approximation for the slope dY/dX, which is equal to the tangent of the slope angle at any point along the beam, is

$$\frac{dY}{dX} = \tan \theta_z = \theta_z \qquad (8.21)$$

This approximation is equivalent to the approximation which was made in reducing Eq. (8.12) to the simpler form shown in Eq. (8.13) during the derivation of the differential equation of bending. Therefore, the symbol θ will be used interchangeably to represent both the slope angle and the slope dY/dX at any point on a beam in the Moment Area Method.

If we now recognize that the integral in Eq. (8.20) is equal to the area under a curve whose ordinates are equal to M_z/EI_z at any location between the two points a and b on the beam (this curve will be defined as the M_z/EI_z *diagram* for the beam), we can now define the First Moment Area Theorem as:

> The change in the slope $\Delta\theta_{(a/b)z}$ of the elastic curve between two points a and b on an initially straight beam, where X_b is greater than X_a, is equal to the area under the M_z/EI_z diagram between the two points.

The sign of the area under the M_z/EI_z diagram will be dictated by the sign of the moment M_z since EI_z is an unsigned quantity. A positive value for $\Delta\theta_{(a/b)z}$ corresponds to a counterclockwise rotation from the tangent to the elastic curve at point a to the tangent to the elastic curve at point b.

The quantity $\Delta\theta_{(a/b)z}$ corresponds to the change in slope of the elastic curve when moving in the positive X direction from point a toward point b along the beam. If we move in the negative X direction, from point b toward point a, the change in slope, which will be designated by the symbol $\Delta\theta_{(b/a)z}$, will have the same magnitude but will be in the opposite direction.

$$\Delta\theta_{(b/a)z} = \theta_{az} - \theta_{bz} = -\Delta\theta_{(a/b)z} \qquad (8.22)$$

An engineer must be very careful to use the correct sign for the change in slope which corresponds to the direction along the X axis between the two points.

Theorem II. The *Second Moment Area Theorem* relates the vertical translation of a point on the elastic curve, at one end of a segment of a beam, from the tangent to the elastic curve at the other end of the segment in terms of the moment on the segment. The relationship can be developed by considering the segment of a beam shown in Figure 8.12. The change in the slope over the distance dX will be $d\theta_z$, and the vertical distance dt at point b between the tangents at the left and the right ends of this elemental segment will be

$$dt = c_b d\theta_z \qquad (8.23)$$

where c_b is the distance from point b to the center of the elemental segment dX. The actual distance dt is an arc length, but for small slope angles, it is essentially equal to the vertical distance. This is again equivalent to the approximation which was made in developing Eq. (8.13). The total vertical distance $t_{b/a}$ of any point b on the elastic curve from the tangent to any other point a on the elastic curve can be found by summing the contributions of each of the elemental lengths of the deformed beam from point a to point b:

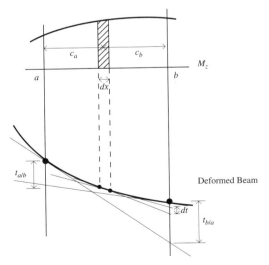

Figure 8.12 Second Moment Area Theorem.

$$t_{b/a} = \int_{X_a}^{X_b} c_b \, d\theta_z \tag{8.24}$$

which, upon substituting the expression in Eq. (8.19) for $d\theta_z$, becomes

$$t_{b/a} = \int_{X_a}^{X_b} c_b \frac{M_z}{EI_z} \, dX \tag{8.25}$$

The quantity $t_{b/a}$ will be called the *tangential de iation* of point b from point a. The physical interpretation of the integral in Eq. (8.25) is the moment of the area under the M_z/EI_z diagram between points a and b about point b. Therefore, we can now define the Second Moment Area Theorem as

> The tangential deviation $t_{b/a}$ of a point b on the elastic curve of an initially straight beam from any other point a on the elastic curve is equal to the moment of the area under the M_z/EI_z diagram between the two points about point b.

The sign of the tangential deviation is dictated by the sign of the moment M_z. The moment arm c_b is considered to be unsigned during the computations. A positive value for $t_{b/a}$ corresponds to point b being up from the tangent to point a.

There is no specific relationship between the tangential deviation $t_{b/a}$ at point b and the tangential deviation $t_{a/b}$ at point a. The tangential deviation $t_{a/b}$ is equal to

$$t_{a/b} = \int_{X_a}^{X_b} c_a \frac{M_z}{EI_z} \, dX \tag{8.26}$$

where c_a is the distance from point a to any point between a and b. As with $t_{b/a}$, the sign of $t_{a/b}$ is dictated by the sign of the moment M_z. The distance c_a is unsigned.

The moment of the M_z/EI_z diagram about either point a or b can be easily computed by multiplying the area under the diagram between the two points by the corresponding distance to its center of gravity. Therefore,

$$t_{a/b} = \overline{c_a} A_{a/b} \tag{8.27}$$

and

$$t_{b/a} = \overline{c_b} A_{a/b} \tag{8.28}$$

where the bar over symbols c_a and c_b corresponds to the distance to the centroid of the total area $A_{a/b}$ under the M_z/EI_z diagram between point a and point b.

Although the Moment Area Theorems do not directly give the actual slope or deflection at any point on a beam, they can be used in combination with the known displacement conditions at the supports to compute the slope and deflection at any point after the bending moment diagram has been generated. The most important factor when using the theorems is the ability of the engineer to draw an approximate deflected shape for the beam from the known values for the bending moment along the beam. For most beams, this is a fairly easy task since the direction of the curvature at any point can be determined by Eq. (8.13) from the sign of the bending moment. The theorems are equally applicable to both statically determinate and statically indeterminate beams as long as the beams meet the restrictions imposed previously during the derivation of the differential equation of bending.

The Moment Area Method can be used to either find numerical values for the slope and deflection at specific points in a beam, by expressing the properties of the beam and the loading as numerical values, or it can be used to determine algebraic expressions for the slope and deflection if the properties of the beam and the loading are expressed as algebraic symbols. It is the opinion of this author that this analysis procedure is best suited for performing a numerical analysis rather than an algebraic analysis. If algebraic expressions are desired for the slope and deflection, then it is usually better to solve the differential equation of bending since a general solution for the entire beam will be obtained. Of course, there can be exceptions to this statement. In this book, we will only consider the application of the Moment Area Method to the solution of numerical problems.

Example Problem 8.8

To demonstrate the use of the Moment Area Theorems we will compute the slope and deflection at the end of a prismatic cantilever beam which is subjected to a concentrated downward force at the free end, as shown in Figure 8.13a. The mathematical model of the beam, with joints at the fixed end and at the free end for reference purposes, is shown in Figure 8.13b. The bending moment diagram, with units of kip-feet, is shown in Figure 8.13c. The curvature of the beam will be negative over the entire length since the bending moment is negative, which will result in the deflected shape shown by the dashed line in Figure 8.13d. This is the only possible deflected shape which is consistent with the direction of the curvature and the boundary conditions of zero slope and zero deflection at the fixed end.

The slope θ_{2z} at the right end of the beam can be expressed in terms of the slope θ_{1z} at the left end and the change in slope $\Delta\theta_{(1/2)z}$ between the two points as

$$\theta_{2z} = \theta_{1z} + \Delta\theta_{(1/2)z}$$

or, since θ_{1z} at the fixed end must be zero

$$\theta_{2z} = \Delta\theta_{(1/2)z}$$

The change in slope between the two ends of the beam can be computed using the First Moment Area Theorem as the area under the M_z/EI_z diagram for the entire beam. Since the quantity EI_z is constant for the beam, this area will be simply equal to the total area under the bending moment diagram divided by EI_z. The area for the triangular bending moment diagram is one half of the product of the base and the height. Although we could substitute numerical values for E and I_z at this point, the quantity EI_z will be carried through the computations as a symbol.

The Moment Area Method

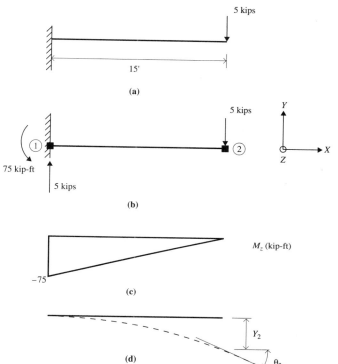

Figure 8.13 Example Problem 8.8.

$$\Delta\theta_{(1/2)z} = \frac{(15)(-75)/2}{EI_z} = -\frac{562.5}{EI_z}$$

from which

$$\theta_{2z} = -\frac{562.5}{EI_z}$$

The slope at the right end is negative, which agrees with the deflected shape of the beam shown in Figure 8.13d.

For this particular beam, the deflection at the right end is exactly equal to the vertical translation of that end from a tangent drawn to the left end since the slope at the left end is zero.

$$Y_2 = t_{2/1}$$

The tangential deviation $t_{2/1}$ can be computed using the Second Moment Area Theorem as the moment of the area under the M_z/EI_z diagram between point 1 and point 2 about point 2. Using Eq. (8.28), and recognizing that the centroid of the triangular M_z/EI_z diagram is 2/3 of the length of the beam from the right end, gives

$$t_{2/1} = \frac{[(2/3)(15)][(15)(-75)/2]}{EI_z} = -\frac{5625}{EI_z}$$

from which

$$Y_2 = -\frac{5625}{EI_z}$$

The deflection at the right end is negative, which also agrees with the deflected shape in Figure 8.12d.

These solutions for the slope and deflection at the right end of the beam can be verified by substituting the values

$$W = -5$$
$$L = 15$$

and

$$X = 15$$

into the following expressions which were derived for the slope and deflection at any point along a cantilever beam by the integration of Eq. (8.13) in Example Problem 8.3:

$$\frac{dY}{dX} = -\frac{WX^2}{2EI_z} + \frac{WLX}{EI_z}$$

$$Y = -\frac{WX^3}{6EI_z} + \frac{WLX^2}{EI_z}$$

from which

$$\frac{dY}{dX} \text{ at } X = 15 \rightarrow -\frac{(-5)(15)^2}{2EI_z} + \frac{(-5)(15)(15)}{EI_z} = -\frac{562.5}{EI_z}$$

$$Y \text{ at } X = 15 \rightarrow -\frac{(-5)(15)^3}{6EI_z} + \frac{(-5)(15)(15)^2}{2EI_z} = -\frac{5625}{EI_z}$$

The two solutions agree.

Numerical values for θ_{2z} and Y_2 can be determined by substituting the numerical values for E and I_z into the previous expressions. Since the units of the moment diagram are kip-feet, care must be taken in substituting these numerical values to ensure that no errors occur in the units. The experience of the author has shown that it is usually better to substitute E and I_z in the units which were used in the original computations rather than trying to find a conversion factor to convert the numbers in the computed expressions into other units before making the substitutions. The moment diagram was originally expressed in kip and feet units since these are the common units which are used by many design engineers. Therefore, E would be substituted as kips per square foot and I_z would be substituted as feet to the fourth. The resulting value for the slope would be dimensionless while the deflection would be in feet. If the bending moment diagram had been expressed in kip-inch or pound inch

The Moment Area Method

units, the numbers which would have to be manipulated would be much larger. It is usually more convenient to work with the smaller numbers. We will not bother to make any numerical substitutions for E and I_z here.

Example Problem 8.9

Find the slope and deflection at the center of a prismatic simply supported beam which is subjected to a downward uniform distributed force, as shown in Figure 8.14a. The mathematical model of the beam, with a joint at each end and at the center for reference purposes, is shown in Figure 8.14b while the bending moment diagram, with the units of kN-m, is shown in Figure 8.14c. The bending moment diagram is a second order curve with a zero slope at the center of the beam. Since the bending moment is positive over the entire beam, the curvature will also be positive. Using this information and the known zero deflections at the pinned supports, we can determine that the deflected shape for the beam must be a downward curve of the shape shown in Figure 8.14d. This is the only possible shape for the beam which is consistent with the support conditions and the positive curvature over the entire length.

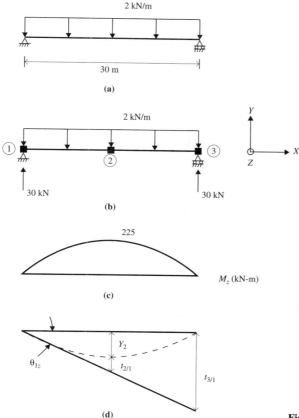

Figure 8.14 Example Problem 8.9.

252   Deflection of Beams   Chap. 8

        Although it is obvious what the deflected shape of this particular beam will be, this will not always be the case. For many beams, with more complicated loadings and support conditions, the exact deflected shape will not be obvious even though the signs of the bending moments are known at all points along the beam. However, an initial assumption must be made for the deflected shape based upon the support conditions and the variation of the signs in the bending moment diagram in order to be able to determine how the Moment Area Theorems might be applied. The assumed deflected shape can be modified if necessary from the additional information which is obtained as the analysis progresses.

        The analysis of this beam by the Moment Area Method is more complicated than the analysis of the cantilever beam in Example Problem 8.8 since the slope is not zero at either of the support points. (The slope does happen to be zero at the center of the beam due to symmetry, but we will assume that we do not recognize this fact at this time.) From the deflected shape of the beam in Figure 8.13d, we can see that the tangential deviation $t_{3/1}$ at joint 3 will be positive since joint 3 is up from the tangent to joint 1. We can also see that the slope of the tangent to the elastic curve θ_{1z} at joint 1 will be negative and can be expressed in terms of $t_{3/1}$ as

$$\theta_{1z} = -\frac{t_{3/1}}{L}$$

As in the previous example problem, the shape of the M_z/EI_z diagram for this prismatic beam will be the same as the shape of the bending moment diagram. For simplicity, the quantity EI_z will be carried through the calculations symbolically. The tangential deviation $t_{3/1}$ can be computed using Eq. (8.28) by recognizing that the area under the parabolic M_z/EI_z diagram, with zero dM_z/dX at the center, is equal to 2/3 of the product of the base and the height, and its centroid is at the center of the beam.

$$t_{3/1} = \frac{(30/2)[(2/3)(30)(225)]}{EI_z} = \frac{67{,}500}{EI_z}$$

from which the slope θ_{1z} will be

$$\theta_{1z} = -\frac{67{,}500/EI_z}{30} = -\frac{2250}{EI_z}$$

With this value known, the slope θ_{2z} at the center of the beam can be computed by

$$\theta_{2z} = \theta_{1z} + \Delta\theta_{(1/2)z}$$

where the change in slope $\Delta\theta_{(1/2)z}$ between point 1 and point 2 can be computed by the First Moment Area Theorem as the area under the M_z/EI_z diagram for the left half of the beam. The area under this parabolic curve will be equal to 2/3 of the product of the base and the height.

$$\Delta\theta_{(1/2)z} = \frac{(2/3)(15)(225)}{EI_z} = \frac{2250}{EI_z}$$

from which

The Moment Area Method

$$\theta_{2z} = -\frac{2250}{EI_z} + \frac{2250}{EI_z} = 0$$

as expected.

From Figure 8.14d we see that the deflection Y_2 at the center of the beam will be negative. One of several possible methods for computing this deflection is to express it in terms of the tangential deviations $t_{3/1}$ and $t_{2/1}$ as

$$Y_2 = -\left(\frac{t_{3/1}}{2} - t_{2/1}\right)$$

The quantity $t_{3/1}/2$ is the vertical distance at the center of the beam from the original undeflected position to the tangent to joint 1 and $t_{2/1}$ is the vertical distance from the center of the deflected beam to the same tangent, as shown in Figure 8.14d. The tangential deviation $t_{2/1}$ can be computed by the Second Moment Area Theorem as the moment of the area under the M_z/EI_z diagram for the left half of the beam about joint 2. Using Eq. (8.28) and recognizing that the centroid of the parabolic area is 3/8 of its length from the right end gives

$$t_{2/1} = \frac{[(3/8)(15)][(2/3)(15)(225)]}{EI_z} = \frac{12{,}656.25}{EI_z}$$

which, upon substituting into the above expression for Y_2, gives

$$Y_2 = -\frac{21{,}093.75}{EI_z}$$

If desired, we can also compute the slope θ_{iz} and deflection Y_i at any other point i on the beam, at a distance X_i from the left end, as

$$\theta_{iz} = \theta_{1z} + \Delta\theta_{(1/i)z}$$

and

$$Y_i = -\left(\frac{X_i}{L}t_{3/1} - t_{i/1}\right)$$

where $\Delta\theta_{(1/i)z}$ is the area under the M_z/EI_z diagram between the left end of the beam and point i and $t_{i/1}$ is the moment of that area about point i. The only difficulties which might be encountered in these computations are the determination of the area and the location of the centroid for the M_z/EI_z area between point i and the left end of the beam. These difficulties will be addressed in the next example problem.

The computed values for θ_{2z} and Y_2 can be verified by substituting

$$\omega = -2$$
$$L = 30$$

and

$$X = 15$$

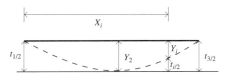

Figure 8.15 Example Problem 8.9, alternate solution.

into the expressions which were derived for the slope and deflection at any point in the simply supported beam in Example Problem 8.2 by integrating Eq. (8.18). This is left as an exercise for the reader.

The computations in this problem could have been significantly simplified if it had been recognized beforehand that the slope at the center of the beam was zero. From Figure 8.15, we can see that the deflection at the center of the beam can be expressed directly in terms of the tangential deviation of either end of the beam from the center as

$$Y_2 = -t_{1/2}$$

or

$$Y_2 = -t_{3/2}$$

where $t_{1/2}$ or $t_{3/2}$ can be computed by the Second Moment Area Theorem. In addition, after either $t_{1/2}$ or $t_{3/2}$ is known, the deflection Y_i at any point i can be computed by either

$$Y_i = -(t_{1/2} - t_{i/2})$$

or

$$Y_i = -(t_{3/2} - t_{i/2})$$

The slope θ_{iz} at the point can be expressed as

$$\theta_{iz} = \theta_{2z} + \Delta\theta_{(2/i)z}$$

or, since the slope θ_{2z} is zero,

$$\theta_{iz} = \Delta\theta_{(2/i)z}$$

The change in slope $\Delta\theta_{(2/i)z}$ can be computed by the First Moment Area Theorem. If the point i is to the right of joint 2 (i.e.; in the positive X direction), this change in slope will be positive, while if the point i is to the left of joint 2 (i.e.; in the negative X direction), this change in slope will be negative. Of course, if we did not know beforehand where the point of zero slope occurred in the beam, this simplified analysis process would be of no value. A structural engineer should always try to take advantage of known conditions in the beam which can help to simplify the analysis.

As shown in the previous example problem, it will often be found that there is more than one way to apply the Moment Area Theorems to compute the slope and deflection at specific points in a beam. An engineer should carefully study the known conditions for the beam to determine the most efficient analysis procedure. The most important factor in deciding how the analysis should be performed is having a realistic initial sketch of the deflected shape of the beam. A realistic

The Moment Area Method

guess for the deflected shape of the beam and careful planning can help to eliminate false starts and unnecessary calculations. It is also extremely important that the sign convention is fully understood so that the directions of the quantities which are computed by the Moment Area Theorems can be interpreted properly.

Bending Moment Diagrams by Cantilever Parts

In the previous example problems, the M_z/EI_z diagrams happened to be in a very convenient shape for computing the numerical values for the changes in slope and the tangential deviations which were needed to determine the slopes and deflections of the beams at the desired points. Unfortunately, for many beams, the shape of the diagrams are much more complicated, and the computation of the area under the curve and the location of the centroid of that area can be become a very tedious operation. However, there is a simple process for drawing the bending moment diagram for any beam which will eliminate these problems. The procedure is to use a modified cantilever mathematical model for the beam during the computation of the bending moment diagrams, which is loaded such that the bending moment diagram for the modified mathematical model is the same as the bending moment diagram for the actual mathematical model. The various loads on the modified mathematical model then can be considered individually to develop a set of bending moment diagrams for which the areas and the centroids can be easily computed.

Example Problem 8.10

To demonstrate this procedure for simplifying the bending moment diagram, we will compute the slope and deflection at the center of the prismatic simply supported beam shown in Figure 8.16a. The beam is subjected to a uniform downward distributed force of 2 kips per foot and moments of 60 kip-feet counterclockwise and 40 kip-feet clockwise at the left and right ends, respectively. The mathematical model for the beam is shown in Figure 8.16b, and the shear force and bending moment diagrams are shown in Figures 8.16c and 8.16d, respectively.

We can see immediately that this bending moment diagram is not going to be as easy to work with as the diagrams in the previous two example problems. The analysis would be greatly simplified if we could replace this diagram with another diagram, or set of diagrams, which represented the same bending moment distribution along the beam but had a shape for which the area and the location of the centroid could be easily computed. This can be accomplished by using the modified mathematical model shown in Figure 8.17a to compute the bending moments for the beam. The bending moment diagram for this cantilever mathematical model will be exactly the same as the bending moment diagram for the original simply supported mathematical model. However, the advantage of this cantilever mathematical model is that the bending moment diagrams for each individual active load on the model are very simple shapes, as shown in Figures 8.17b through 8.17d. Since the sum of the ordinates of these three bending moment diagrams at any point is equal to the ordinate of the original bending moment diagram at the same point, we also can sum the areas or the moments of the areas for any zone in these three diagrams and have the same

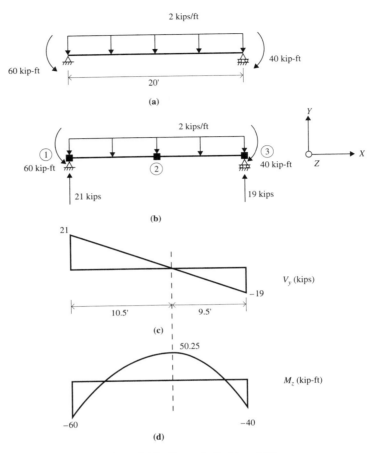

Figure 8.16 Example Problem 8.10.

result that we would obtain by performing the same operation on the original diagram. Essentially, what we have done is replace the original bending moment diagram by a set of cantilever parts.

Note that each of the cantilever bending moment diagrams is a curve of the form

$$M = aX^n \tag{8.29}$$

where X is the distance from the free end of the cantilever beam, a is a numeric coefficient, and n is the order of the curve. The values of n for the three curves in Figures 8.17b through 8.17d are 0, 1 and 2, respectively. The advantage of curves of this type is that it is possible to express the area under the curve and the location of the centroid by the set of simple equations:

$$A = \frac{bh}{n+1} \tag{8.30}$$

The Moment Area Method

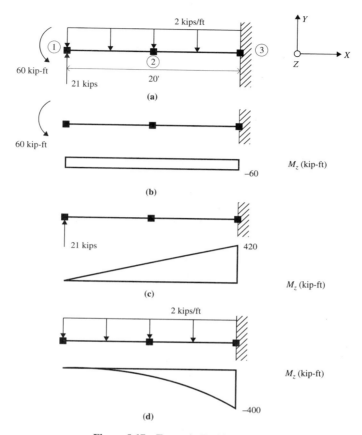

Figure 8.17 Example Problem 8.10.

$$c = \frac{b}{n+2} \qquad (8.31)$$

where A is the area under the curve and b, c and h are the distances defined in Figure 8.18. It will be found that the cantilever bending moment diagrams for triangular, parabolic and higher order distributed forces have these same properties. This technique allows the bending moment diagram for almost any beam loading to be transformed into a set of diagrams which are well suited for use in the Moment Area Method. However, the analyst must be careful when using this technique to not fall into the trap of using the boundary conditions for the cantilever mathematical model when computing the slopes and deflections of the beam. The cantilever mathematical model is only to be used to compute the modified bending moment diagram. The boundary conditions for the original mathematical model must be used to compute the slopes and deflections of the beam.

Now that we have transformed the bending moment diagram for this beam into a form that is easy to work with, our next step is to decide how the Moment Area

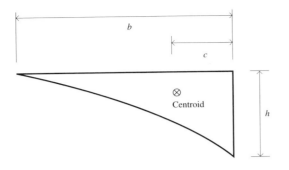

Figure 8.18 Cantilever bending moment diagram.

Theorems can be used to compute the slope and deflection at the center of the beam. For this, we need to know the deflected shape of the beam. If the beam were subjected only to the two end moments, it would bow upward with negative curvature over the entire length, while if it were subjected only to the downward distributed force, it would bow downward with positive curvature. From the original bending moment diagram in Figure 8.16d, we can see that the actual deformed shape falls between these two extremes since the curvature is negative adjacent to the supports and positive in the interior. The slope θ_{1z} of the beam at the left end could be positive or negative, or it could even be zero. Figure 8.19 shows four possible cases for the deflected shape of the beam which will satisfy the curvature requirements imposed by the bending moment diagram. The table at the bottom of the figure shows the signs for the quantities $t_{2/1}$, $t_{3/1}$, θ_{1z} and Y_2 for each of these cases. For each of these deflected shapes, the slope θ_{2z} and the deflection Y_2 at the center of the beam can be

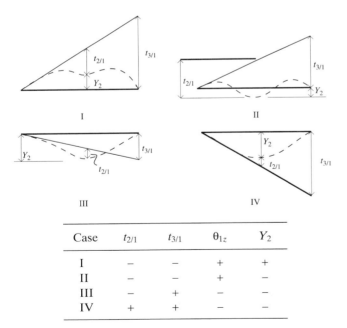

Case	$t_{2/1}$	$t_{3/1}$	θ_{1z}	Y_2
I	−	−	+	+
II	−	−	+	−
III	−	+	−	−
IV	+	+	−	−

Figure 8.19 Example Problem 8.10.

The Moment Area Method

computed by the same expressions which were used in the analysis of the beam in Example Problem 8.8 as long as the correct signs are inserted for $\Delta\theta_{(1/2)z}$, $t_{2/1}$ and $t_{3/1}$.

$$\theta_{2z} = \theta_{1z} + \Delta\theta_{(1/2)z}$$

where

$$\theta_{1z} = -\frac{t_{3/1}}{L}$$

and

$$Y_2 = -\left(\frac{t_{3/1}}{2} - t_{2/1}\right)$$

The change in slope $\Delta\theta_{(1/2)z}$ can be computed by the First Moment Area Theorem by adding the contributions from each of the three cantilever bending moment diagrams in Figures 8.17b to 8.17d:

$$\Delta\theta_{(1/2)z} = \frac{(10)(-60)}{EI_z} + \frac{(10)(210)/2}{EI_z} + \frac{(10)(-100)/3}{EI_z} = \frac{116.67}{EI_z}$$

The tangential deviations $t_{2/1}$ and $t_{3/1}$ can be computed by the Second Moment Area Theorem in a similar manner as

$$t_{2/1} = \frac{(10/2)\,[(10)(-60)]}{EI_z} + \frac{(10/3)\,[(10)(210)/2]}{EI_z} + \frac{(10/4)\,[(10)(-100)/3]}{EI_z} = -\frac{333.33}{EI_z}$$

and

$$t_{3/1} = \frac{(20/2)\,[(20)(-60)]}{EI_z} + \frac{(20/3)\,[(20)(420)/2]}{EI_z} + \frac{(20/4)\,[(20)(-400)/3]}{EI_z} = \frac{2666.67}{EI_z}$$

The positive value for $t_{3/1}$ shows that the actual deformed shape for the beam must be either Case III or Case IV. Therefore, the final computed values for the slope θ_{1z} and the deflection Y_2 should both be negative. Using the previous expression to compute θ_{1z}, with the positive value for $t_{3/1}$, gives

$$\theta_{1z} = -\frac{2666.67/EI_z}{20} = -\frac{133.33}{EI_z}$$

from which

$$\theta_{2z} = -\frac{133.33}{EI_z} + \frac{116.67}{EI_z} = -\frac{16.67}{EI_z}$$

The negative value for $t_{2/1}$ shows that the deformed shape for the beam corresponds to Case III. Using the previous expression to compute Y_2, with the positive value for $t_{3/1}$ and the negative value for $t_{2/1}$, gives

$$Y_2 = -\left[\frac{2.666.67/EI_z}{2} - \left(-\frac{333.33}{EI_z}\right)\right] = -\frac{1666.67}{EI_z}$$

Therefore, both the slope and the deflection are negative, as expected.

The negative value for the slope θ_{2z} shows that the point of maximum downward deflection for the beam occurs to the right of the center. The location of the point of maximum deflection at a distance X_i from the left end of the beam can be determined by setting the slope θ_{iz} at that point to zero in the expression

$$\theta_{iz} = \theta_{1z} + \Delta\theta_{(1/i)z}$$

from which

$$\Delta\theta_{(1/i)z} = -\theta_{1z}$$

If we now develop an algebraic expression for $\Delta\theta_{(1/i)z}$ in terms of the distance X_i, we can use this relationship to determine the point of maximum deflection. Using the first moment area theorem to compute $\Delta\theta_{(1/i)z}$ from each of the cantilever bending moment diagrams gives

$$\Delta\theta_{(1/i)z} = \frac{(X_i)(-60)}{EI_z} + \frac{(X_i)[(X_i/20)(420)]/2}{EI_z} + \frac{(X_i)[(X_i/20)^2(-400)]/3}{EI_z}$$

which, after equating to $-\theta_{1z}$ and dividing out the symbol EI_z, gives

$$-60X_i + 10.5X_i^2 - 0.333X_i^3 = 133.33$$

The three roots for this polynomial equation are

$$X_i = -1.6934; \ 10.3325; \ 22.8609$$

from which we can determine that the point of maximum deflection is located 10.3325 feet from the left end of the beam since the other two roots correspond to points which are beyond the limits of the beam ends. Note that the maximum deflection does not occur at the point of maximum positive bending moment, which is 10.5 feet from the left end of the beam. There is no relationship between the point of maximum bending moment and the point of maximum deflection for a simply supported beam subjected to an arbitrary loading. The maximum deflection $Y_{\max}$ now can be computed using the value for X_i by either

$$Y_{\max} = -t_{1/i}$$

or

$$Y_{\max} = -t_{3/i}$$

where $t_{1/i}$ or $t_{3/i}$ can be computed by the Second Moment Area Theorem. Due to the form of the cantilever bending moment diagrams, it is much easier to compute $t_{1/i}$ for this particular beam.

$$t_{1/i} = \frac{[X_i/2][(X_i)(-60)]}{EI_z} + \frac{[2X_i/3][(X_i)[(X_i/20)(420)]/2]}{EI_z}$$
$$+ \frac{[3X_i/4][(X_i)[(X_i/20)^2(-400)]/3]}{EI_z}$$

from which, after substituting the computed value for X_i, gives

The Moment Area Method

$$t_{1/i} = \frac{1669.45}{EI_z}$$

Therefore, the maximum deflection is

$$Y_{max} = -t_{1/i} = -\frac{1669.45}{EI_z}$$

This value can be verified by computing $t_{3/i}$. The value of $t_{3/i}$ should be equal to the value of $t_{1/i}$ within expected roundoff error. This is left as an exercise for the reader. It will be found that it will be much more convenient to compute $t_{3/i}$ if the bending moment diagram is redrawn by cantilever parts about the left end of the beam. An engineer should try to plan ahead when drawing the bending moment diagram by cantilever parts to ensure that it will be in the most convenient form for the calculations which must be performed. The modified mathematical model for computing the cantilever parts can have the fixed support at any desired location.

Example Problem 8.11

Compute the slope and deflection at the right end of the beam shown in Figure 8.20a. The mathematical model for the beam is shown in Figure 8.20b and the bending moment diagram is shown in Figure 8.20c. This is another situation where the total bending moment diagram is not convenient to work with. Therefore, we will draw the bending moment diagram by cantilever parts for use in the computations. However, since the fixed

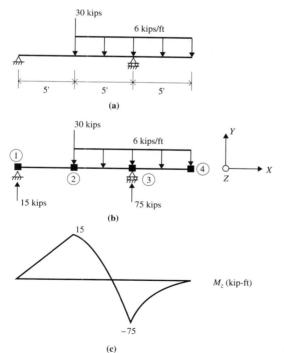

Figure 8.20 Example Problem 8.11.

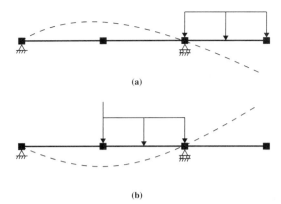

Figure 8.21

support for the modified mathematical model for drawing the cantilever bending moment diagrams may be located at any point, we must first decide how the Moment Area Theorems will be applied to compute the deflection at the right end of this beam so that the cantilever bending moment diagrams will be in the most convenient form.

If the beam were only subjected to that portion of the loading to the right of joint 3, the curvature would be negative over the entire beam and joint 4 would deflect down, as shown in Figure 8.21a. If the beam were only subjected to that portion of the load between joint 1 and joint 3, the curvature would be positive between the supports and zero to the right of joint 3. Joint 4 would deflect up for this loading, as shown in Figure 8.21b, with the portion of the beam between joints 3 and 4 being a straight line. From the bending moment diagram for the actual loading we can see that the curvature will change from positive to negative while moving from left to right along the beam. It is not possible to determine from this diagram whether joint 4 deflects up or down. Figure 8.22 shows four possible cases for the deflected shape for the beam which are consistent with the required directions for the curvatures and the support conditions.

One method which can be used to compute the deflection Y_4 at the right end for each of these deflected shapes is to express it in terms of the tangential deviations $t_{3/1}$ and $t_{4/1}$ as

$$Y_4 = -(1.5 t_{3/1} - t_{4/1})$$

as shown by the tangent to the left end of the beam for each of these cases. Also, the slope θ_{4z} at the right end can be computed by

$$\theta_{4z} = \theta_{1z} + \Delta\theta_{(1/4)z}$$

where

$$\theta_{1z} = -\frac{t_{3/1}}{L_{1/3}}$$

where $L_{1/3}$ is the length of the beam from joint 1 to joint 3. The correct signs for Y_4 and θ_{1z} will be obtained from these expression if the values of $t_{3/1}$ and $t_{4/1}$ are substituted with the correct signs. The table in Figure 8.22 shows the signs for $t_{3/1}$, $t_{4/1}$, θ_{1z} and Y_4 for each of these cases.

The Moment Area Method

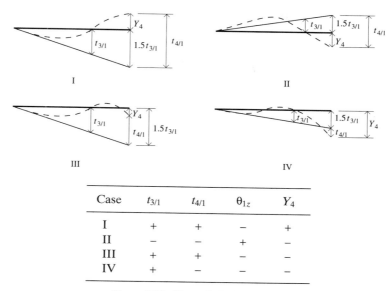

Figure 8.22 Example Problem 8.11.

Now that we know that we must compute $t_{3/1}$, $t_{4/1}$ and $\Delta\theta_{(1/4)z}$ to analyze this problem, we can see that the most convenient form for the bending moment diagram will be obtained by drawing it by cantilever parts about the right end of the beam, as shown in Figure 8.23. By using these four cantilever diagrams, the values for the tangential deviations $t_{3/1}$ and $t_{4/1}$ can be computed by the Second Moment Area Theorem as

$$t_{3/1} = \frac{(10/3)[(10)(150)/2]}{EI_z} + \frac{(5/3)[(5)(-150)/2]}{EI_z} + 0 + \frac{(5/4)[(5)(-75)/3]}{EI_z}$$

from which

$$t_{3/1} = \frac{1718.75}{EI_z}$$

and

$$t_{4/1} = \frac{(15/3)[(15)(225)/2]}{EI_z} + \frac{(10/3)[(10)(-300)/2]}{EI_z} + \frac{(5/3)[(5)(375)/2]}{EI_z}$$
$$+ \frac{(10/4)][(10)(-300)/3]}{EI_z}$$

from which

$$t_{4/1} = \frac{2500}{EI_z}$$

The change in slope $\Delta\theta_{(1/4)z}$ can be computed by the First Moment Area Theorem as

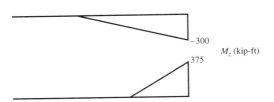

Figure 8.23 Example Problem 8.11, moment diagram by cantilever parts.

$$\Delta\theta_{(1/4)z} = \frac{(15)(225)/2}{EI_z} + \frac{(10)(-300)/2}{EI_z} + \frac{(5)(375)/2}{EI_z} + \frac{(10)(-300)/3}{EI_z}$$

from which

$$\Delta\theta_{(1/4)z} = \frac{125}{EI_z}$$

Since both $t_{3/1}$ and $t_{4/1}$ are positive, the deflected shape for the beam must correspond to either Case I or Case III. Substituting these values into the previous expression for Y_4, θ_{1z} and θ_{4z} gives

$$Y_4 = -\left[1.5\left(\frac{1718.75}{EI_z}\right) - \frac{2500}{EI_z}\right] = -\frac{78.125}{EI_z}$$

and

$$\theta_{1z} = -\frac{1718.75/EI_z}{10} = -\frac{171.875}{EI_z}$$

from which

$$\theta_{4z} = -\frac{171.875}{EI_z} + \frac{125}{EI_z} = -\frac{46.875}{EI_z}$$

The Moment Area Method

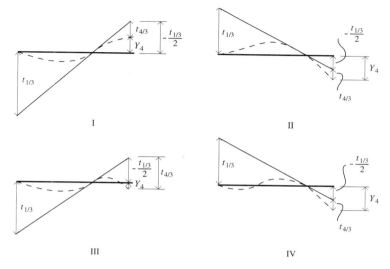

Figure 8.24 Example Problem 8.11.

The deflected shape corresponds to Case III since Y_4 is negative.

An alternate approach for computing Y_4 is to draw a tangent to the elastic curve at joint 3 for each of the previous deflected shape cases, as shown in Figure 8.24. By using this tangent, the deflection Y_4 can be expressed in terms of the tangential deviations $t_{1/3}$ and $t_{4/3}$ as

$$Y_4 = -\left(-\frac{t_{1/3}}{2} - t_{4/3}\right)$$

and the slope θ_{4z} can be computed by

$$\theta_{4z} = \theta_{3z} + \Delta\theta_{(3/4)z}$$

where

$$\theta_{3z} = \frac{t_{1/3}}{L_{1/3}}$$

For this approach, a convenient form for the cantilever bending moment diagrams will be obtained by considering the fixed support in the modified mathematical model to be located at joint 3, as shown in Figure 8.25. The values of $t_{1/3}$, $t_{4/3}$ and $\Delta\theta_{(3/4)z}$ can be easily computed by the Moment Area Theorems from these diagrams as

$$t_{1/3} = \frac{781.25}{EI_z}$$

$$t_{4/3} = -\frac{468.75}{EI_z}$$

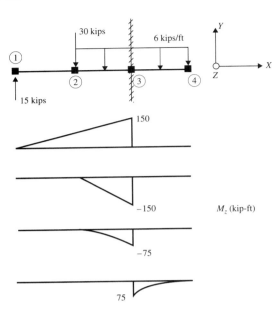

Figure 8.25 Example Problem 8.11.

$$\Delta\theta_{(3/4)z} = -\frac{125}{EI_z}$$

By using these values, the deflection at the right end of the beam can now be computed as

$$Y_4 = -\left[-\left(\frac{781.25/EI_z}{2}\right) - \left(-\frac{468.75}{EI_z}\right)\right] = -\frac{78.125}{EI_z}$$

while the slope at joint 3 can be computed as

$$\theta_{3z} = \frac{781.25/EI_z}{10} = \frac{78.125}{EI_z}$$

from which the slope at the right end is

$$\theta_{4z} = \frac{78.125}{EI_z} + \left(-\frac{125}{EI_z}\right) = -\frac{46.875}{EI_z}$$

These values agree with the previous solution. Using an alternate calculation procedure such as this can be very valuable in checking the results of an analysis before proceeding to other stages in the design process.

Note that the deflection Y_4 and the slope θ_{3z} happen to have the same numerical values with opposite signs for this beam. This is merely a coincidence and the reader should not try to draw any general conclusions from it. The numerical values will be different for other beam loadings.

The Moment Area Method

Nonprismatic Beams

The Moment Area Method also can be used for the analysis of nonprismatic beams. The only difference which will occur in the analysis, compared to the analysis of a prismatic beam, is that the M_z/EI_z diagram will no longer be the same shape as the bending moment diagram. The shape of the diagram will now depend on the variation of EI_z over the length of the beam.

Example Problem 8.12

Compute the slopes at the two ends and the slope and deflection at the center of the nonprismatic simply supported beam which was analyzed previously in Example Problem 8.7 by direct integration of Eq. (8.13). The mathematical model for the beam is shown in Figure 8.26a and the M_z/EI_z diagram, which was obtained by dividing the bending moment by EI_z in the left half of the beam and by $2EI_z$ in the right half, is shown in Figure 8.26b.

There is no particular advantage in drawing the bending moment diagram by cantilever parts for this beam. The diagram shown in Figure 8.26b will be used in the analysis.

Since the moment is positive over the entire length of the beam, the curvature will also be positive over the entire length which will result in the deflected shape shown in Figure 8.26c. Since this is the same deflected shape as the simply supported beam in Example Problem 8.9, the same expressions which were used to compute θ_{1z} and Y_2 in that analysis also can be used here.

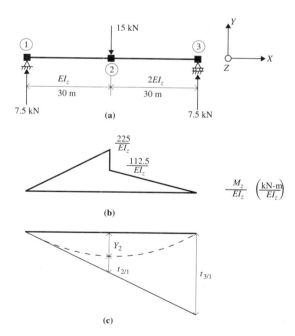

Figure 8.26 Example Problem 8.12.

$$\theta_{1z} = -\frac{t_{3/1}}{L}$$

$$Y_2 = -\left(\frac{t_{3/1}}{2} - t_{2/1}\right)$$

The tangential deviations $t_{2/1}$ and $t_{3/1}$ can be computed by the Second Moment Area Theorem as

$$t_{2/1} = \frac{33{,}750}{EI_z}$$

$$t_{3/1} = \frac{168{,}750}{EI_z}$$

from which

$$\theta_{1z} = -\frac{2812.5}{EI_z}$$

$$Y_2 = -\frac{50{,}625}{EI_z}$$

The slopes θ_{2z} and θ_{3z} can be computed by

$$\theta_{2z} = \theta_{1z} + \Delta\theta_{(1/2)z}$$

$$\theta_{3z} = \theta_{1z} + \Delta\theta_{(1/3)z}$$

where $\Delta\theta_{(1/2)z}$ and $\Delta\theta_{(1/3)z}$ can be computed by the First Moment Area Theorem as

$$\Delta\theta_{(1/2)z} = \frac{3375}{EI_z}$$

$$\Delta\theta_{(1/3)z} = \frac{5062.5}{EI_z}$$

from which

$$\theta_{2z} = \frac{562.5}{EI_z}$$

$$\theta_{3z} = \frac{2250}{EI_z}$$

The computed values for θ_{1z}, θ_{2z}, θ_{3z} and Y_4 can be verified by evaluating the expressions which were developed for the slope and deflection in Example Problem 8.7. This is left as an exercise for the reader.

The positive value for θ_{2z} shows that the maximum downward deflection in the beam will occur to the left of the center. The location of this maximum deflection

The Conjugate Beam Method

and its magnitude can be computed by the same process used in Example Problem 8.10.

THE CONJUGATE BEAM METHOD

In some situations, during an engineering analysis, it will be found that it is possible to determine the solution to a particular problem by solving a totally different problem, which might even have a different physical meaning, whose solution happens to have the same numerical value as the solution of the desired problem. This other problem is known as an analogy. We can use this procedure to solve the beam deflection equations for the slope and deflection at a specific point in a beam by solving an analogous problem using the beam equilibrium equations. Some structural engineers prefer this procedure to compute the slope and deflection in a beam over the previously described procedures since they feel more comfortable solving an equilibrium problem rather than a beam deflection problem. In reality, the use of this procedure does not reduce the computational effort required to solve the problem since the number and type of calculations which must be performed are exactly the same for both the original problem and the analogous problem. The only advantage of the process is that it allows the engineers to use an analysis procedure with which they are more familiar.

To develop an analogy for solving beam deflection problems, we will compare the forms of the beam equilibrium equations and the beam deflection equations. The beam equilibrium equations, which were presented previously in Chapter 5, can be summarized as

$$\Delta V_y = \int q_y \, dX \quad \text{or} \quad q_y = \frac{dV_y}{dX} = \frac{d^2 M_z}{dX^2} \qquad (8.32)$$

$$\Delta M_z = \int V_y \, dX \quad \text{or} \quad V_y = \frac{dM_z}{dX} \qquad (8.33)$$

while the beam deflection equations, which were shown previously in this chapter, are

$$\Delta \theta_z = \int \frac{M_z}{EI_z} \, dX \quad \text{or} \quad \frac{M_z}{EI_z} = \frac{d\theta_z}{dX} = \frac{d^2 Y}{dX^2} \qquad (8.34)$$

$$\Delta Y = \int \theta_z \, dX \quad \text{or} \quad \theta_z = \frac{dY}{dX} \qquad (8.35)$$

A comparison of these two sets of equations shows that they have exactly the same form. In fact, if we make the following substitutions for the symbols in the beam deflection equations

$$Y = M_z^* \qquad (8.36)$$

$$\theta_z = V_y^* \qquad (8.37)$$

and

$$\frac{M_z}{EI_z} = q_y^* \qquad (8.38)$$

they become

$$\Delta V_y^* = \int q_y^* dX \qquad \text{or} \qquad q_y^* = \frac{dV_y^*}{dX} = \frac{d^2 M_z^*}{dX^2} \qquad (8.39)$$

$$\Delta M_z^* = \int V_y^* dX \qquad \text{or} \qquad V_y^* = \frac{dM_z^*}{dX} \qquad (8.40)$$

which look exactly like the beam equilibrium equations. This indicates that the solution of the beam deflection equations and the beam equilibrium equations should involve the same type of calculations. If this is correct, then the same calculation process can be used to solve for the slope and deflection at a specific point in a beam in terms of the quantity M_z/EI_z as would be used to compute the shear force V_y^* and bending moment M_z^* at that point in an analogous beam in terms of an analogous distributed force q_y^*. The magnitude of the analogous distributed force at each point on the analogous beam would be equal to M_z/EI_z for the real beam. The analogous beam will be called the *conjugate beam* and the analogous distributed force will be called the *conjugate distributed force*, while the procedure to perform the equilibrium analysis of the conjugate beam will be known as the *Conjugate Beam Method*.

Example Problem 8.13

To demonstrate the Conjugate Beam Method and how it compares to the Moment Area Method, we will compute the slopes at the two ends and the center and the deflection at the center of the nonprismatic simply supported beam which was analyzed by the Moment Area Method in Example Problem 8.12. The conjugate beam and the conjugate distributed force are shown in Figure 8.27a. The conjugate distributed force q_y^* at each point along the conjugate beam is equal to the ordinate of the M_z/EI_z diagram for the real beam shown previously in Figure 8.26b. The force acts up since the bending moment in the real beam is positive. (For reference purposes during the analysis the left and right ends of the conjugate beam will be designated as points 1 and 3 respectively while the center will be designated as point 2. This is consistent with the numbering scheme used for the joints in the mathematical model of the beam in the previous example problem.)

The reaction at the left end of the conjugate beam can be computed by summing moments about point 3:

$$\Sigma_3 M_z \rightarrow R_{1Y}^* = -\frac{2812.5}{EI_z} = \frac{2812.5}{EI_z} \downarrow$$

The Conjugate Beam Method

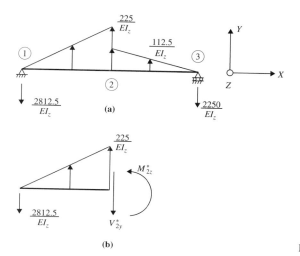

Figure 8.27 Example Problem 8.13.

from which the shear force at the left end is

$$V^*_{1y} = -\frac{2812.5}{EI_z}$$

The shear forces V^*_{2y} and V^*_{3y} at the center and the right end of the conjugate beam can be computed by Eq. (5.12) from Chapter 5 as

$$V^*_{2y} = V^*_{1y} + \int_{X_1}^{X_2} q^*_y dX$$

and

$$V^*_{3y} = V^*_{1y} + \int_{X_1}^{X_3} q^*_y dX$$

in which the integral on the right side is equal to the magnitude of the conjugate force between the two points. This is equivalent to the area under the load diagram which was used in Chapter 5 when computing the change in the shear force in a beam. Performing these operations gives

$$V^*_{2y} = \frac{562.5}{EI_z}$$

and

$$V^*_{3y} = \frac{2250}{EI_z}$$

The moment at the center of the conjugate beam can be computed by summing moments about point 2 for the free body diagram of the left half of the conjugate beam shown in Figure 8.27b.

$$\Sigma_2 M_Z \rightarrow M_{2z}^* = -\frac{50{,}625}{EI_z}$$

The slopes at the two ends and at the center of the real beam now can be determined from the shear forces at these points in the conjugate beam by using the analogy shown in Eq. (8.37):

$$\theta_{1z} = V_{1y}^* = -\frac{2812.5}{EI_z}$$

$$\theta_{2z} = V_{2y}^* = \frac{562.5}{EI_z}$$

$$\theta_{3z} = V_{3y}^* = \frac{2250}{EI_z}$$

while the deflection at the center can be determined by using the analogy shown in Eq. (8.36):

$$Y_2 = M_{2z}^* = -\frac{50{,}625}{EI_z} = \frac{50{,}625}{EI_z} \downarrow$$

Note that these values agree exactly with the values obtained by the analysis of this beam by the Moment Area Method.

If the individual calculations are compared for Example Problem 8.12 for the analysis by the Moment Area Method and for Example Problem 8.13 for the analysis by the Conjugate Beam Method, it will be found that exactly the same numbers were added, subtracted, multiplied and divided during the computations.

1. The calculations in Example Problem 8.13 to compute V_{1y}^*, by summing moments for the conjugate beam about the right end to determine R_{1Y}^*, were the same calculations which were performed in Example Problem 8.12 to compute θ_{1z} by the expression

$$\theta_{1z} = -\frac{t_{3/1}}{L}$$

2. The calculations in Example Problem 8.13 to compute V_{2y}^* and V_{3y}^*, after V_{1y}^* had been determined, were the same calculations which were performed in Example Problem 8.12 to compute θ_{2z} and θ_{3z}, after θ_{1z} had been determined, by the expressions

$$\theta_{2z} = \theta_{1z} + \Delta\theta_{(1/2)z}$$
$$\theta_{3z} = \theta_{1z} + \Delta\theta_{(1/3)z}$$

3. The calculations in Example Problem 8.13 to compute M_{2z}^* by summing moments about the center of the beam for the left segment were the same cal-

The Conjugate Beam Method

culations which were performed in Example Problem 8.12 to compute Y_2 by the expression

$$Y_2 = -\left(\frac{t_{3/1}}{2} - t_{2/1}\right)$$

Therefore, there were no savings in the computational effort for one method over the other. The primary advantage of the Conjugate Beam Method over the Moment Area Method is that it is not necessary to draw and later verify the deflected shape of the beam from the signs of the curvatures and the known displacements at the supports for complicated beam geometries and loadings. Many engineers are more comfortable with an equilibrium analysis than with a geometric analysis, particularly in determining the signs of the computed results.

Support Conditions for the Conjugate Beam

In the previous example problem, the support conditions for the real beam and the conjugate were identical. However, this will not always be the case. For many beams, the support conditions for the conjugate beam will be entirely different than those for the real beam. The important factors for the supports in the real beam are how they affect the slope and the deflection of the beam. However, the important factors for the supports in the conjugate beam are how they affect the conjugate shear force and the conjugate bending moment. These effects are independent of each other. A beam can have a zero shear force or bending moment at a point where the slope or deflection is nonzero or a nonzero shear force or bending moment at a point where the slope or deflection is zero.

Example Problem 8.14

To demonstrate the importance of having the correct support conditions for the conjugate beam, we will use the Conjugate Beam Method to compute the slope and deflection at the free end of a prismatic cantilever beam which is subjected to a positive concentrated force W at the free end, as shown in Figure 8.28a. The bending moment diagram for this beam will be a first order curve with an ordinate of WL at the fixed end and an ordinate of zero at the free end. Figure 8.28b shows the conjugate beam subjected to the conjugate loading which will be used in the first attempt at analyzing this beam. The support conditions for this conjugate beam are the same as the support conditions for the real beam. The left end will be designated as point 1 and the right end will be designated as point 2 for reference. The shear force and the bending moment at the fixed end of the conjugate beam can be computed by

$$\Sigma F_Y^* \to R_{1Y}^* = -\frac{WL^2}{2EI_z} \to V_{1y}^* = \frac{WL^2}{2EI_z}$$

$$\Sigma_1 M_Z^* \to M_{1z}^* = \frac{WL^3}{6EI_z}$$

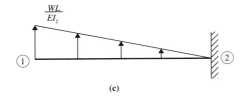

Figure 8.28 Example Problem 8.14.

while both the shear force and bending moment at the free end must be zero:

$$V_{2y}^* = 0$$
$$M_{2z}^* = 0$$

If we now use the analogies shown in Eqs. (8.36) and (8.37), the slopes and deflections at the two ends of the real beam will be

$$\theta_{1z} = -\frac{WL^2}{2EI_z}$$

$$\theta_{2z} = 0$$

$$Y_1 = \frac{WL^3}{6EI_z}$$

$$Y_2 = 0$$

However, these slopes and deflections are not consistent with the support conditions for the real beam. Something is wrong. The problem is in the support conditions which were used for the conjugate beam.

Figure 8.28c shows the conjugate beam which will be used in the second attempt to analyze the real beam. Note that the support conditions have been reversed at the two ends. For this beam, the shear force and the bending moment at the fixed end will be

$$\Sigma F_Y^* \to R_{2Y}^* = -\frac{WL^2}{2EI_z} \to V_{2y}^* = \frac{WL^2}{2EI_z}$$

The Conjugate Beam Method

$$\Sigma_2 M_Z^* \to M_{2z}^* = \frac{WL^3}{3EI_z}$$

while the shear force and bending moment at the free end will be

$$V_{1y}^* = 0$$

$$M_{1z}^* = 0$$

If we now use the analogies in Eqs. (8.36) and (8.37) for this conjugate beam, the slopes and deflections at the two ends of the real beam will be

$$\theta_{1z} = 0$$

$$\theta_{2z} = \frac{WL^2}{2EI_z}$$

$$Y_1 = 0$$

$$Y_2 = \frac{WL^3}{3EI_z}$$

which are consistent with the support conditions for the real beam. These expressions for the slope and deflection at the free end agree with the expressions which was obtained for this same beam in Example Problem 8.3 by the integration of Eq. (8.13).

The previous example problem shows that extreme care must be exercised when choosing the support conditions for the conjugate beam. If the boundary conditions are not correct, the computed slopes and deflections will be meaningless. The supports for the conjugate beam must be chosen so that the values for the shear force and the bending moment in the conjugate beam at the supports are consistent with the values for the slope and the deflection in the real beam at the supports. Figure 8.29 shows a number of different support conditions for a real beam and the corresponding support conditions which must exist in the conjugate beam.

Example Problem 8.15

Compute the slope and deflection by the Conjugate Beam Method at the right end of the beam in Figure 8.20 which was analyzed in Example Problem 8.11 by the Moment Area Method. By using the rules shown in Figure 8.29, the conjugate beam for this analysis will have a pinned support at joint 1, an interior pin at joint 3 and a fixed support at joint 4, as shown in Figure 8.30a. The conjugate distributed force will have the same shape as the bending moment diagram shown in Figure 8.20c. The desired slope and deflection can be determined by computing the shear force V_{4y}^* and the bending moment M_{4z}^* for this beam.

Although the operations to perform the static analysis of this conjugate beam are fairly straight forward, the actual computations will be complicated due to the form of the conjugate distributed force. However, this difficulty can be easily overcome by using the procedure which was used when analyzing this same beam by the

276   Deflection of Beams   Chap. 8

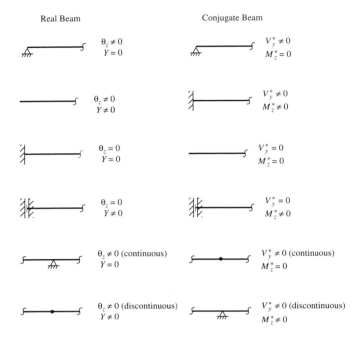

Figure 8.29 Conjugate beam support conditions.

Moment Area Method by resolving the bending moment diagram for the real beam into a set of cantilever parts, as shown previously in Figure 8.23. By using this procedure, the static analysis for the original conjugate distributed force can be achieved by summing the results of the static analyses for the four conjugate distributed forces shown in Figures 8.30b through 8.30e. The magnitude and the location of the resultant force for each of these distributed forces can be computed by Eqs. 8.30 and 8.31.

The shear force V^*_{4y} and the bending moment M^*_{4z} now can be easily determined using these simplified conjugate loadings. The interior pin in the conjugate beam will supply an equation of condition which can be used in the analysis.

$$\Sigma_{3/\text{left}} M^*_Z \rightarrow R^*_{1Y} = -\frac{171.875}{EI_z}$$

$$\Sigma F^*_Y \rightarrow R^*_{4Y} = \frac{46.875}{EI_z} \rightarrow V^*_{4Y} = -\frac{46.875}{EI_z} \rightarrow \theta_{4z} = -\frac{46.875}{EI_z}$$

$$\Sigma_4 M^*_Z \rightarrow M^*_{4z} = -\frac{78.125}{EI_z} \rightarrow Y_4 = -\frac{78.125}{EI_z} = \frac{78.125}{EI_z} \downarrow$$

As expected, the slope and deflection at the right end agree exactly with the values obtained by the Moment Area Method. As with the previous example problem, a

The Conjugate Beam Method

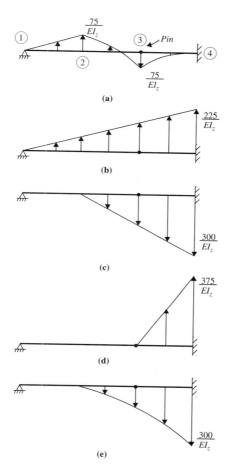

Figure 8.30 Example Problem 8.15.

comparison of the analyses by the Moment Area Method and the Conjugate Beam Method will show that same computations were performed in both analyses.

Example Problem 8.16

Compute the slope and deflection by the Conjugate Beam Method at the point of application of the downward concentrated force for the beam shown in Figure 8.31a. The mathematical model for the beam is shown in Figure 8.31b and the bending moment diagram is shown in Figure 8.31c. By using the rules shown in Figure 8.29, the conjugate beam for this analysis will have a free end at joint 1, an interior pinned support at joint 2 and a roller support at joint 4, as shown in Figure 8.31d.

The first step in the analysis will be to compute the reactions for the conjugate beam.

$$\Sigma_4 M_Z^* \rightarrow R_{2Y}^* = \frac{7530}{EI_z}$$

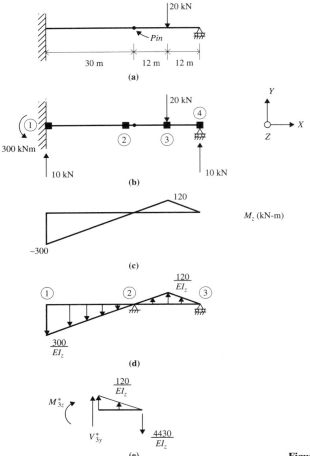

Figure 8.31 Example Problem 8.16.

$$\Sigma F_Y^* \to R_{4Y}^* = -\frac{4430}{EI_z}$$

The next step is to compute V_{3y}^* and M_{3z}^* using the free body diagram of the portion of the beam between points 3 and 4 shown in Figure 8.31e

$$\Sigma F_Y^* \to V_{3y}^* = \frac{2990}{EI_z} \to \theta_{3z} = \frac{2990}{EI_z}$$

$$\Sigma M_Z^* \to M_{3z}^* = \frac{117,840}{EI_z} \to Y_3 = \frac{117,840}{EI_z}$$

The reader will find that the computation of the slope and deflection at any point to the right of the interior pin in the beam in the previous example problem

The Conjugate Beam Method

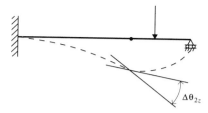

Figure 8.32 Kink angle in real beam.

is much easier by the Conjugate Beam Method than by the Moment Area Method. The interior pin at point 3 will cause the beam to deflect into a shape similar to that shown in Figure 8.32, with a kink angle $\Delta\theta_{2z}$ at the location of the pin. Although the deflection is continuous across the pin, the slope will be discontinuous, which will cause difficulty in the application of the Moment Area Theorems. This difficulty is not encountered in the Conjugate Beam Method since the interior knife edge support in the conjugate beam automatically compensates for the slope discontinuity. The vertical reactive force at this interior support, which corresponds to the discontinuity in the shear force ΔV_{2y}^* in the conjugate beam, is exactly equal to the kink angle in the real beam.

$$\Delta\theta_{2z} = \Delta V_{2y}^* = R_{2Y}^* = \frac{7530}{EI_z}$$

Since the change in the shear force in the conjugate beam is positive, the change in the slope across the pin in the real beam is counterclockwise, as shown in Figure 8.32.

Analysis of Statically Indeterminate Beams by the Conjugate Beam Method

The Conjugate Beam Method can be used to compute the slopes and deflections for both statically determinate and statically indeterminate beams after the bending moments at all points along the beam have been determined. The only difference that will be encountered when applying this method to the analysis of a statically indeterminate beam is that the conjugate beam will appear to be unstable. However, it will still be possible to perform the static analysis of the conjugate beam to compute the shear force and bending moment at any point since the conjugate distributed force will be such that the beam will be in a state of *unstable equilibrium*.

Example Problem 8.17

To demonstrate the application of the Conjugate Beam Method to the analysis of a statically indeterminate beam, we will compute the slope and deflection at a point 60 meters from the left end of the prismatic fixed-fixed beam, whose mathematical model is shown in Figure 8.33a. The bending moment diagram for the beam, which was developed by another engineer, is shown in Figure 8.33b. Although we cannot veri-

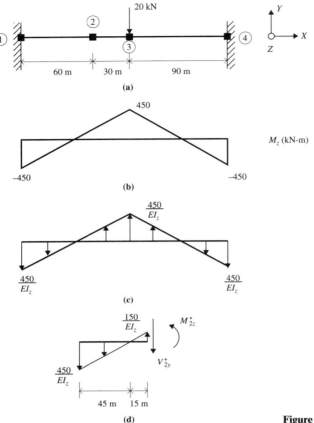

Figure 8.33 Example Problem 8.17.

fy it at this time, this is the correct diagram for this beam. (Procedures for analyzing statically indeterminate beams will be presented in Chapters 11 through 15.)

In order to satisfy the requirements of zero slope and zero deflection at the ends of the real beam, the conjugate beam must have zero shear force and zero bending moment at its ends. The only way that these conditions can be satisfied is for the conjugate beam to have free ends, as shown in Figure 8.33c. This is consistent with the previously defined relationships between the real beam and conjugate beam supports shown in Figure 8.29. Although it appears that this unstable conjugate beam will cause difficulty in the computation of the desired conjugate shear force and bending moment, a quick calculation will show that ΣF_Y is satisfied and ΣM_Z about any point is satisfied for the conjugate distributed force even though there are no supports for the conjugate beam. Therefore, the conjugate beam is in a state of unstable equilibrium.

The slope and deflection at joint 2 in the real beam can be determined by computing the shear force V_{2y}^* and the bending moment M_{2z}^* in the conjugate beam using the free body diagram shown in Figure 8.33d.

$$\Sigma F_Y^* \rightarrow V_{2y}^* = -\frac{9000}{EI_z} \rightarrow \theta_{2z} = -\frac{9000}{EI_z}$$

$$\Sigma_2 M_Z^* \rightarrow M_{2z}^* = -\frac{292{,}500}{EI_z} \rightarrow Y_2 = -\frac{292{,}500}{EI_z}$$

The Conjugate Beam Method can be used to compute the slope and deflection at any point for any statically indeterminate beam. The combination of the supports for the conjugate beam and the conjugate distributed force will always result in a system which is in unstable equilibrium.

COMPUTER PROGRAM BEAMVM

The computer program BEAMVM, which was used in Chapter 5 to plot the shear force and bending moment diagrams for a beam, also can be used to compute the deflections and plot the deflected shape for a linear elastic stable beam. The instructions for using the program are in the ASCII disk file BEAMVM.DOC.

Figure 8.34 shows the input data file for BEAMVM for the analysis of the simply supported beam with an overhang which was analyzed in Example Problem 8.11. The material modulus of elasticity E and the cross section moment of inertia I_z have been entered with unit values so that the computed values for the deflections are consistent with the manual analysis in which the quantity EI_z was carried through symbolically. Figure 8.35 shows plots of the shear force and bending moment diagrams, while Figure 8.36 shows a plot of the deflected shape of the beam. The computed values for the slopes and deflections at the joints are listed to the right of the plot. If the program user desires a value for the slope and de-

```
Example Problem 8.11 - Analysis by Program BEAMVM
4,1,1,0,2,1,2
Joint Coordinates
1,0.0
2,5.0
3,10.0
4,15.0
Material Data
1,1.0
Cross Section Data
1,1.0
Support Restraints
1,1,1,0
3,0,1,0
Joint Loads
2,-30.0,0.0
Distributed Member Loads
2,-6.0,-6.0
3,-6.0,-6.0
```

Figure 8.34 Example Problem 8.11, BEAMVM input.

Example Problem 8.11 - Analysis by Program BEAMVM
Shear Force and Bending Moment Diagrams

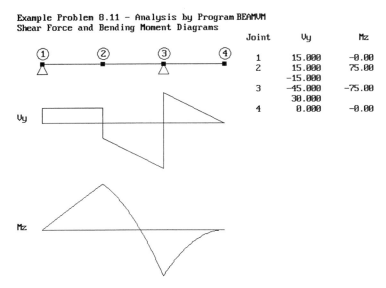

Figure 8.35 Example Problem 8.11, V_y and M_z diagrams from BEAMVM.

Example Problem 8.11 - Analysis by Program BEAMVM
Deflected Shape of Beam - Multiplication Factor = .005

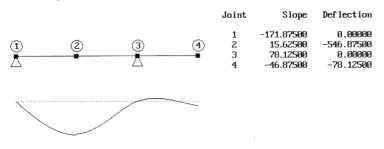

Figure 8.36 Example Problem 8.11, deflected shape from BEAMVM.

flection at some specific point in the beam, then a joint must be inserted in the mathematical model at that point. The computed values for the slopes at joints 1 and 4 and the deflection at joint 4 agree with the values computed previously in Example Problem 8.11 by the Moment Area Method and in Example Problem 8.15 by the Conjugate Beam Method. Note that the overhanging portion of the beam actually deflects upward adjacent to the right support, but then deflects down at the right end due to the negative curvature. This deformed shape is consistent with the listed positive value for the slope of the beam at joint 3. Figures 8.37a through 8.37d show a series of plots of the deflected shape of this same beam with magnitudes of the downward concentrated force at joint 2 of zero, 15, 30 and 45 kips, respectively. The deflection at the right end changes from a negative to a

Computer Program BEAMVM

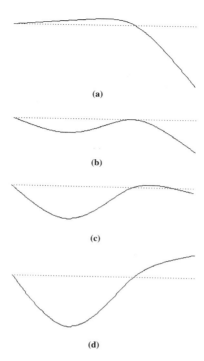

Figure 8.37

positive value as the magnitude of the downward force increases. Plots of this type can be very valuable in understanding how beams behave for various support and loading conditions.

The computed deflections for the beam have been multiplied by a user defined multiplication factor in order to obtain a realistic plot. For the plots in Figures 8.36 and 8.37, a factor of 0.005 was used since the values of the computed deflections are very large numbers due to the unit values for E and I_z. The program user must be careful when specifying the multiplication factor. Too large of a value will result in an unrealistic plot for the deflected shape of the beam, which might even extend beyond the limits of the screen, as shown in Figure 8.38, while if the value is too small, the deflections will not be noticeable on the plot. The program will print out the maximum computed deflection for the beam when requesting the value for the multiplication factor, so that the user has some basis for specifying an acceptable value. For most beams, for which realistic values are specified for E and I_z, the multiplying factor will probably be of the order of several hundred or more since the numerical values for the deflections will be very small numbers compared to the overall length of the beam. It might be necessary to try several different values for the multiplying factor before an acceptable plot is obtained.

284 Deflection of Beams   Chap. 8

```
Example Problem 8.11 - Analysis by Program BEAMVM
Deflected Shape of Beam - Multiplication Factor = .05
```

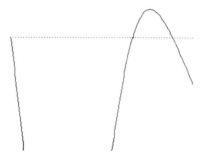

Joint	Slope	Deflection
1	−171.87500	0.00000
2	15.62500	−546.87500
3	78.12500	0.00000
4	−46.87500	−78.12500

Figure 8.38 Example Problem 8.11, deflected shape from BEAMVM.

SUGGESTED PROBLEMS

SP8.1 to **SP8.4** Derive expressions for the slope and deflection for each of the prismatic beams shown in Figures SP8.1 to SP8.4 by integrating the equation

$$\frac{d^4Y}{dX^4} = \frac{q_y}{EI_z}$$

The origin of the global coordinate system is at the left end of each beam. The global X axis extends toward the right and the global Y axis is up.

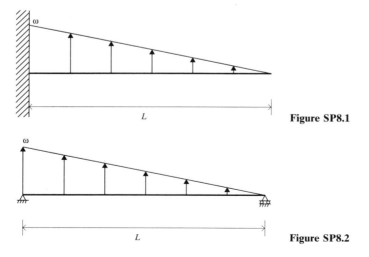

Figure SP8.1

Figure SP8.2

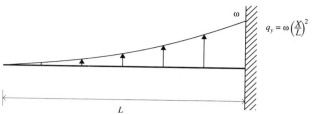

Figure SP8.3

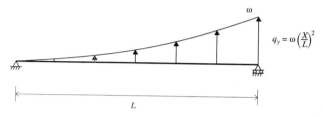

Figure SP8.4

SP8.5 to **SP8.10** Derive expressions for the slope and deflection for each of the prismatic beams shown in Figures SP8.5 to SP8.10 by integrating the equation

$$\frac{d^2Y}{dX^2} = \frac{M_z}{EI_z}$$

The origin of the global coordinate system is at the left end of each beam. The global X axis extends toward the right and the global Y axis is up.

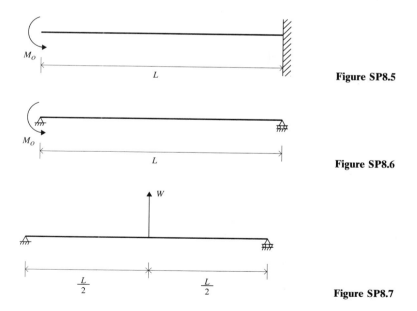

Figure SP8.5

Figure SP8.6

Figure SP8.7

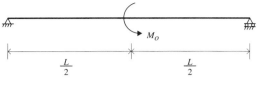

Figure SP8.8

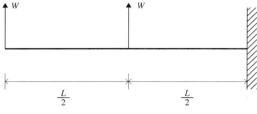

Figure SP8.9

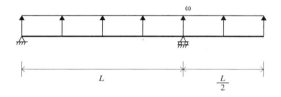

Figure SP8.10

SP8.11 and **SP8.12** Derive expressions for the slope and deflection for each of the nonprismatic beams shown in Figures SP8.11 and SP8.12 by integrating the equation

$$\frac{d^2Y}{dX^2} = \frac{M_z}{EI_z}$$

The origin of the global coordinate system is at the left end of each beam. The global X axis extends toward the right and the global Y axis is up.

Figure SP8.11

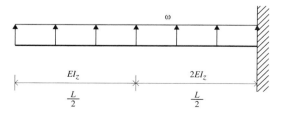

Figure SP8.12

Suggested Problems

SP8.13 Derive expressions for the shear force, bending moment, slope and deflection for the prismatic statically indeterminate beam shown in Figure SP8.13 by integrating the equation

$$\frac{d^4 Y}{dX^4} = \frac{q_y}{EI_z}$$

The origin of the global coordinate system is at the left end of the beam. The global X axis extends toward the right and the global Y axis is up.

Evaluate the expressions for the shear force and the bending moment at each end of the beam and compare the resulting expressions to Eqs. (6.15) through (6.18) in Chapter 6.

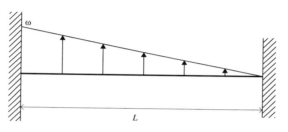

Figure SP8.13

SP8.14 to SP8.20 Compute the slope and vertical deflection at point a for each of the beams shown in Figures SP8.14 to SP8.20 by the Moment Area Method. Verify your solutions with the program BEAMVM. Hint: Perform the analysis treating the quantity EI_z as a symbol and then substitute the numerical values for E and I_z into the final expressions for the slope and deflection.

Figure SP8.14

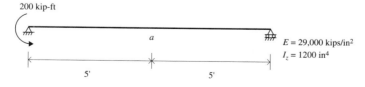

Figure SP8.15

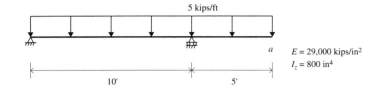

Figure SP8.16

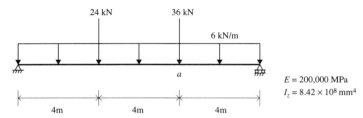

Figure SP8.17

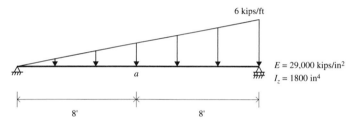

Figure SP8.18

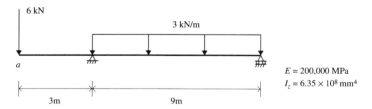

Figure SP8.19

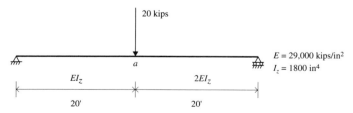

Figure SP8.20

SP8.21 to SP8.23 Determine the location and the magnitude of the maximum vertical translation for each of the beams shown in Figures SP8.21 to SP8.23 using the Moment Area Theorems. Treat the quantity EI_z as a symbol in the calculations.

See Suggested Problem SP8.32 for a procedure for verifying your solutions with the program BEAMVM.

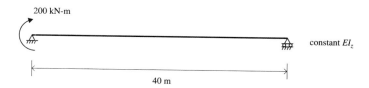

Figure SP8.21

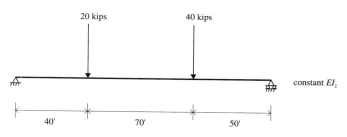

Figure SP8.22

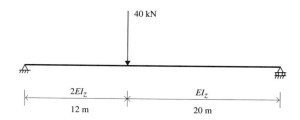

Figure SP8.23

SP8.24 to SP8.30 Compute the indicated slopes and deflections for Suggested Problems SP8.14 through SP8.20 by the Conjugate Beam Method.

SP8.31 Compute the indicated quantities for the beam shown in Figure SP8.31 by the Conjugate Beam Method
  (a) The kink angle in the pin at point a
  (b) The slope and vertical deflection at point b
  (c) The location and the magnitude of the maximum vertical deflection

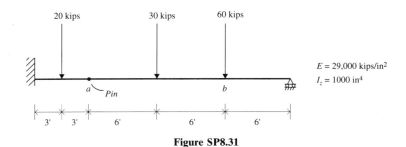

Figure SP8.31

SP8.32 Use the program BEAMVM to plot the shear force and the bending moment diagrams for the conjugate beam for Suggested Problem SP8.31. Note that these diagrams correspond to the slope diagram and the deflection diagram for the real beam. After the bending moment diagram has been generated for the whole conjugate beam, the location and magnitude of the maximum deflection can be determined by plotting the bending moment diagram for the individual member in the mathematical model which contains the maximum bending moment. (Use values of 1.0 for E and I_z for the conjugate beam in the input data since the deflections for the conjugate beam are not needed.)

This same procedure also can be used to verify the solutions for Suggested Problems SP8.21 through SP8.23.

9

Deflection of Trusses and Frames

DEFLECTION COMPUTATIONS BY CONSERVATION OF ENERGY

In the previous chapter, several procedures were discussed for computing the slopes and deflections of beams. Although these same procedures are also applicable for computing the local slopes and deflections of an individual member in a frame, relative to its two end joints, they are not well suited for computing the global deflection of a specific point in a frame since the deformations of all of the members in the frame, which affect that particular deflection, must be combined by some type of geometric relationship. Even for small frames, with only a few members and simple geometries, this can be a difficult task, while for large frames, it can result in an almost impossible set of computations. What is needed at this point is an analysis procedure which can be used to compute the deflection at any point in any direction for any type of structural system. Fortunately, a procedure which meets this requirement can be developed based upon a simple concept which the reader should be familiar with from an earlier physics course. This concept is the Principle of the Conservation of Energy from which we can conclude that the total work performed by the external loads on the structure to deform the structure must be equal to the total work performed by the internal stresses in the members to deform the members plus any work lost in the form of heat, such as due to friction at the connections between the members. For our present applications of this procedure, we will assume that these losses are small and can be ignored.

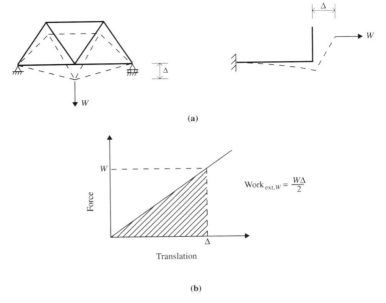

Figure 9.1

External Work

As the external loads are applied to a structure, the points where the loads are applied will displace. The *external work* is equal to the work performed by the external loads as they move through the displacements at their points of application. These external loads may consist of both forces and moments. The work performed by each of these types of loads will be considered separately.

External Force. If a concentrated force W is applied to a structure whose members are composed of a linear elastic material, as shown by the typical truss and frame in Figure 9.1a, the point where the force is applied will translate through a distance Δ in the direction of the force. The translation at the loaded joint in the truss occurs due to the axial deformation in the members, whereas the translation in the frame occurs due to a combination of axial, bending and shear deformations in the members. For either case, the relationship between the magnitude of the applied force, as the force increases from zero to W, and the translation of the point of application of the force in the direction of the force can be assumed to be linear, as shown in Figure 9.1b, as long as the material stress at all points in the structure is less than the yield stress for the material and the geometry changes which occur in the structure are small. From our basic physics course, we know that the external work performed by the force W, as it translates through the distance Δ, will be equal to the shaded area under the force-translation curve:

Deflection Computations by Conservation of Energy

$$\text{Work}_{\text{ext, }W} = \frac{W\Delta}{2} \tag{9.1}$$

A simple explanation for the factor of one half is that the average force which is acting during the time that the translation Δ occurs is $W/2$ since the force W is causing the translation.

External Moment. If a moment M is applied to a point on a structure whose members are composed of a linear elastic material, as shown by the frame in Figure 9.2a, the point where the moment is applied will rotate through an angle θ in the direction of the moment due to the bending deformation in the members. The relationship between the magnitude of the moment, as it increases from zero to M, and the rotation can be assumed to be linear, as shown in Figure 9.2b, as long as the conditions concerning the material stresses and the geometry changes meet the same assumptions described previously for the force W. Therefore, since a moment will perform work as it rotates, the external work performed by the moment M will be

$$\text{Work}_{\text{ext, }M} = \frac{M\theta}{2} \tag{9.2}$$

which is equal to the shaded area under the moment-rotation curve in Figure 9.2b. The rotation θ must be expressed in radians in this relationship. The point where the moment is applied might also translate, but the moment will perform no work as it moves through that translation. Only the rotation of the point of application of a moment must be considered in computing the work performed by the moment.

Internal Work

As the external loads are applied to the structure, stresses will be introduced into the members, which will cause the material in the members to deform. The type and magnitude of the member deformations will depend upon the type and mag-

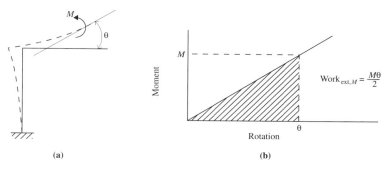

Figure 9.2

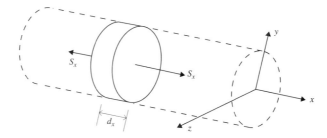

Figure 9.3

nitude of the stresses in the members. As the members deform, the internal forces and moments acting on the member cross sections will displace. The *internal work* in the members is equal to the work performed by these internal forces and moments as they displace due to the internal member deformations. We will consider four different types of internal member cross section loads: axial forces, bending moments, shear forces and twisting moments (i.e.; torsion).

Axial Force. An expression which represents the work performed by the axial force as a member deforms can be developed by considering an infinitesimal segment from the member of length dx, which is subjected to an axial force S_x acting along the longitudinal axis of the member, as shown in Figure 9.3. The internal work which is performed as the segment deforms elastically due to the equal and opposite forces S_x at the ends of the segment will be

$$d\text{Work}_{\text{int}, S_x} = \frac{S_x dL_x}{2} \tag{9.3}$$

where the change in length dL_x of the segment along the longitudinal local x axis of the member will be equal to the strain ε_x multiplied by the length of the segment:

$$dL_x = \varepsilon_x dx \tag{9.4}$$

The strain ε_x for an elastic material can be expressed in terms of the axial stress σ_x and the modulus of elasticity E of the material as

$$\varepsilon_x = \frac{\sigma_x}{E} \tag{9.5}$$

while the stress σ_x on the cross section of the element is equal to

$$\sigma_x = \frac{S_x}{A_x} \tag{9.6}$$

where A_x is the area of the cross section perpendicular to the longitudinal axis of the member. Combining Eqs. (9.4) through (9.6) results in

Deflection Computations by Conservation of Energy

Figure 9.4

$$dL_x = \frac{S_x}{A_x E} dx \tag{9.7}$$

which, upon substituting into Eq. (9.3) and integrating over the length of the member, gives the total internal work to deform the member as

$$\text{Work}_{\text{int}, S_x} = \int \frac{S_x^2}{2 A_x E} dx \tag{9.8}$$

For the specific case of a prismatic member of length L, which is subjected to equal and opposite end forces, as shown in Figure 9.4, the quantities S_x, A_x and E will be constant over the length of the member. Therefore, they can be moved outside of the integral before the integration is performed and Eq. (9.8) will become

$$\text{Work}_{\text{int}, S_x} = \frac{S_x^2 L}{2 A_x E} \tag{9.9}$$

If the member is nonprismatic, such that the quantity $A_x E$ varies over the length of the member, or if S_x is not constant due to either intermediate concentrated axial forces between the ends of the member or a distributed force on the member with an axial component, the variation of these quantities must be considered during the integration of Eq. (9.8).

The total internal work in the structure, due to the axial forces in the members, can be determined by evaluating either Eq. (9.8) or Eq. (9.9) for each member, depending upon which equation is applicable to that particular member, and then summing the individual values for all members.

Example Problem 9.1

Compute the horizontal translation of joint 3 in the statically determinate plane truss whose mathematical model is shown in Figure 9.5a. All members in the truss are prismatic.

The first step in the analysis is to compute the reactions and the member forces for the truss. The results of this analysis are shown in Figure 9.5b. The verification of these values is left as an exercise for the reader.

An expression which can be used to determine the horizontal translation Δ at the point of application of a concentrated force W can be developed by equating the external work, as expressed by Eq. (9.1), and the total internal work which is performed in deforming the NM prismatic members in the truss

$$\frac{W\Delta}{2} = \sum_{n=1}^{NM} \left(\frac{S_x^2 L}{2 A_x E} \right)_n$$

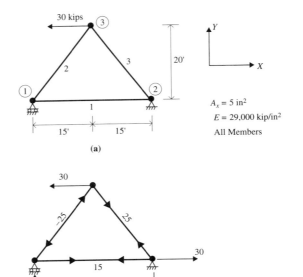

Figure 9.5 Example Problem 9.1.

from which, upon eliminating the common factor of 2 on each side and solving for Δ, we obtain

$$\Delta = \frac{1}{W}\left[\sum_{n=1}^{NM}\left(\frac{S_x^2 L}{A_x E}\right)_n\right] \quad (9.10)$$

This expression can be used to compute the deflection for any plane truss or space truss with prismatic members which is subjected to a single concentrated force. The deflection will be at the point of application of the force in the direction of the force.

The quantity inside the square brackets on the right side of Eq. (9.10) can be easily evaluated in a table, as shown in Figure 9.5c.

$$\sum_{n=1}^{NM}\left(\frac{S_x^2 L}{A_x E}\right)_n = 3.145$$

This value now can be substituted into Eq. (9.10) to obtain the horizontal translation of joint 3:

$$\Delta_{3X} = \frac{1}{W}(3.145) = \frac{3.145}{30} = 0.105 \text{ inch}$$

The positive value for Δ_{3X} signifies that the horizontal force at joint 3 will perform positive work as the structure deforms. Therefore, since this force acts toward the left, Δ_{3X} is also toward the left.

$$\Delta_{3X} = 0.105 \text{ inch} \leftarrow$$

An engineer must be very careful in any analysis of this type to ensure that the units of all quantities used in the calculations are consistent. In this particular problem, all quantities were expressed in units of kips and inches. Therefore, the computed displacement Δ_{3X} is in inches. The positive direction for the displacement is defined by the positive direction of the applied load on the structure rather than by the directions of the global coordinate axes as in the beam problems in Chapter 8.

Bending Moment. An expression which represents the internal work performed by the bending moment as a plane frame member deforms can be developed by considering an infinitesimal segment of length dx from a member in pure bending as shown in Figure 9.6, where the local x axis extends along the length of the member and the local y and z axes are the principal axes of the cross section. The internal work which is performed as the segment deforms elastically due to the two equal and opposite moments M_z at the ends of the segment will be

$$d\text{Work}_{\text{int}, M_z} = \frac{M_z d\theta_z}{2} \qquad (9.11)$$

which, upon substituting the expression for $d\theta_z$ which was shown previously in Eq. (8.19) in Chapter 8, gives.

$$d\text{Work}_{\text{int}, M_z} = \frac{M_z^2}{2EI_z} dx \qquad (9.12)$$

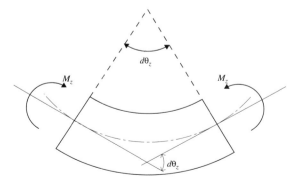

Figure 9.6

The total work to deform any member can be determined by integrating Eq. (9.12) over the length of the member:

$$\text{Work}_{\text{int}, M_z} = \int \frac{M_z^2}{2EI_z} dx \qquad (9.13)$$

The variation of the bending moment M_z over the length of the member can be determined from the bending moment diagram. If the member is nonprismatic the variation of EI_z also must be considered during the integration. This expression can now be used in conjunction with Eq. (9.1) to compute the translation in a frame at the point of application of a concentrated force or with Eq. (9.2) to compute the rotation at the point of application of a concentrated moment.

For a space frame member, the components of the bending moment about both principal axes of the cross section must be considered. The total internal work due to bending for this case will be

$$\text{Work}_{\text{int}, M} = \text{Work}_{\text{int}, M_y} + \text{Work}_{\text{int}, M_z} = \int \frac{M_y^2}{2EI_y} dx + \int \frac{M_z^2}{2EI_z} dx \qquad (9.14)$$

where M_y and I_y are the bending moment and the moment of inertia, respectively, about the local y axis for the cross section.

Example Problem 9.2

Compute the rotation at the point of application of the external moment for the plane frame shown in Figure 9.7a. The value of EI_z is constant throughout the frame.

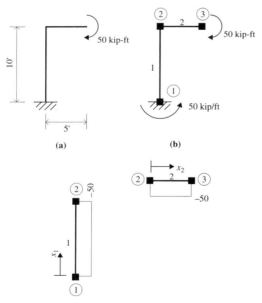

Figure 9.7 Example Problem 9.2.

Deflection Computations by Conservation of Energy

The mathematical model for the frame is shown in Figure 9.7b and an exploded view of the members, which shows the individual bending moment diagrams in kip and feet units, is shown in Figure 9.7c. Note that the bending moments are constant along the length of each member due to the type of loading for this frame. They are plotted on the compression side of the members to be consistent with the procedure described in Chapter 6 for drawing the bending moment diagrams for plane frame members. Since the axial force and the shear force are zero in each member, the only contributions to the rotation at the point of application of the external moment will be the bending deformations in the two members.

The rotation angle θ can be computed by equating the external work, as expressed by Eq. (9.2), and the total internal work due to the bending in each member, as expressed by Eq. (9.13). Performing this operation, and solving for θ results in the expression

$$\theta = \frac{1}{M} \left[\sum_{n=1}^{NM} \left(\int \frac{M_z^2}{EI_z} dx \right)_n \right] \tag{9.15}$$

which, upon substituting the specific values for this frame, gives

$$\theta = \frac{1}{50} \left[\int_0^{10} \frac{(-50)^2}{EI_z} dx_1 + \int_0^5 \frac{(-50)^2}{EI_z} dx_2 \right]$$

from which

$$\theta = \frac{750}{EI_z}$$

Note that the signs of the bending moments which are substituted into Eq. (9.15) are not important since their values are squared during the computations. Any desired bending moment sign convention may be used in this analysis procedure.

The actual value of the rotation angle in radians can be obtained by substituting the values for E and I_z in kip and feet units, since all of the values which were used to evaluate Eq. (9.15) were in these units. For example, for typical values of E and I_z of

$$E = 29{,}000 \text{ kips/in}^2 = 4{,}176{,}000 \text{ kips/ft}^2$$

and

$$I_z = 1000 \text{ in}^4 = 0.04823 \text{ ft}^4$$

the rotation will be

$$\theta = 0.003724 \text{ radians} = 0.2133 \text{ degrees}$$

The positive value for θ means that the moment will do positive work as the frame deforms. Therefore, since the applied moment is clockwise, the rotation will be clockwise.

Shear Force. From our previous sophomore level mechanics course we know that at any point in a bending member, which is subjected to a vertical shear

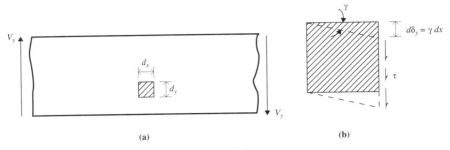

Figure 9.8

force V_y, the vertical shear stress and the longitudinal shear stress have equal magnitudes τ and the variation of the shear stress over the depth of the section is given by

$$\tau = \frac{V_y Q_z}{I_z b} \qquad (9.16)$$

where Q_z is the moment about the neutral axis of that portion of the cross section above the point where τ is being evaluated and b is the width of the cross section. The shear stress is assumed to be constant over the width of the section. The vertical and longitudinal shear stresses on any infinitesimal element in the member with dimensions dx, dy and dz along the local member x, y and z axes, as shown in Figure 9.8a, will cause the element to deform into a rhombus, as shown in Figure 9.8b, due to the shear strain γ, which occurs in the element. The elemental vertical shear force dV_y acting on each vertical face of the element will be equal to the shear stress multiplied by the area of the face:

$$dV_y = \tau\, dy\, dz \qquad (9.17)$$

while the relative vertical translation $d\delta_y$ between the two vertical faces will be

$$d\delta_y = \gamma\, dx \qquad (9.18)$$

Using these values, the internal work which will be performed as the element deforms can be expressed as

$$d\text{Work}_{\text{int}, \tau} = \frac{dV_y\, d\delta_y}{2} \qquad (9.19)$$

or, on substituting the expressions in Eqs. (9.17) and (9.18),

$$d\text{Work}_{\text{int}, \tau} = \frac{\tau \gamma}{2}\, dx\, dy\, dz \qquad (9.20)$$

An expression which can be used to compute the total internal work in a member due to the shear stress can be obtained from Eq. (9.20) by substituting the linear shear stress-shear strain relationship

Deflection Computations by Conservation of Energy

$$\gamma = \frac{\tau}{G} \tag{9.21}$$

and integrating over the volume of the member

$$\text{Work}_{\text{int}, \tau} = \iiint \frac{\tau^2}{2G} \, dx \, dy \, dz \tag{9.22}$$

The shear modulus of the material G, which is equal to the slope of the shear stress-shear strain curve in the elastic range, can be expressed in terms of the modulus of elasticity E and Poisson's Ratio ν for the material as

$$G = \frac{E}{2(1 + \nu)} \tag{9.23}$$

It is assumed that the reader has encountered this relationship previously in a sophomore level course dealing with engineering materials. A typical value for the dimensionless quantity ν for many structural materials, such as steel or concrete, is approximately 0.3.

Example Problem 9.3

To demonstrate the effect of the shear deformation in a beam on the vertical deflection, we will compute the vertical deflection at the end of an elastic cantilever beam which is subjected to a positive vertical force at the free end, as shown in Figure 9.9a. The beam is prismatic with a rectangular cross section of width b and depth d. The mathematical model for the beam is shown in Figure 9.9b and the shear force and bending moment diagrams are shown in Figure 9.9c.

The deflection at the point of application of the concentrated force W can be computed by equating the external work performed by the force, as expressed by Eq. (9.1), and the internal work performed by the bending moment M_z and the shear force V_y, as expressed by Eqs. (9.13) and (9.22), respectively. Performing this operation and solving for Δ gives

$$\Delta = \frac{1}{W} \int \frac{M_z^2}{EI_z} \, dx + \frac{1}{W} \iiint \frac{\tau^2}{G} \, dx \, dy \, dz \tag{9.24}$$

or

$$\Delta = \Delta_{M_z} + \Delta_{V_y}$$

where Δ_{M_z} is the contribution of the bending deformation in the member to the vertical deflection and Δ_{V_y} is the contribution of the shear deformation. Equation (9.24) can be used for any elastic beam which is subjected to a single concentrated force as long as expressions for M_z and τ can be determined at all points in the beam. If the mathematical model for the structure has more than one member, the right side of the equation would have to be modified to sum the effects for each member.

The contribution of the bending deformation to the deflection can be computed by evaluating the expression for Δ_{M_z} in Eq. (9.24) by the following operation:

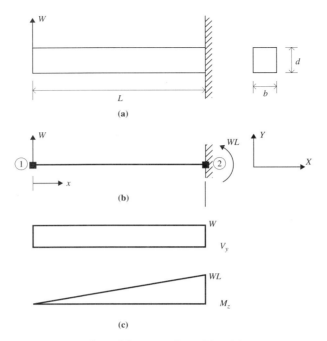

Figure 9.9 Example Problem 9.3.

$$\Delta_{Mz} = \frac{1}{W} \int \frac{M_z^2}{EI_z} dx = \frac{1}{W} \int_0^L \frac{(Wx)^2}{EI_z} dx = \frac{WL^3}{3EI_z}$$

in which the origin of the local coordinate system for the single member in the mathematical model of the beam has been considered to be at the left end of the member.

Before the contribution of the shear distortion can be computed, an expression for the shear stress at all points in the beam must be developed using Eq. (9.16). An expression for Q_z can be determined as the moment of the shaded area in the upper portion of the cross section, as shown in Figure 9.10

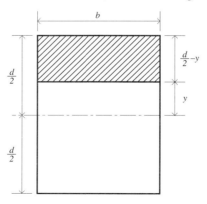

Figure 9.10 Example Problem 9.3.

Deflection Computations by Conservation of Energy

$$Q_z = b\left(\frac{d}{2} - y\right)\left(y + \frac{d/2 - y}{2}\right) = b\left(\frac{d^2}{8} - \frac{y^2}{2}\right)$$

which, upon substituting into Eq. (9.16) along with the constant value of W for V_y, gives

$$\tau = \frac{W(d^2/8 - y^2/2)}{I_z}$$

at any point in the beam. Note that this expression does not contain the variable x since the shear stress at any level y in the cross section is constant along the length of the beam. Substituting this expression into the expression for Δ_{V_y} in Eq. (9.24) and integrating gives

$$\Delta_{V_y} = \frac{1}{W}\iiint \frac{\tau^2}{G}\,dx\,dy\,dz$$

$$= \frac{1}{W}\int_{-b/2}^{b/2}\int_{-d/2}^{d/2}\int_0^L \frac{[-W(d^2/8 - y^2/2)/I_z]^2}{G}\,dx\,dy\,dz = \frac{WLbd^5}{120GI_z^2}$$

Therefore, the total vertical deflection at the end of the beam is

$$\Delta = \frac{WL^3}{3EI_z} + \frac{WLbd^5}{120GI_z^2}$$

Since this value for Δ is positive, it will be in the same direction as the force W. Although this expression can be used as it stands, it can be put into a more useful form for comparing the relative magnitudes of the deflections Δ_{M_z} and Δ_{V_y} by substituting

$$I_z = \frac{bd^3}{12}$$

for one of the I_z symbols in the denominator of the second term and

$$G = \frac{E}{2.6}$$

to correspond to a typical structural material according to Eq. (9.23). The resulting expression, after some manipulation, is

$$\Delta = \frac{WL^3}{3EI_z}\left[1 + 0.78\left(\frac{d}{L}\right)^2\right] = \Delta_{M_z}\left[1 + 0.78\left(\frac{d}{L}\right)^2\right]$$

Note that the expression outside of the brackets agrees with the expression for the deflection which was derived for this same beam in Example Problem 8.3 in Chapter 8 by integrating the differential equation of bending. The expression inside the brackets represents the factor by which the bending deflection must be multiplied to include the effect of shear deformation in the beam. This multiplying factor can be expressed as

$$\frac{\Delta}{\Delta_{Mz}} = 1 + 0.78\left(\frac{d}{L}\right)^2$$

It is obvious from this expression that as the length of the beam increases compared to its depth, the relative contribution of the shear deformations to the overall deflection of the beam decreases. The following table shows the value of the multiplying factor Δ/Δ_{Mz} for various depth to length ratios of the beam

d/L	Δ/Δ_{M_z}
0.01	1.0001
0.10	1.0078
0.20	1.0312
0.50	1.1950

For beams with a depth to length ratios of 0.1 or smaller, the increase in the deflection due to the shear distortion is less than 1%. Although these specific numbers are only valid for this particular prismatic cantilever beam, similar results will be obtained for other beam geometries and other loadings. Therefore, the assumption which was made in Chapter 8 in which the effects of shear distortion in long slender members were considered to be negligible in most deflection computations is valid. It is usually only necessary to consider this effect in short stubby members where deflection is a critical factor in the overall serviceability of the beam.

An alternate approach for computing the shear deflection in any bending member can be developed by expressing the internal work which is performed by the shear force, as a small segment of the member deforms, in the form

$$d\text{Work}_{\text{int}, V_y} = \frac{V_y d\alpha_y}{2} \tag{9.25}$$

where the quantity $d\alpha_y$ is the overall effective vertical shear distortion over an infinitesimal length dx in the member, as shown in Figure 9.11. This shear distortion can be expressed as

$$d\alpha_y = \frac{V_y}{A_y G} dx \tag{9.26}$$

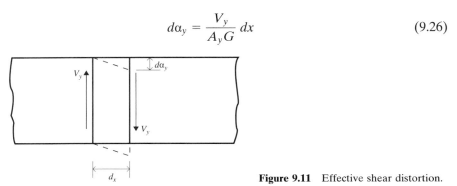

Figure 9.11 Effective shear distortion.

Deflection Computations by Conservation of Energy

in which A_y is known as the *effective shear area* of the cross section in the local y direction. The total internal work in the member due to the shear forces can be obtained by combining Eqs. (9.25) and (9.26) and integrating over the length of the member:

$$\text{Work}_{\text{int}, V_y} = \int \frac{V_y^2}{2A_y G} dx \qquad (9.27)$$

For a space frame member, the components of the shear force acting along both principal axes of the cross section must be considered. The total internal work due to the shear forces for this case will be

$$\text{Work}_{\text{int}, V} = \text{Work}_{\text{int}, V_y} + \text{Work}_{\text{int}, V_z} = \int \frac{V_y^2}{2A_y G} dx + \int \frac{V_z^2}{2A_z G} dx \qquad (9.28)$$

where V_z and A_z are the shear force and the effective shear area, respectively, for the cross section in the local z direction.

The effective shear area depends on the specific shape and dimensions of the cross section. Several common cross section shapes and the corresponding values for A_y are shown in Figure 9.12. Additional information on the concept of the effective shear area for a cross section and the derivation of Eq. (9.26) can be found in Gere and Timoshenko (1972) and Weaver and Gere (1980). The derivation of this equation will not be presented here.

Example Problem 9.3 Revisited

To demonstrate the use of Eq. (9.27), we will use it to recompute the shear deflection for the cantilever beam shown previously in Figure 9.9a. For this case, the expression for computing the shear deflection, after equating the external work to the internal work and solving for Δ_{V_y}, will be

$$\Delta_{V_y} = \frac{1}{W} \int \frac{V_y^2}{A_y G} dx \qquad (9.29)$$

which, upon substituting the expression for V_y corresponding to the shear force diagram in Figure 9.9c, and integrating, gives

$$\Delta_{V_y} = \frac{1}{W} \int_0^L \frac{W^2}{A_y G} dx = \frac{WL}{A_y G}$$

Although this expression does not resemble the expression for Δ_{V_y} which was obtained previously, it can be converted into the same form by several algebraic manipulations. Substituting

$$A_y = \frac{5}{6} bd$$

for the rectangular section, as specified in Figure 9.12, and multiplying the numerator and the denominator by I_z^2 gives

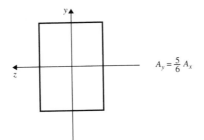

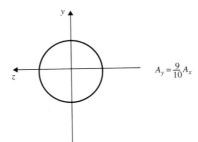

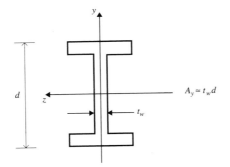

Figure 9.12 Effective shear area for various cross-sections.

$$\Delta_{V_y} = \frac{WLI_z^2}{(5/6)bdGI_z^2}$$

which, upon substituting

$$I_z^2 = \left(\frac{bd^3}{12}\right)^2$$

in the numerator and simplifying, becomes

$$\Delta_{V_y} = \frac{WLbd^5}{120GI_z^2}$$

which is the same expression obtained in the previous analysis.

It is obvious from this second analysis that the use of Eq. (9.27) to represent the internal work due to the shear force in a member not only simplifies the computations for computing the shear deflection in a beam or a plane frame, but also results in a much simpler final expression for the deflection compared to that obtained by using Eq. (9.22). We will be using the effective shear area later in this chapter and in future chapters to consider the effect of shear deformation during the analysis of beams and frames by other methods. The effective shear area is also required as input in computer programs which have the capability to consider the effects of shear deformation during the analysis of beams and frames. We will be discussing such a program later in this chapter.

Twisting Moments. Most sophomore level mechanics courses present a brief discussion of the torsion equations for circular shafts subjected to a twisting moment M_x. One of the equations which is usually developed is the following expression for the total twist angle θ_x over the length of a prismatic shaft which is subjected to a constant twisting moment over its entire length

$$\theta_x = \frac{M_x L}{J_x G} \tag{9.30}$$

where: L is the length of the shaft; J_x is the polar moment of inertia of the circular cross section; and G is the shear modulus of the material. The polar moment of inertia is given by

$$J_x = \frac{\pi r^4}{2} \tag{9.31}$$

where r is the radius of the shaft. Equation (9.30) can also be used to compute the twist angle for a noncircular member in a space frame by substituting the value for the torsional constant of the cross section for J_x. The value for this torsional constant varies with both the shape and the size of the cross section. It is only equal to the polar moment of inertia for the specific case of a circular cross section. For example, for a wide flange section an approximate value for the torsional constant, which is accurate enough for most structural computations, can be computed by the expression

$$J_x = \frac{2 t_f^3 b}{3} + \frac{t_w^3 h}{3} \tag{9.32}$$

where: t_f is the thickness of the flanges; b is the width of the flanges; t_w is the thickness of the web; and h is the height of the cross section. Values for the torsional constant are listed for common structural shapes in a number of structural engineering design manuals (American Institute of Steel Construction, 1986, 1989) and reference books (Young, 1989).

The internal work which is performed during the deformation of an elastic member by a twisting moment M_x will be equal to

$$\text{Work}_{\text{int}, M_x} = \frac{M_x \theta_x}{2} \tag{9.33}$$

which, on substituting Eq. (9.30) for θ_x, gives

$$\text{Work}_{\text{int}, M_x} = \frac{M_x^2 L}{2 J_x G} \tag{9.34}$$

By equating the total internal work in the members, due to the twisting moments, to the external work, due to a concentrated load, the contribution of the torsion in the members to the displacement at the point of application of the load can be determined.

We will not carry the topic of torsion any further at this time since the primary application of Eq. (9.34) would be for the computation of torsional deformations in space frame structures. Except for very trivial cases, for which the use of this equation is obvious, these computations can become very complex and tedious and are beyond the scope of what we are trying to accomplish here. The torsional constant J_x is required as input in any computer program which computes displacements in space frame structures.

DEFLECTION COMPUTATIONS BY CASTIGLIANO'S THEOREM

Although the previous procedure of equating the external work to the internal work can be used to compute deflections and rotations in essentially any type of structure, it is very limited in its applications since the structure only can be subjected to a single concentrated force or moment. If more than one force or moment acts on the structure, the expression for the external work on the left side of the equation will contain a term corresponding to the displacement at the point of application of each of these loads. We will then have the situation of one equation with multiple unknowns, which cannot be solved for the individual displacements. We need a more versatile analysis procedure which can consider the effects of multiple loads acting on the structure.

To develop such a procedure, consider the case of a structure which is subjected to an active load system **W**, which may consist of any combination of concentrated forces and moments at any points and any type of distributed forces on any of the members. The concentrated forces and moments will be designated as W_1 through W_{NW}, where NW is the total number of forces and moments, as shown in Figure 9.13. A generic representation of the structure is shown in this figure since the structure may have any form. It may be a statically determinate or statically indeterminate truss or frame in two or three dimensions. The supports may be of any number and type as long as the structure is stable. We will now consider two specific cases for the loading on the structure:

Deflection Computations by Castigliano's Theorem

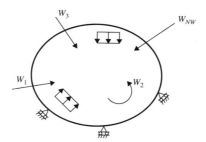

Figure 9.13

Case I. Assume that after the load system **W** is applied, one of the concentrated forces or moments W_i is increased by an infinitesimal amount dW_i. The total work which will be performed for this case as the structure assumes its final deformed shape can be expressed as

$$\text{Work}_{\text{Case I}} = \text{Work}_W + \frac{\partial \text{Work}_W}{\partial W_i}(dW_i) + \text{Work}_{dW_i} \qquad (9.35)$$

where Work_W is the work which is performed by the load system **W** and Work_{dW_i} is the work which would be performed by the infinitesimal load dW_i acting by itself. The middle term on the right side of the expression corresponds to the change in the work which would be performed by the load system **W** due to the change in the force W_i.

Case II. Assume that the infinitesimal load dW_i is applied first and then the load system **W** is applied. The total work which will be performed for this case can be expressed as

$$\text{Work}_{\text{Case II}} = \text{Work}_{dW_i} + \text{Work}_W + dW_i(\Delta_i) \qquad (9.36)$$

where Δ_i is the displacement at load W_i due to the load system **W**. The term $dW_i(\Delta_i)$ corresponds to the work which will be performed by the infinitesimal load dW_i as it moves through the displacement Δ_i which is caused by the load system **W** at the point of application of W_i. A factor of one half is not present since the infinitesimal load dW_i is acting during the entire time that the displacement Δ_i is produced.

Since the total work which will be performed by a set of loads must be independent of the order in which the loads are applied, we can equate the two expressions in Eqs. (9.35) and (9.36). Performing this operation and canceling like terms on each side gives

$$\Delta_i = \frac{\partial \text{Work}_W}{\partial W_i} \qquad (9.37)$$

This relationship was originally published by the Italian engineer Alberto Castigliano in 1879 as one of a set of relationships dealing with the equilibrium and deflection of structural systems. This particular relationship is usually called *Castigliano's Second Theorem*, which can be stated as follows:

For a stable elastic structure, which is subjected to any load system, the displacement Δ_i at the point of application of any concentrated load W_i, in the direction of that load, will be equal to the partial derivative of the total work performed by the load system with respect to the concentrated load W_i.

If the load W_i is a concentrated force, the displacement Δ_i will be a translation, whereas, if W_i is a moment, then Δ_i will be a rotation. If Δ_i is positive, it will be in the direction that the load is acting, whereas, if Δ_i is negative, then it will be in the opposite direction. The advantage of this procedure for computing displacements in a structure, compared to the procedure described in the previous section of this chapter, is that there may be any number of concentrated forces and moments, and also distributed forces, acting on the structure. The only disadvantage to the procedure is that there must be a concentrated load acting at the point, of the correct type and in the correct direction, where the displacement is to be computed. However, as we will see later in this chapter, there is a simple procedure for getting around this restriction.

For completeness in our discussion, the relationship corresponding to *Castigliano's First Theorem* is

$$W_i = \frac{\partial \text{Work}_W}{\partial \Delta_i} \qquad (9.38)$$

Although this relationship can be very useful for some types of analysis, we will not be considering any applications for it at this time. We will only be concerned with the application of Castigliano's Second Theorem, as expressed by Eq. (9.37), for the computation of structure displacements.

Example Problem 9.4

Find the vertical deflection at the end of the prismatic cantilever beam shown in Figure 9.14a by Castigliano's Second Theorem. The beam is subjected to a positive concentrated force W at its end and a positive uniform distributed force ω over its entire length. Include the effect of both bending and shear deformation in the beam. The mathematical model for the beam is shown in Figure 9.14b and the shear force and bending moment diagrams are shown in Figure 9.14c.

The first step in the analysis is to develop expressions for the shear force V_y and the bending moment M_z at any point x along the beam from the free end. The origin of the local coordinate system will be considered to be at the left end of the single member in the mathematical model of this beam.

$$V_y = W + \omega x$$

$$M_z = Wx + \frac{\omega x^2}{2}$$

The work performed by the load system as the structure deforms can be expressed in terms of M_z and V_y as

$$\text{Work} = \text{Work}_{M_z} + \text{Work}_{V_y}$$

Deflection Computations by Castigliano's Theorem

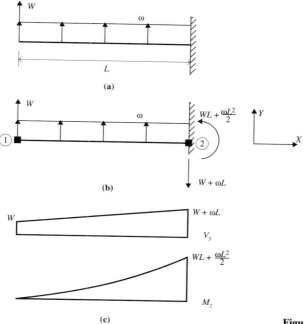

Figure 9.14 Example Problem 9.4.

where

$$\text{Work}_{M_z} = \int \frac{M_z^2}{2EI_z}\, dx$$

and

$$\text{Work}_{V_y} = \int \frac{V_y^2}{2A_y G}\, dx$$

The displacement Δ at the point of application of the concentrated force W can be computed by Castigliano's Second Theorem as

$$\Delta = \Delta_{M_z} + \Delta_{V_y}$$

where

$$\Delta_{M_z} = \frac{\partial \text{Work}_{M_z}}{\partial W}$$

and

$$\Delta_{V_y} = \frac{\partial \text{Work}_{V_y}}{\partial W}$$

There are two approaches which can be used to compute the displacements Δ_{M_z} and Δ_{V_y}. The difference in the two approaches is the order in which the integration and the differentiation is performed.

Approach I. The differentiation will be performed after the integration. The internal work due to the bending moment M_z is

$$\text{Work}_{M_z} = \int_0^L \frac{(Wx + \omega x^2/2)^2}{2EI_z} dx = \frac{W^2 L^3}{6EI_z} + \frac{W\omega L^4}{8EI_z} + \frac{\omega^2 L^5}{40EI_z}$$

from which the deflection ΔM_z is

$$\Delta_{M_z} = \frac{\partial \text{Work}_{M_z}}{\partial W} = \frac{WL^3}{3EI_z} + \frac{\omega L^4}{8EI_z}$$

The internal work due to the shear force V_y is

$$\text{Work}_{V_y} = \int_0^L \frac{(W + \omega x)^2}{2A_y G} dx = \frac{W^2 L}{2A_y G} + \frac{W\omega L^2}{2A_y G} + \frac{\omega^2 L^3}{6A_y G}$$

from which the deflection ΔV_y is

$$\Delta_{V_y} = \frac{\partial \text{Work}_{V_y}}{\partial W} = \frac{WL}{A_y G} + \frac{\omega L^2}{2A_y G}$$

Approach II. The differentiation will be performed before the integration. The deflection due to the bending moment M_z is

$$\Delta_{M_z} = \frac{\partial \text{Work}_{M_z}}{\partial W} = \frac{\partial \text{Work}_{M_z}}{\partial M_z} \frac{\partial M_z}{\partial W} = \int \frac{M_z}{EI_z} \frac{\partial M_z}{\partial W} dx$$

which, on substituting the expression for M_z, gives

$$\Delta_{M_z} = \int_0^L \left(\frac{Wx + \omega x^2/2}{EI_z}\right)(x) dx = \frac{WL^3}{3EI_z} + \frac{\omega L^4}{8EI_z}$$

The deflection due to the shear force V_y is

$$\Delta_{V_y} = \frac{\partial \text{Work}_{V_y}}{\partial W} = \frac{\partial \text{Work}_{V_y}}{\partial V_y} \frac{\partial V_y}{\partial W} = \int \frac{V_y}{A_y G} \frac{\partial V_y}{\partial W} dx$$

which, on substituting the expression for V_y, gives

$$\Delta_{V_y} = \int_0^L \left(\frac{W + \omega x}{A_y G}\right)(1) dx = \frac{WL}{A_y G} + \frac{\omega L^2}{2A_y G}$$

The final results for the deflections are the same for both approaches. Note that the bending deflection and the shear deflection due to the concentrated force agree with the bending deflection and the shear deflection obtained in Example Problem 9.3 for a cantilever beam with only a concentrated force at the free end.

As we can see from the two analyses in Example Problem 9.4, the order in which the integration and the differentiation are performed is not important when using Castigliano's Theorem. However, it will usually be more convenient to perform the differentiation before performing the integration.

Deflection Computations by Castigliano's Theorem

Superposition: A Word of Warning

In Chapter 8, it was shown that the computations involving the bending moment diagrams in both the Moment Area Method and the Conjugate Beam Method could be significantly simplified by considering the bending moment diagrams for each load on the structure separately and then superimposing the results at the end of the calculations. It would seem logical that the same approach might be used for computing the work performed by the loads on a structure when using Castigliano's Theorem. However, there is a problem which will be encountered when attempting to use the concept of superposition which might not be obvious at first glance. To demonstrate this problem, we will compute the internal work performed by the bending moments produced by the concentrated force and the distributed force separately for the cantilever beam from Example Problem 9.4.

The work due to the concentrated force W acting alone is

$$\text{Work}_{M_z, W} = \int \frac{(Wx)^2}{2EI_z} dx = \frac{W^2 L^3}{6EI_z}$$

while the work due to the distributed force acting alone is

$$\text{Work}_{M_z, \omega} = \int \frac{(\omega x^2/2)^2}{2EI_z} dx = \frac{\omega^2 L^5}{40EI_z}$$

which, on superimposing, gives

$$\text{Work}_{M_z} = \text{Work}_{M_z, W} + \text{Work}_{M_z, \omega} = \frac{W^2 L^3}{6EI_z} + \frac{\omega^2 L^5}{40EI_z}$$

However, comparing this expression to the correct expression for Work_{M_z} which was obtained in Example Problem 9.4

$$\text{Work}_{M_z} = \frac{W^2 L^3}{6EI_z} + \frac{W \omega L^4}{8EI_z} + \frac{\omega^2 L^5}{40EI_z}$$

shows that a term is missing in the superimposed results. This additional term occurs due to the interaction of the two loads as the structure deforms. A simple explanation is that one of the loads performs work while moving through the displacement produced by the other load. This interaction is not represented if the work due to the bending moment for each load is computed separately, but it is introduced by the cross product term when the expression for M_z, which includes the effects of both the concentrated force and the distributed force, is squared. Without this interaction term, the effect of the distributed force will be lost when taking the derivative of Work_{M_z} with respect to W. Therefore, when computing the external work for any system, all active loads must be considered to act on the structure simultaneously. It is possible to consider the effects of the axial deformations, the bending deformations and the shear deformations separately since they are not normally considered to interact.

DEFLECTION COMPUTATIONS BY THE PRINCIPLE OF VIRTUAL WORK

The *Principle of Virtual Work*, which was originally developed by Johann Bernoulli in 1717, is one of the most powerful tools available for structural analysis. In this chapter we will only be concerned with its application to the computation of displacements in trusses and frames. However, as we will see in later chapters, it also has other important applications.

To develop the basic concept of the Principle of Virtual Work we will consider a structure under three different load cases. For the types of problems which we will be considering here, the structure may be a statically determinate or statically indeterminate truss or frame in two or three dimensions. However, it might actually consist of any type of structural system composed of essentially any type of structural components.

Case I. Assume that the structure is subjected to a load system **Q**, which may consist of any combination of concentrated forces and moments acting at any points on the structure, as shown in Figure 9.15. As in the development of Castigliano's Theorem, the structure is shown in generic form in this figure. The loads will be designated as Q_1 through Q_{NQ}, where NQ is the number of loads. This load system will consist of both the active and the reactive loads for the structure. Therefore, for any stable structure, the load system **Q** will automatically be a system which is in equilibrium (i.e.; both the resultant force and the resultant moment for the load system will be zero). The external work performed by the load system **Q** while the structure deforms will be designated as $\text{Work}_{\text{ext}, Q}$ while the internal work which is performed by the member axial forces, shear forces and bending and twisting moments as the mem-

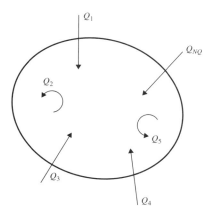

Figure 9.15 Virtual Load System.

Deflection Computations by the Principle of Virtual Work

bers deform will be designated as Work$_{int, Q}$. The load system **Q** will be known as the *irtual load system*.

Case II. Assume that the structure is subjected to a load system **W** which may consist of any combination of concentrated forces and moments at any points on the structure and distributed forces on any members. As with the load system **Q**, the loads in the load system **W** will consist of both the active and reactive loads. Therefore, the load system will be in equilibrium for a stable structure. The external work performed by the load system **W** while the structure deforms will be designated as Work$_{ext, W}$ while the internal work which is performed by the member axial forces, shear forces and bending and twisting moments as the members deform will be designated as Work$_{int, W}$. The load system **W** will be known as the *real load system*.

Case III. Assume that the structure is first subjected to the load system **Q** and then, with the load system **Q** on the structure, it is subjected to the load system **W**. The total external work for this combined loading case will be

$$\text{Work}_{ext, QW} = \text{Work}_{ext, Q} + \text{Work}_{ext, W} + \sum_{i=1}^{NQ} Q_i \Delta_{i, W} \qquad (9.39)$$

where the quantity $Q_i \Delta_{i, W}$ represents the work performed by any load Q_i in the load system **Q** while moving through the displacement $\Delta_{i, W}$ at its point of application, which is caused by the load system **W**. A factor of one half is not present since the load Q_i is acting during the entire time that the displacement $\Delta_{i, W}$ occurs.

The total internal work for this case will be

$$\text{Work}_{int, QW} = \text{Work}_{int, Q} + \text{Work}_{int, W}$$

$$+ \sum_{n=1}^{NM} \left(\int S_{x, Q} dL_{x, W} + \int V_{y, Q} d\alpha_{y, W} + \int V_{z, Q} d\alpha_{z, W} \right. \qquad (9.40)$$

$$\left. + \int M_{x, Q} d\theta_{x, W} + \int M_{y, Q} d\theta_{y, W} + \int M_{z, Q} d\theta_{z, W} \right)_n$$

where the integral expressions on the right side represent the work performed by the axial forces $S_{x, Q}$, the shear forces $V_{y, Q}$ and $V_{z, Q}$, the twisting moments $M_{x, Q}$ and the bending moments $M_{y, Q}$ and $M_{z, Q}$ in the members due to the load system **Q** while moving through the member deformations $dL_{x, W}$, $d\alpha_{y, W}$, $d\alpha_{z, W}$, $d\theta_{x, W}$, $d\theta_{y, W}$ and $d\theta_{z, W}$, which are caused by the load system **W**. Again, a factor of one half is not present since the member axial forces, shear forces and moments due to **Q** are present during the entire time that the member deformations due to **W** occur.

The external work and the internal work for each of the separate load cases must be equal. Therefore, by equating Eqs. (9.39) and (9.40) and eliminating equal terms on each side corresponding to load Cases I and II, we obtain

$$\sum_{i=1}^{NQ} Q_i \Delta_{i,W} = \sum_{n=1}^{NM} \left(\int S_{x,Q} dL_{x,W} + \int V_{y,Q} d\alpha_{y,W} + \int V_{z,Q} d\alpha_{z,W} \right.$$
$$\left. + \int M_{x,Q} d\theta_{x,W} + \int M_{y,Q} d\theta_{y,W} + \int M_{z,Q} d\theta_{z,W} \right)_n \quad (9.41)$$

This equation represents the basic concept of the Principle of Virtual Work, which can be expressed as:

> If a structure, which is subjected to a load system **Q** which is in equilibrium, undergoes a set of external displacements and geometrically compatible internal member deformations, the external work performed by the load system **Q** moving through the external displacements will be equal to the internal work performed by the member cross section loads corresponding to the load system **Q** moving through the internal member deformations.

For the particular structure being considered here, the external displacements and the internal member deformations were produced by an external load system **W**. For this specific case, Eq. (9.41), after substituting the appropriate expressions for the member deformations due to the real load system, becomes

$$\sum_{i=1}^{NQ} Q_i \Delta_{i,W} = \sum_{n=1}^{NM} \left(\int \frac{S_{x,Q} S_{x,W}}{A_x E} dx + \int \frac{V_{y,Q} V_{y,W}}{A_y G} dx + \int \frac{V_{z,Q} V_{z,W}}{A_z G} dx \right.$$
$$\left. + \int \frac{M_{x,Q} M_{x,W}}{J_x G} dx + \int \frac{M_{y,Q} M_{y,W}}{EI_y} dx + \int \frac{M_{z,Q} M_{z,W}}{EI_z} dx \right)_n \quad (9.42)$$

However, the external displacements and the member deformations also can be due to any other cause as long as they are geometrically compatible. The only requirement for the load system **Q** is that it is a system of concentrated loads which are in equilibrium.

One of the applications of the Principle of Virtual Work is to compute the displacements which will be produced in a structure due to an external or internal action. This action might be the application of an external load system, but it could also consist of other actions such as an internal temperature change, which causes one or more members to deform, or a support settlement which produces an external displacement. The following example problems will demonstrate the use of the Principle of Virtual Work for computing structural displacements.

Example Problem 9.5

Compute the horizontal and vertical translation of joint 3 in the statically determinate truss shown previously in Figure 9.5a. This is the same truss for which the horizontal translation was computed by equating the external work to the internal work in Example Problem 9.1.

Deflection Computations by the Principle of Virtual Work

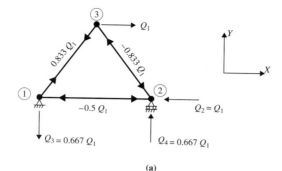

Member	$S_{x,Q}$	$S_{x,w}$ (kips)	L (in)	A_x (in²)	E (kips/in²)	$\dfrac{S_{x,Q}S_{x,w}L}{A_xE}$
1	−0.5	15	360	5	29,000	−0.019
2	0.833	−25	300	5	29,000	−0.043
3	−0.833	25	300	5	29,000	−0.043

$$\Sigma = -0.105$$

(b)

Figure 9.16 Example Problem 9.5.

Since only axial forces exist in the members in a truss, Eq. (9.42) will reduce to

$$\sum_{i=1}^{NQ} Q_i \Delta_{i,W} = \sum_{n=1}^{NM} \left(\int \frac{S_{x,Q} S_{x,W}}{A_x E} dx \right)_n$$

which, upon recognizing that all of the members in the truss are prismatic, can be expressed as

$$\sum_{i=1}^{NQ} Q_i \Delta_{i,W} = \sum_{n=1}^{NM} \left(\frac{S_{x,Q} S_{x,W} L}{A_x E} \right)_n \qquad (9.43)$$

This expression now can be used to compute the translation of any joint in the truss in any direction by choosing the appropriate virtual load system **Q**.

Horizontal Translation at Joint 3. Equation (9.43) can be used to compute the horizontal translation at joint 3 by choosing the virtual load system shown in Figure 9.16a. This load system consists of a horizontal force Q_1 of arbitrary magnitude in the positive global X direction at joint 3 and a set of additional forces Q_2 through Q_4 at joints 1 and 2 acting in the directions of the support restraints at these joints. The magnitudes and directions of Q_2 through Q_4 have been chosen so that the load system **Q** is in equilibrium. (Note that the magnitudes and directions of Q_2 through Q_4 are exactly equal to the reactive forces which would occur at the support joints if the truss, with the support conditions shown in Figure 9.5a, were subjected to a horizontal active force Q_1 at joint 3.) The member forces for the load system **Q** are shown adjacent to each member.

If we now evaluate Eq. (9.43) for this load system, we obtain

$$Q_1\Delta_{1,\,W} + Q_2\Delta_{2,\,W} + Q_3\Delta_{3,\,W} + Q_4\Delta_{4,\,W} = \sum_{n=1}^{3}\left(\frac{S_{x,Q}S_{x,W}L}{A_xE}\right)_n$$

which, on recognizing that $\Delta_{2,\,W}$ through $\Delta_{4,\,W}$, which correspond to the displacements in the directions of the support restraints for the truss due to the real load system, must be zero, gives

$$Q_1\Delta_{1,\,W} = \sum_{n=1}^{3}\left(\frac{S_{x,Q}S_{x,W}L}{A_xE}\right)_n$$

which can be solved for $\Delta_{1,\,W}$, which is equal to the horizontal displacement Δ_{3X} of joint 3 due to the real load system. Since the magnitude of the force Q_1 is arbitrary, the arithmetic can be simplified by choosing a unit value for this force, for which Eq. (9.43) reduces to the following simple expression for computing Δ_{3X}

$$\Delta_{3X} = \sum_{n=1}^{3}\left(\frac{S_{x,Q}S_{x,W}L}{A_xE}\right)_n$$

The right hand side summation for this equation can be easily evaluated in a table, as shown in Figure 9.16b, from which

$$\Delta_{3X} = -0.105 \text{ inch}$$

The negative value for Δ_{3X} means that the force Q_1 does negative work. Therefore, the horizontal translation of joint 3 is toward the left.

$$\Delta_{3X} = 0.105 \text{ inch} \leftarrow$$

If Q_1 had been chosen to act in the opposite direction, the computed value of Δ_{3X} would be positive, which would still correspond to a translation toward the left. Note that this computed translation agrees with the value computed in Example Problem 9.1.

Vertical Translation at Joint 3. The vertical translation of joint 3 can be determined by using the virtual load system, shown in Figure 9.17a. For this load system, a unit vertical force Q_1 acts at joint 3 in the positive direction of the global Y axis, while the forces Q_2 and Q_3 are equal to the vertical reactive forces which would occur at joints 1 and 2. Substituting into Eq. (9.43) and recognizing that the displacements $\Delta_{2,\,W}$ and $\Delta_{3,\,W}$ for the real load system are zero and that the displacement $\Delta_{1,\,W}$ is equal to the displacement Δ_{3Y} results in the following expression for computing Δ_{3Y}:

$$\Delta_{3Y} = \sum_{n=1}^{3}\left(\frac{S_{x,Q}S_{x,W}L}{A_xE}\right)_n$$

Evaluation of this expression in the table in Figure 9.17b gives

$$\Delta_{3Y} = -0.014 \text{ inch}$$

Since the translation is negative, it is down to correspond to the opposite direction of the upward force Q_1.

$$\Delta_{3Y} = 0.014 \text{ inch} \downarrow$$

Deflection Computations by the Principle of Virtual Work

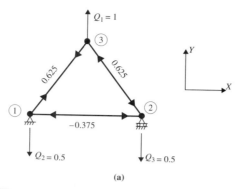

Member	$S_{x,Q}$	$S_{x,w}$ (kips)	L (in)	A_x (in^2)	E (kips/in^2)	$\dfrac{S_{x,Q} S_{x,w} L}{A_x E}$
1	−0.375	15	360	5	29,000	−0.014
2	0.625	−25	300	5	29,000	−0.032
3	0.625	25	300	5	29,000	0.032

$$\Sigma = -0.014$$

(b)

Figure 9.17 Example Problem 9.5.

Equation (9.43) can be used to compute the translation in any direction at any joint in a truss after the member forces have been computed. The technique of assigning a unit value to the virtual force at the point where the displacement is to be computed is often known as the *Dummy Unit Load Method*. Since only one equation can be developed for any virtual load system, it must be chosen so that either zero displacements, or known nonzero displacements, occur at all of the virtual loads except the load acting at the joint in the direction where the desired displacement is to be computed. The situation where nonzero displacements can exist at the points of application of one or more of the virtual loads is demonstrated in the next example problem.

Example Problem 9.6

In a previous analysis by another engineer, it was found that the vertical translations of joints 2 and 4 of the truss shown in Figure 9.18a are 0.128 inches down, and 0.240 inches down, respectively. It is now necessary to also compute the vertical translation of joint 6.

The first step in the analysis is to compute the reactive forces and the member forces for the truss due to the real load system. The results of this analysis are shown in Figure 9.18b

The obvious approach to compute the vertical translation of joint 6 is to use the virtual load system shown in Figure 9.19a, where a unit upward vertical force Q_1 is applied at joint 6 and balancing forces Q_2 and Q_3 are applied in the directions of the vertical restraints at the support joints.

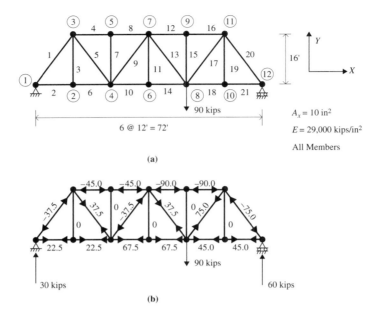

Figure 9.18 Example Problem 9.6.

Since the translations $\Delta_{2,W}$ and $\Delta_{3,W}$ must be zero, the translation Δ_{6Y} can be found by Eq. (9.43) as

$$\Delta_{6Y} = \Delta_{1,W} = \sum_{n=1}^{21}\left(\frac{S_{x,Q}S_{x,W}L}{A_XE}\right)_n$$

which, upon evaluation of the right side summation in the table in Figure 9.19b, gives

$$\Delta_{6Y} = -0.318 \text{ inch} = 0.318 \text{ inch} \downarrow$$

Although the previous analysis was simple and straightforward, it did involve a considerable amount of computation to obtain the final result. The computational effort for this problem can be considerably reduced by using the known translations at joints 2 and 4 in the application of Eq. (9.43). For this set of computations, the virtual load system shown in Figure 9.20a will be used, which consists of a downward force Q_2 of magnitude 2 at joint 4 and unit upward balancing vertical forces Q_1 and Q_3 at joints 6 and 2 respectively.

For this loading, Eq. (9.43) will become

$$Q_1\Delta_{1,W} + Q_2\Delta_{2,W} + Q_3\Delta_{3,W} = \sum_{n=1}^{21}\left(\frac{S_{x,Q}S_{x,W}L}{A_xE}\right)_n$$

for which the right side summation can be evaluated in the table in Figure 9.20b. Note that the evaluation of this table required considerably fewer computations than were required for the table in Figure 9.19b, since it was not necessary to include any member for which the virtual member force $S_{x,Q}$ is zero. If we now substitute this

Deflection Computations by the Principle of Virtual Work

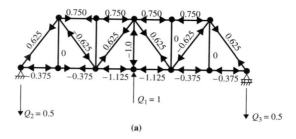

Member	$S_{x,Q}$	$S_{x,W}$ (kips)	L (in)	A_x (in²)	E (kips/in²)	$\dfrac{S_{x,Q} S_{x,W} L}{A_x E}$
1	0.625	−37.5	240.0	10.0	29,000	−0.01940
2	−0.375	22.5	144.0	10.0	29,000	−0.00419
3	0	0	192.0	10.0	29,000	0
4	0.750	−45.0	144.0	10.0	29,000	−0.01676
5	−0.625	37.5	240.0	10.0	29,000	−0.01940
6	−0.375	22.5	144.0	10.0	29,000	−0.00419
7	0	0	192.0	10.0	29,000	0
8	0.750	−45.0	144.0	10.0	29,000	−0.01676
9	0.625	−37.5	240.0	10.0	29,000	−0.01940
10	−1.125	67.5	144.0	10.0	29,000	−0.03771
11	−1.0	0	192.0	10.0	29,000	0
12	0.750	−90.0	144.0	10.0	29,000	−0.03352
13	0.625	37.5	240.0	10.0	29,000	0.01940
14	−1.125	67.5	144.0	10.0	29,000	−0.03771
15	0	0	192.0	10.0	29,000	0
16	0.750	−90.0	144.0	10.0	29,000	−0.03352
17	−0.625	75.0	240.0	10.0	29,000	−0.03879
18	−0.375	45.0	144.0	10.0	29,000	−0.00838
19	0	0	192.0	10.0	29,000	0
20	0.625	−75.0	240.0	10.0	29,000	−0.03879
21	−0.375	45.0	144.0	10.0	29,000	−0.00838

$$\Sigma = -0.3175$$

(b)

Figure 9.19 Example Problem 9.6.

value, and the known values for the translations $\Delta_{2,W}$ and $\Delta_{3,W}$ and the loads Q_1, Q_2 and Q_3 into the preceding equation, we obtain

$$(1.0)\Delta_{1,W} + (2.0)(0.240) + (1.0)(-0.128) = 0.0336$$

from which the desired translation at joint 6 can be computed as

$$\Delta_{6Y} = \Delta_{1,W} = -0.318 \text{ inch} = 0.318 \text{ inch} \downarrow$$

The displacement $\Delta_{2,W}$ was considered to be positive in this computation since it is in the same direction as Q_2, while the displacement $\Delta_{3,W}$ was considered to be negative

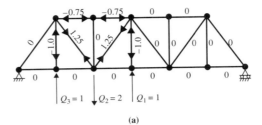

Member	$S_{x,Q}$	$S_{x,w}$ (kips)	L (in)	A_x (in²)	E (kips/in²)	$\dfrac{S_{x,Q} S_{x,w} L}{A_x E}$
3	−1.0	0	192.0	10.0	29,000	0
4	−0.75	−45.0	144.0	10.0	29,000	0.0168
5	1.25	37.5	240.0	10.0	29,000	0.0388
8	−0.75	−45.0	144.0	10.0	29,000	0.0168
9	1.25	−37.5	240.0	10.0	29,000	−0.0388
11	−1.0	0	192.0	10.0	29,000	0

$\Sigma = 0.0336$

(b)

Figure 9.20 Example Problem 9.6.

since it is in the opposite direction to Q_3. The negative value for $\Delta_{1,w}$ indicates that it is opposite to Q_1.

There is often more than one virtual load system which can be used to compute the displacement at a specific point in a structure. The only requirement for the virtual load system is that it must consist of a set of concentrated forces and moments which are in equilibrium. Other than that, there are no restrictions on the number of virtual loads and where they may be placed on the structure. Of course, as explained previously, only one virtual work equation can be developed for each virtual load system. Therefore, the virtual load system must be chosen so that only one unknown displacement exists in the equation.

Example Problem 9.7

Compute the vertical translation at joint 3 for the statically indeterminate truss shown in Figure 9.21a. The reactive forces and the member forces, which were obtained in a previous analysis by another engineer, are shown in Figure 9.21b.

We will now consider several different virtual load systems which may be used for this analysis.

Virtual Load System I. At first glance it would appear that the virtual load system which must be used for this analysis will consist of a unit upward vertical force at joint 3 and the reactive forces at the supports which would occur for this statically indeterminate truss, as represented by the forces Q_1 through Q_5 in Figure 9.22a. (These reactive forces and the corresponding member forces were computed using the analy-

Deflection Computations by the Principle of Virtual Work

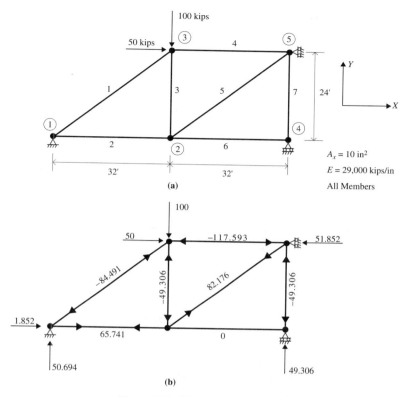

Figure 9.21 Example Problem 9.7.

sis procedure which will be presented later in Chapter 11. For now, we will merely accept these values as being correct.)

Since the translations $\Delta_{2,\,W}$ through $\Delta_{5,\,W}$ must be zero due to the support restraints for this truss, Eq. (9.43) will reduce to

$$\Delta_{3Y} = \Delta_{1,\,W} = \sum_{n=1}^{7}\left(\frac{S_{x,Q}S_{x,W}L}{A_x E}\right)_n$$

which, upon evaluating the right side summation in the table in Figure 9.22b, gives

$$\Delta_{3Y} = -0.440 \text{ inch} = 0.440 \text{ inch} \downarrow$$

The previous virtual load system was chosen so that the displacements which it would produce in the truss would be consistent with the support restraints. However, if we look carefully at the general virtual work relationship, which is represented by Eq. (9.41), we see that the displacements due to the virtual load system do not enter into the relationship. The only requirement for the virtual load system is that it must be in equilibrium. Therefore, the support conditions for the structure can be ignored for the virtual load system, as long as only one unknown displacement occurs on the left side of the virtual work equation.

324 Deflection of Trusses and Frames   Chap. 9

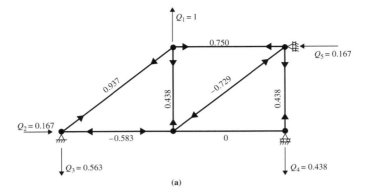

Member	$S_{x,Q}$	$S_{x,w}$ (kips)	L (in)	A_x (in²)	E (kips/in²)	$\dfrac{S_{x,Q} S_{x,w} L}{A_x E}$
1	0.937	−84.891	480.0	10.0	29,000	−0.131
2	−0.583	65.741	384.0	10.0	29,000	−0.051
3	0.438	−49.306	288.0	10.0	29,000	−0.021
4	0.750	−117.593	384.0	10.0	29,000	−0.117
5	−0.729	82.176	480.0	10.0	29,000	−0.099
6	0	0	384.0	10.0	29,000	0
7	0.438	−49.306	288.0	10.0	29,000	−0.021

$\Sigma = -0.440$

(b)

Figure 9.22   Example Problem 9.7, Virtual Load System I.

Virtual Load Systems II and III.   Figures 9.23a and 9.24a show two additional virtual load systems which can be used to compute the vertical translation of joint 3 for this truss. Each of these load systems consists of a unit upward vertical force at joint 3 and additional forces corresponding to the reactions which would occur if selected support restraints were removed from the original truss to leave a statically determinate stable structure. Since these additional forces occur at points which do not translate in the original truss, the left side of Eq. (9.43) reduces to the same form shown previously for Virtual Load System I. The tables in Figures 9.23b and 9.24b show that the computed translation of joint 3 for each load system agrees with the previous solution within expected roundoff error.

Since there are an infinite number of combinations of forces at the four support restraints which will balance a unit upward vertical force at joint 3 for this statically indeterminate truss, there are an infinite number of virtual load systems which can be used to compute the desired translation. The virtual load system shown in Figure 9.24 results in the simplest possible set of computations, whereas the system shown in Figure 9.22 is the most complicated since a considerable amount of computation was required to compute the reactive forces for the statically indeterminate truss subjected to a unit virtual force at joint 3.

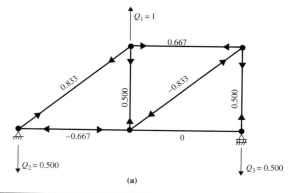

Member	$S_{x,Q}$	$S_{x,w}$ (kips)	L (in)	A_x (in^2)	E (kips/in^2)	$\dfrac{S_{x,Q}S_{x,w}L}{A_xE}$
1	0.833	−84.891	480.0	10.0	29,000	−0.116
2	−0.667	65.741	384.0	10.0	29,000	−0.058
3	0.500	−49.306	288.0	10.0	29,000	−0.024
4	0.667	−117.593	384.0	10.0	29,000	−0.104
5	−0.833	82.176	480.0	10.0	29,000	−0.113
6	0	0	384.0	10.0	29,000	0
7	0.500	−49.306	288.0	10.0	29,000	−0.024

$\Sigma = -0.439$

(b)

Figure 9.23 Example Problem 9.7, Virtual Load System II.

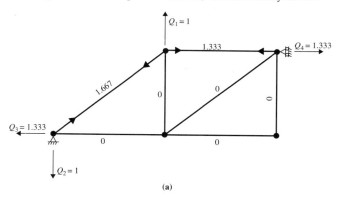

Member	$S_{x,Q}$	$S_{x,w}$ (kips)	L (in)	A_x (in^2)	E (kips/in^2)	$\dfrac{S_{x,Q}S_{x,w}L}{A_xE}$
1	1.667	−84.891	480.0	10.0	29,000	−0.233
2	0	65.741	384.0	10.0	29,000	0
3	0	−49.306	288.0	10.0	29,000	0
4	1.333	−117.593	384.0	10.0	29,000	−0.208
5	0	82.176	480.0	10.0	29,000	0
6	0	0	384.0	10.0	29,000	0
7	0	−49.306	288.0	10.0	29,000	0

$\Sigma = -0.441$

(b)

Figure 9.24 Example Problem 9.7, Virtual Load System III.

Example Problem 9.8

Compute the vertical translation and the rotation at the center of the prismatic beam shown in Figure 9.25a. Include the effects of both shear and bending deformations in the beam.

The mathematical model of the beam, which consists of three joints and two members, is shown in Figure 9.25b. The real load system **W** for the beam consists of the active distributed force and the concentrated reactive forces at the support joints. Each member has its own local coordinate system with the local axes x_1 and x_2 corresponding to the longitudinal position along each member measured from the left end of the member. The shear force and bending moments in each member due to the real load system can be expressed in terms of these local coordinates as follows:

Member 1 ($0 \leq x_1 \leq 20$):
$$V_{y,W} = 30 - 2x_1$$
$$M_{z,W} = 30x_1 - x_1^2$$

Member 2 ($0 \leq x_2 \leq 20$):
$$V_{y,W} = -10$$
$$M_{z,W} = 200 - 10x_2$$

Since only shear deformations in the local y direction and bending deformations about the local z axis must be considered for this beam, Eq. (9.42) will reduce to

$$\sum_{i=1}^{NQ} Q_i \Delta_{i,W} = \sum_{n=1}^{NM} \left(\int \frac{V_{y,Q} V_{y,W}}{A_y G} dx + \int \frac{M_{z,Q} M_{z,W}}{EI_z} dx \right)_n \quad (9.44)$$

This expression now can be used to compute the translation or rotation at any point along the beam by choosing the appropriate virtual load system **Q**.

Vertical Translation at Center of the Beam. The vertical translation at the center of the beam can be computed by using the virtual load system shown in Figure 9.25c, for which the virtual shear forces and bending moments are as follows.

Member 1 ($0 \leq x_1 \leq 20$):
$$V_{y,Q} = -0.5$$
$$M_{z,Q} = -0.5x_1$$

Member 2 ($0 \leq x_2 \leq 20$):
$$V_{y,Q} = 0.5$$
$$M_{z,Q} = 0.5x_2 - 10$$

Since the displacements $\Delta_{2,W}$ and $\Delta_{3,W}$ are zero at the support joints, Eq. (9.44) will become

$$\Delta_{2Y} = \Delta_{1,W} = \sum_{n=1}^{2} \left(\int \frac{V_{y,Q} V_{y,W}}{A_y G} dx + \int \frac{M_{z,Q} M_{z,W}}{EI_z} dx \right)_n$$

Deflection Computations by the Principle of Virtual Work

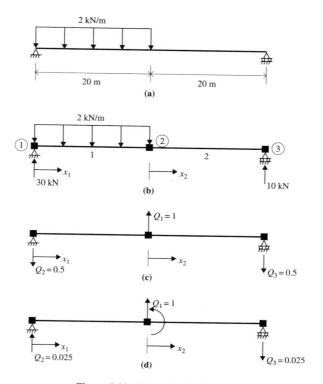

Figure 9.25 Example Problem 9.8.

which, on substituting the preceding values for $V_{y,Q}$, $V_{y,w}$, $M_{z,Q}$ and $M_{z,w}$, becomes

$$\Delta_{2Y} = \int_0^{20} \frac{(-0.5)(30 - 2x_1)}{A_y G} dx_1 + \int_0^{20} \frac{(0.5)(-10)}{A_y G} dx_2$$

$$+ \int_0^{20} \frac{(-0.5x_1)(30x_1 - x_1^2)}{EI_z} dx_1 + \int_0^{20} \frac{(0.5x_2 - 10)(200 - 10x_2)}{EI_z} dx_2$$

from which

$$\Delta_{2Y} = -\frac{200}{A_y G} - \frac{33{,}333.33}{EI_z}$$

Since the sign of the translation is negative, it will be in the opposite direction of Q_1. Therefore, Δ_{2Y} is down.

Rotation at Center of Beam. The rotation at the center of the beam can be computed by using the virtual load system shown in Figure 9.25d, for which the virtual shear forces and bending moments are as follows.

Member 1 ($0 \leq x_1 \leq 20$):

$$V_{y,Q} = 0.025$$

$$M_{z,Q} = 0.025 x_1$$

Member 2 ($0 \leq x_2 \leq 20$):

$$V_{y,Q} = 0.025$$

$$M_{z,Q} = -0.5 + 0.025 x_2$$

For this virtual load system, Eq. (9.44) will become

$$\theta_{2Z} = \Delta_{1,w} = \sum_{n=1}^{2} \left(\int \frac{V_{y,Q} V_{y,W}}{A_y G} dx + \int \frac{M_{z,Q} M_{z,W}}{EI_z} dx \right)_n$$

which, on substituting the expressions for $V_{y,Q}$, $V_{y,W}$, $M_{z,Q}$ and $M_{z,W}$, becomes

$$\theta_{2Z} = \int_0^{20} \frac{(0.025)(30 - 2x_1)}{A_y G} dx_1 + \int_0^{20} \frac{(0.025)(-10)}{A_y G} dx_2$$

$$+ \int_0^{20} \frac{(0.025 x_1)(30 x_1 - x_1^2)}{EI_z} dx_1 + \int_0^{20} \frac{(-0.5 + 0.025 x_2)(200 - 10 x_2)}{EI_z} dx_2$$

from which

$$\theta_{2Z} = \frac{10}{A_y G} + \frac{333.33}{EI_z}$$

Since the sign of the rotation is positive, it is counterclockwise.

The portion of the translation and the rotation produced by the bending deformations in the beam can be verified by using the expressions for Y and dY/dX which were derived in Example Problem 8.4 in Chapter 8 for a beam subjected to this same type of loading. The magnitudes of the translation in meters and the rotation in radians can be determined by substituting specific values for A_y, I_z, G, and E in kN and meter units.

Example Problem 9.9

Compute the horizontal translation of joint 5 for the plane bridge truss shown in Figure 9.26a if the following errors were made during the construction of the truss:

1. Support joint 6 was located 0.75 inches down from its correct elevation due to an error during the construction of the right abutment.
2. Member 5 was fabricated 0.5 inches too short and member 6 was fabricated 0.25 inches too long due to an error in reading the fabrication drawings

Since no active loads are being considered for this statically determinate truss, the forces in the members will be zero and the only member deformations will be those

Deflection Computations by the Principle of Virtual Work

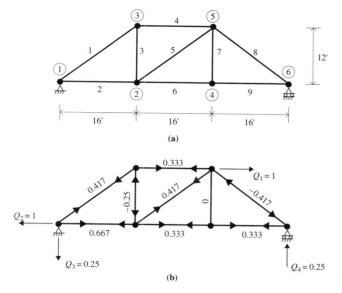

Figure 9.26 Example Problem 9.9.

which are caused by the fabrication errors. Therefore, the virtual work relationship represented by Eq. (9.41) should be expressed in the form

$$\sum_{i=1}^{NQ} Q_i \Delta_i = \sum_{n=1}^{NM} (S_{x,Q} \Delta L_x)_n$$

where Δ_i represents the translation at any concentrated force Q_i in the virtual load system and ΔL_x represents the change in length of any member in the truss due to the fabrication errors.

The horizontal translation of joint 5 can be determined by using the virtual load system shown in Figure 9.26b. Substituting the quantities from this loading into the previous virtual work relationship, and recognizing that the only nonzero joint displacements are Δ_1 and Δ_4 and the only nonzero member deformations are ΔL_5 and ΔL_6 gives

$$Q_1 \Delta_1 + Q_4 \Delta_4 = S_{5x,Q} \Delta L_5 + S_{6x,Q} \Delta L_6$$

or

$$(1.0)\Delta_{5X} + (0.25)(-0.75) = (0.417)(-0.5) + (0.333)(0.25)$$

from which

$$\Delta_{5X} = 0.062 \text{ inch} \rightarrow$$

The change in length of member 5 is considered to be negative since the member shortens and member 5 is in tension due to the virtual load system. The change in length of member 6 is considered to be positive since the member elongates and member 6 is in tension due to the virtual load system.

As we can see from the previous example problems, the Principle of Virtual Work is an extremely versatile tool for computing displacements in any type of structural system. By choosing an appropriate virtual load system, the translation at any joint in a truss or the translation or rotation at any point in a beam or frame can be determined due to any combination of external and/or internal effects in the structure. As we will see in later chapters, the Principle of Virtual Work also has other applications in the field of structural analysis.

CASTIGLIANO'S THEOREM REVISITED

In the previous discussion of Castigliano's Theorem for computing displacements in structures, it appeared that this theorem only could be used if a concentrated load of the correct type in the correct direction was present at the point on the structure where the displacement was to be computed. However, this restriction on the application of this theorem can be eliminated by using a simple trick during the analysis. To demonstrate this analysis procedure, consider the prismatic cantilever beam whose mathematical model is shown in Figure 9.27a, which is subjected to a positive distributed force ω over its length. If we try to apply Castigliano's Theorem to compute the vertical translation at the free end of this beam, we encounter difficulty since there is no vertical concentrated force acting at this point. However, we would be able to apply this procedure if the beam were subjected to the loading, shown in Figure 9.27b. For this loading, the bending moment at any point is equal to

$$M_z = Wx + \frac{\omega x^2}{2}$$

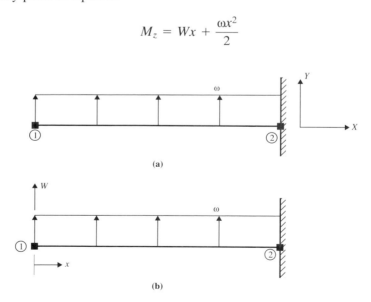

Figure 9.27

Castigliano's Theorem Revisited

and the desired translation can be determined by

$$\Delta_{1Y} = \frac{\partial \text{Work}}{\partial W} = \frac{\partial \text{Work}}{\partial M_z}\frac{\partial M_z}{\partial W} = \int \frac{M_z}{EI_z}\frac{\partial M_z}{\partial W}\,dx$$

which gives

$$\Delta_{1Y} = \int_0^L \left(\frac{Wx + \omega x^2/2}{EI_z}\right)(x)\,dx = \frac{WL^3}{3EI_z} + \frac{\omega L^4}{8EI_z}$$

Although this is not the desired solution, we can easily transform it into the desired result by taking the limit of the final expression for the translation as the magnitude of the concentrated force W goes to zero. Therefore, the deflection for the beam in Figure 9.27a will be

$$\Delta_{1Y} = \lim_{W \to 0}\left(\frac{WL^3}{3EI_z} + \frac{\omega L^4}{8EI_z}\right) = \frac{\omega L^4}{8EI_z}$$

This same process can be used for any type of structure under any type of loading. However, the algebra involved in performing the analysis usually can be reduced by a slight change in the order of the calculations by taking the limit as W goes to zero before performing the integration.

$$\Delta_{1Y} = \lim_{W \to 0} \int \frac{M_z}{EI_z}\frac{\partial M_z}{\partial W}\,dx \tag{9.45}$$

For this case, the expression for M_z used in the integration will reduce to

$$M_z = \frac{\omega x^2}{2}$$

while the expression for $\partial M_z/\partial W$ will not change. Performing the integration with these values gives

$$\Delta_{1Y} = \int_0^L \left(\frac{\omega x^2/2}{EI_z}\right)(x)\,dx = \frac{\omega L^4}{8EI_z}$$

which agrees with the previous analysis.

If we now compare this last analysis to the analysis which would be performed if we solved this same problem using the Dummy Unit Load form of the Principle of Virtual Work, we will see that they involve exactly the same set of calculations. The quantity M_z in Eq. (9.45) is equivalent to the moment $M_{z,w}$ in the virtual work relationship in Eq. (9.44) while the quantity $\partial M_z/\partial W$ is equivalent to the moment $M_{z,Q}$ since the rate of change of the moment with respect to W is equal to the moment which would be produced by a unit value of W. The load W is equivalent to the unit virtual load which would be used when performing the analysis by the Principle of Virtual Work.

FATPAK II - STUDENT EDITION

A general purpose structural analysis computer package named FATPAK II (Frame And Truss PAcKage, Version II) is supplied for this book. This package is the student version of the commercial structural analysis package FATPAK V (Structural Software Systems, 1996). It consists of a set of programs which will analyze both statically determinate and statically indeterminate trusses and frames for a variety of conditions. The individual programs which are included in the package are: T2DII for the analysis of plane trusses; T3DII for the analysis of space trusses; F2DII for the analysis of plane frames; and F3DII for the analysis of space frames. The input to each of the programs consists of a description of the properties of the mathematical model for the structure and the conditions for which it is to be analyzed, while the output consists of the displacements of the joints, the member end loads and the support reactions. In addition, the program user may plot both the original geometry of the mathematical model before the analysis is performed and the final deformed shape after the analysis has been completed. The shear force and bending moment diagrams may also be plotted for any member in a plane frame or space frame. A beam can be analyzed by the program F2DII since a beam is merely a special case of a plane frame in which all of the members are collinear. The joint loads and the joint displacements for any structure are expressed as components in a global right hand orthogonal coordinate system, as shown previously in Figure 1.3 in Chapter 1, while the member distributed forces and member end loads are expressed as components in the local coordinate system for each member. A description of the orientation of the global and the local member coordinate axes and instructions for using each of the individual programs in FATPAK II are contained in the ASCII disk file FATPAKII.DOC.

The analysis of any structure is performed by the Stiffness Method in which a set of linear simultaneous equations is formed by summing the stiffnesses of the members in the structure at each joint. The unknowns in these equations are the joint displacements. We will not discuss the details of this analysis procedure at this time. It will be discussed in Chapter 15 along with two computer programs which demonstrate its implementation on the computer for the analysis of plane trusses and plane frames. The source code for the programs in FATPAK II will not be discussed in this book.

No unit conversions are made at any step during the analysis of a structure by FATPAK II. Therefore, the input data may be in any desired units as long as all units are consistent. The output will be expressed in the same units as the input. Any numbers in the input data may be entered in either decimal or exponential format. In addition, the program will give the option during the analysis of listing the output in either of these formats. It will usually be found that the decimal format will be adequate when working in either kip and feet or kip and inch units, but the exponential format might be needed when working in SI units due to the magnitude of some of the numbers for some combinations of force and length units. The calculations for the displacements are performed in double precision arithmetic in FATPAK II to reduce roundoff error, since the analysis of large structures might result in the solution of several hundred linear simultaneous equations.

The programs in FATPAK II may be used by the reader to verify the manual solutions for the suggested problems at the end of this chapter and all following chapters. They also may be used for a variety of educational purposes to gain a better understanding of the behavior of structures under various loading and support conditions. The reader is licensed to use the programs for any educational purpose, either in a college level structural engineering course or for self study. Any commercial use of FATPAK II is a violation of the license under which it is being distributed. The program FATPAK V is licensed for commercial purposes.

Computer Solution for Example Problems 9.1 and 9.5

Figure 9.28 shows a listing of the input data in kip and inch units for the plane truss shown in Figure 9.5 for the program T2DII in FATPAK II. The primary difference between this input data and that for the program SDPTRUSS, which was used previously, is that information now must be supplied concerning the material properties and the cross section properties for each member. This information is needed by T2DII in order to perform the analysis by the Stiffness Method. This information was not needed by SDPTRUSS since it generated a set of equilibrium equations at the joints which were only dependent upon the truss geometry. This equilibrium analysis was restricted to statically determinate trusses, whereas T2DII can analyze both statically determinate and statically indeterminate trusses.

```
Example Problems 9.1 and 9.5 - Analysis by FATPAK II
3,3,1,1,2,0,1,0,0
Joint Coordinates
1,0.0,0.0
2,360.0,0.0
3,180.0,240.0
Material Data
1,29000.0
Cross Section Data
1,5.0
Member Data
1,1,2
2,1,3
3,2,3
Support Restraints
1,0,1
2,1,1
Joint Loads
3,-30.0,0.0
```

Figure 9.28 Example Problem 9.5, T2DII input.

Example Problems 9.1 and 9.5 - Analysis by FATPAK II

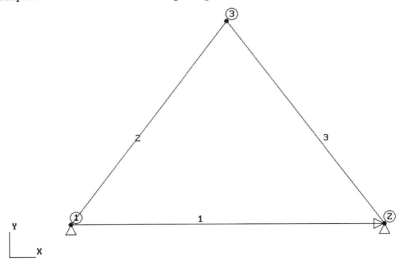

Figure 9.29 Example Problem 9.5, geometry plot from T2DII.

```
FATPAK II -- Frame and Truss Package -- Student  Edi
(C) Copyright 1995 by Structural Software Systems
Program T2DII - Plane Truss Analysis


Example Problems 9.1 and 9.5 - Analysis by FATPAK II

Joint Displacements

Joint          X-Tran              Y-Tran

  1            -0.037              0.000
  2             0.000              0.000
  3            -0.105             -0.014

Member Forces (Tension Positive)

Member          Force

  1            15.000
  2           -25.000
  3            25.000

Reactions

Joint            RX                  RY

  1             0.000              20.000
  2            30.000             -20.000
```

Figure 9.30 Example Problem 9.5, T2DII output.

Figure 9.29 shows a plot of the geometry of the truss by T2DII before the analysis was performed and Figure 9.30 shows a listing of the output from the program. The primary difference between this output and the output from the program SDPTRUSS is that in addition to a listing of the member forces and the support reaction force components, it also contains a listing of the global displacement components of the joints. These displacements could not be computed by SDPTRUSS. The X and Y translations of joint 3 of -0.105 inches and -0.014 inches, respectively, agree with the values which were computed using the Principle of Virtual Work in Example Problem 9.5.

Figure 9.31 on page 336 shows a plot of the deformed shape of the truss after the analysis was completed. The solid lines represent the location of the members in the deformed position, while the dotted lines correspond to their original undeformed positions. The displacements have been multiplied by a factor of 500 so that they can be clearly seen in the plot. The program user may specify any desired magnitude for the displacement multiplication factor.

Computer Solution for Example Problem 9.2

Figure 9.32 on page 336 shows a listing of the input data in kip and inch units for the plane frame shown in Figure 9.7 for the program F2DII in FATPAK II. As with the previous input data listing for T2DII, the primary difference between this input data and that for the program SDPFRAME is the material and cross section information.

Figure 9.33 on page 337 shows a listing of the output from the program and Figure 9.34 shows the deformed shape of the frame which was generated after the analysis was completed. A displacement multiplication factor of 50 was specified for this plot. The results of the analysis show that the rotation at joint 3 is 0.00372 radians clockwise, which agrees with the value which will be obtained from the final expression for θ in Example Problem 9.2 for the material and cross section properties which were specified in the input data. Finally, Figure 9.35 on page 338 shows a plot of the shear force and bending moment diagrams for member 2 in the mathematical model of the frame. Although this particular plot is not very interesting since the shear force is zero and the bending moment is constant for each member in the frame, plots of the shear force and bending moment diagrams can be extremely valuable for understanding the results of the analysis for frames with more complicated geometries and loadings.

Computer Solution for Example Problem 9.9

Figure 9.36 on page 338 shows a listing of the input data for the plane truss shown in Figure 9.26 for the program T2DII in FATPAK II. The errors in the member lengths are specified under the heading Initial Member Deformations while the error in elevation of the right support joint is specified under the heading Displaced Support Joints. Figure 9.37 on page 339 shows a listing of the results of

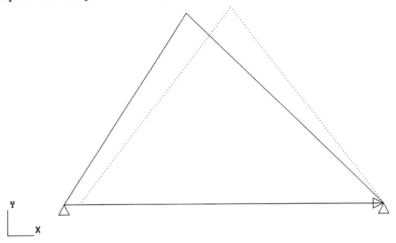

Figure 9.31 Example Problem 9.5, deformed geometry plot from T2DII.

```
Example Problem 9.2 - Analysis by FATPAK II
3,2,1,1,0,1,0,1,0,0,0
Joint Coordinates
1,0.0,0.0
2,0.0,120.0
3,60.0,120.0
Material Data
1,29000.0,0.0
Cross Section Data
1,1.0,0.0,1000.0
Member Data
1,1,2
2,2,3
Support Restraints
1,1,1,1
Joint Loads
3,0.0,0.0,-600.0
```

Figure 9.32 Example Problem 9.2, F2DII input.

```
FATPAK II -- Frame and Truss Package -- Student Edition
(C) Copyright 1995 by Structural Software Systems
Program F2DII -- Plane Frame Analysis

Example Problem 9.2 - Analysis by FATPAK II

Joint Displacements

Joint       X-Tran          Y-Tran          Z-Rot

  1         0.000           0.000           0.00000
  2         0.149          -0.000          -0.00248
  3         0.149          -0.186          -0.00372

Member End Loads

Member   Joint        Sx              Vy              Mz

  1        1        0.000          -0.000           600.00
           2       -0.000           0.000          -600.00
  2        2        0.000           0.000           600.00
           3        0.000          -0.000          -600.00

Reactions

Joint                RX              RY              MZ

  1                 0.000           0.000           600.00
```

Figure 9.33 Example Problem 9.2, F2DII output.

Example Problem 9.2 - Analysis by FATPAK II
Displacement Multiplication Factor = 50

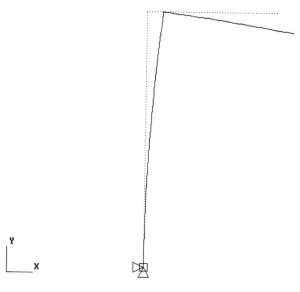

Figure 9.34 Example Problem 9.2, deformed geometry plot from F2DII.

337

```
Example Problem 9.2 - Analysis by FATPAK II
Member 2  -- Shear and Moment Diagrams
```

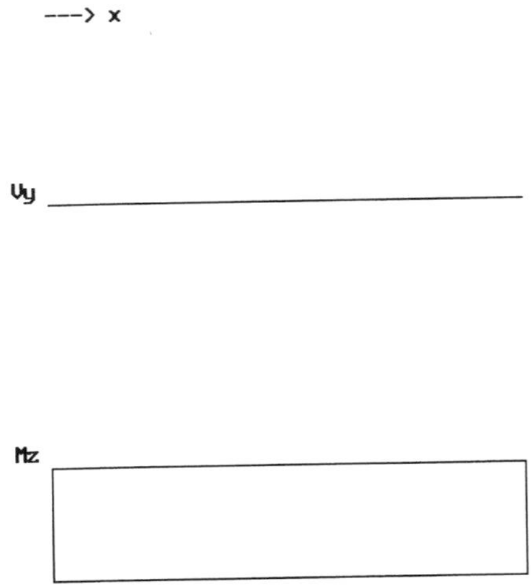

```
---> x                              x        Vy         Mz

                                  0.0L      0.000     -600.00
                                  0.1L      0.000     -600.00
                                  0.2L      0.000     -600.00
                                  0.3L      0.000     -600.00
                                  0.4L      0.000     -600.00
Vy                                0.5L      0.000     -600.00
                                  0.6L      0.000     -600.00
                                  0.7L      0.000     -600.00
                                  0.8L      0.000     -600.00
                                  0.9L      0.000     -600.00
                                  1.0L      0.000     -600.00

                                  Left end is joint 2
                                  Right end is joint 3

Mz
```

Figure 9.35 Example Problem 9.2, V_y and M_z diagram from F2DII.

```
Example Problem 9.9 - Analysis by FATPAK II
6,9,1,1,2,0,0,2,1
Joint Coordinats
1,0.0,0.0
2,192.0,0.0
3,192.0,144.0
4,384.0,0.0
5,384.0,144.0
6,576.0,0.0
Material Data
1,29000.0
Cross Section Data
1,10.0
Member Data
1,1,3
2,1,2
3,2,3
4,3,5
5,2,5
6,2,4
7,4,5
8,5,6
9,4,6
Support Restraints
1,1,1
6,0,1
Initial Member Deformations
5,-0.5
6,0.25
Displaced Support Joints
6,0.0,-0.75
```

Figure 9.36 Example Problem 9.9, T2DII input.

```
FATPAK II -- Frame and Truss Package -- Student  Edition
(C) Copyright 1995 by Structural Software Systems
Program T2DII - Plane Truss Analysis

Example Problem 9.9 - Analysis by FATPAK II

Joint Displacements

Joint        X-Tran          Y-Tran

  1          0.000           0.000
  2         -0.000          -0.083
  3          0.062          -0.083
  4          0.250          -1.000
  5          0.063          -1.000
  6          0.250          -0.750

Member Forces (Tension Positive)

Member       Force

  1          0.000
  2         -0.000
  3          0.000
  4          0.000
  5          0.000
  6         -0.000
  7          0.000
  8         -0.000
  9         -0.000

Reactions

Joint        RX              RY

  1          0.000          -0.000
  6          0.000           0.000
```

Figure 9.37 Example Problem 9.9, T2DII output.

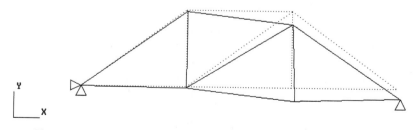

Figure 9.38 Example Problem 9.9, deformed geometry plot from T2DII.

the analysis. The horizontal translation of joint 5 of 0.063 inches agrees with the result obtained in Example Problem 9.9 within expected roundoff error. The computer solution is probably the more accurate of the two results. Figure 9.38 on page 339 shows a plot of the deformed shape of the truss with a displacement multiplication factor of 25.

Deformed Plots of Space Structures

Figure 9.39 shows the deformed plot of an electric transmission tower by the program T3DII in FATPAK II for a set on unbalanced wire tension forces along the transmission line, while Figure 9.40 shows the deformed plot of a two story space frame by the program F3DII in FATPAK II for a horizontal wind load. Plots of this type can be extremely useful in determining how a structure behaves for a particular loading condition. It is often difficult to mentally form a picture of the deformed shape of a structure from the printed joint displacements in the program output. There is an old saying, "A picture is worth a thousand words". For a structural engineer, it can be stated as "A deformed geometry plot is worth a thousand printed joint displacements".

BETTI'S LAW AND MAXWELL'S THEOREM

Castigliano's Theorem and the Principle of Virtual Work are only two of a large number of the applications which can be made of the internal and external work which is performed as a structure deforms under a set of loads. In fact, there are a number of books which are devoted exclusively to this subject. Before we end this chapter, we will consider two more useful relationships which can be developed using this concept. These relationships are known as *Betti's Law* and *Maxwell's Theorem*. Although we will not be using these relationships to compute displacements in trusses and frames, as we did with the previous relationships developed in this chapter, we will find that they will be very useful to us in several of the following chapters.

Betti's Law

Betti's Law is a relationship which shows that the work performed by two totally different load systems on a structure are not completely independent. To develop this relationship, consider a structure subjected to two different load systems W_A and W_B, as shown in Figures 9.41a and 9.41b. Load system W_A consists of a set of concentrated loads $W_{A,1}$ through $W_{A,NWA}$, where NWA is the number of loads in load system W_A, while load system W_B consists of a set of concentrated loads $W_{B,1}$ through $W_{B,NWB}$, where NWB is the number of loads in load system W_B. The loads in the two load systems may act at different points on the structure. Each load system consists of both the active and reactive loads on the struc-

Transmission Tower for Program T3DII
Displacement Multiplication Factor = 50

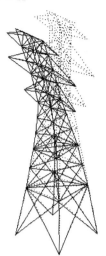

Figure 9.39 Deformed geometry plot from T3DII.

ture. Therefore, each load system will automatically be in equilibrium. The work performed by load system **W$_A$** as the structure deforms will be designated as Work$_A$, while the work performed by load system **W$_B$** will be designated as Work$_B$. We will now consider two different loading cases for the structure.

Case I. Assume that load system **W$_A$** is first applied to the structure and then load system **W$_B$** is applied. The total work performed as the structure deforms under the combined effects of the two load systems is

$$\text{Work}_{AB} = \text{Work}_A + \text{Work}_B + \sum_{i=1}^{NWA} W_{A,i} \Delta_{(A,i),B} \qquad (9.46)$$

where $\Delta_{(A,i),B}$ is the displacement at load $W_{A,i}$ due to load system **W$_B$**. The third term on the right side of the equation corresponds to the work performed by the loads in load system **W$_A$** as they move through the displacements due to load system **W$_B$**. A factor of one half is not present since the loads in load system **W$_A$** are acting during the entire time that the displacements due to load system **W$_B$** are produced.

Case II. Assume that load system **W$_B$** is first applied to the structure and then load system **W$_A$** is applied. The total work performed as the structure deforms under the combined effects of the two load systems is

Two Story Space Frame for Program F3DII
Displacement Multiplication Factor = 150

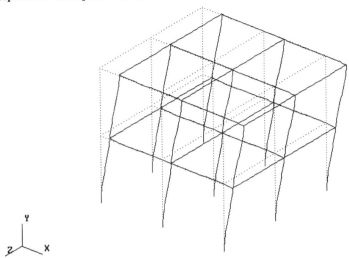

Figure 9.40 Deformed geometry plot from F3DII.

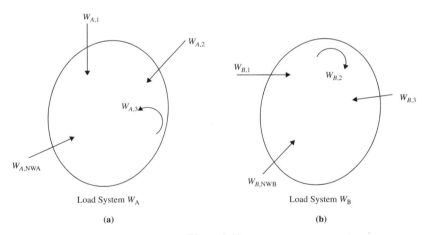

Figure 9.41

$$\text{Work}_{BA} = \text{Work}_B + \text{Work}_A + \sum_{j=1}^{\text{NWB}} W_{B,j}\Delta_{(B,j),\,A} \qquad (9.47)$$

where $\Delta_{(B,j),\,A}$ is the displacement at load $W_{B,j}$ due to the load system $\mathbf{W_A}$. In a similar manner to Eq. (9.46), the third term on the right side of Eq. (9.47) corresponds to the work performed by the loads in load system $\mathbf{W_B}$ as they move through the displacements due to load system $\mathbf{W_A}$.

Betti's Law and Maxwell's Theorem

The total work performed for the two situations must be equal. Therefore, equating Eqs. (9.46) and (9.47) and canceling out like terms on either side gives

$$\sum_{i=1}^{NWA} W_{A,i}\Delta_{(A,i),B} = \sum_{j=1}^{NWB} W_{B,j}\Delta_{(B,j),A} \qquad (9.48)$$

from which Betti's Law can be stated as

> The work performed by a load system W_A, which is in equilibrium, acting on a linear elastic structure, moving through the displacements caused by a load system W_B, which is also in equilibrium, is equal to the work performed by load system W_B moving through the displacements caused by load system W_A.

This relationship can be very useful for several different applications in the analysis of various types of structures. We will now use it to develop Maxwell's Theorem.

Maxwell's Theorem

Maxwell's Theorem shows that the displacements produced by loads at two different points in a structure are not completely independent. To develop this relationship consider a structure subjected to two different load systems W_A and W_B, as shown in Figures 9.42a and 9.42b. Load system W_A consists of a single active concentrated load W at point i and the corresponding reactive loads at the supports, while load system W_B consists of a single active concentrated load W at point j and the associated reactive loads at the supports. The displacement at point j due to load system W_A, in the direction of the force in load system W_B, will be designated as $\Delta_{j,A}$ while the displacement at point i due to load system W_B, in the direction of the force in load system W_A, will be designated $\Delta_{i,B}$. The type of supports for the structure are not important since it may be either statically deter-

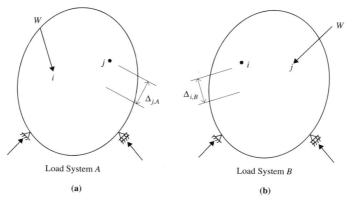

Figure 9.42

minate or statically indeterminate. The only requirements are that the material must be linear elastic and the structure must be stable.

If we now apply Betti's Law, while recognizing that the displacements at the reactive loads must be zero for each load case, we obtain

$$W\Delta_{i,B} = W\Delta_{j,A} \tag{9.49}$$

from which

$$\Delta_{i,B} = \Delta_{j,A} \tag{9.50}$$

This relationship is the basis of Maxwell's Theorem, which can be stated in its general form as

> The displacement at point i, in a linear elastic structure, due to a concentrated load at point j is equal to the displacement at point j due to a concentrated load of the same magnitude at point i.

The displacement at each point will be measured in the direction of the concentrated load at that point. The only other restriction on this statement, in addition to the structure being linear elastic and stable, is that the displacement at either point must be consistent with the type of load at that point. If the load at a point is a concentrated force, then the displacement at that point will be a translation, while if the load is a moment, then the displacement will be a rotation. The displacement at any point will be in the same direction as the load at that point and its positive direction will be in the same direction as the load. This theorem is often referred to as *Maxwell's Reciprocal Displacement Theorem*.

This relationship can be put into a more useful form for use in later chapters by expressing the displacements at points i and j as

$$\Delta_{i,B} = f_{i,j} W \tag{9.51}$$

and

$$\Delta_{j,A} = f_{j,i} W \tag{9.52}$$

where $f_{i,j}$ is the displacement at point i due to a unit load at point j and $f_{j,i}$ is the displacement at point j due to a unit load at point i. If we now substitute these expressions into Eq. (9.50) and cancel out the term W on each side, we obtain

$$f_{i,j} = f_{j,i} \tag{9.53}$$

from which a more specific form of Maxwell's Theorem can be stated as

> The displacement at point i, in an elastic structure, due to a unit load at point j is equal to the displacement at point j due to a unit load at point i.

As in the previous statement for Maxwell's Theorem, the displacement at each point will be measured in the direction of the unit load at that point. The quanti-

Suggested Problems

ties $f_{i,j}$ and $f_{j,i}$ are known as *flexibility coefficients*. They will be very useful to us later in the analysis of statically indeterminate structures.

REFERENCES

AMERICAN INSTITUTE OF STEEL CONSTRUCTION (1986). *Manual of Steel Construction—Allowable Stress Design*. Chicago: AISC.

AMERICAN INSTITUTE OF STEEL CONSTRUCTION (1989). *Manual of Steel Construction—Load and Resistance Factor Design*. Chicago: AISC.

GERE, JAMES M., and STEVEN P. TIMOSHENKO (1972). *Mechanics of Materials*. Brooks/Cole Publishers.

STRUCTURAL SOFTWARE SYSTEMS (1996). FATPAK V-Professional Edition. Monroeville, PA.: Structural Software Systems.

WEAVER, WILLIAM, JR., and JAMES M. GERE (1980). *Matrix Analysis of Framed Structures*. New York: D. Van Nostrand.

YOUNG, WARREN, C. (1989). *Roarke's Formulas for Stress and Strain*. New York: McGraw-Hill.

SUGGESTED PROBLEMS

SP9.1 Compute the horizontal translation at the point of application of the concentrated force for the plane truss shown in Figure SP9.1 by equating the external work to the internal work.

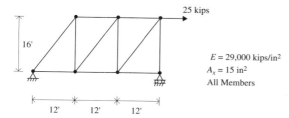

Figure SP9.1

SP9.2 Compute the vertical translation at the center of the beam shown in Figure SP9.2 by equating the external work to the internal work. Include the effects of bending and shear deformation in the beam. Develop the final expression for the translation in a form similar to the final expression in Example Problem 9.3.

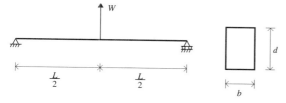

Cross-Section   Figure SP9.2

346   Deflection of Trusses and Frames   Chap. 9

(a) Compute the effect of the shear deformation by using the expression for the shear stress in Eq. (9.16).

(b) Compute the effect of the shear deformation by using the effective shear area A_y of the cross section.

SP9.3 to **SP9.5** Develop an expression for the vertical translation at point a for each of the beams shown by Figures SP9.3 to SP9.5 by using Castigliano's Second Theorem.

(a) Perform the differentiation after the integration.

(b) Perform the differentiation before the integration.

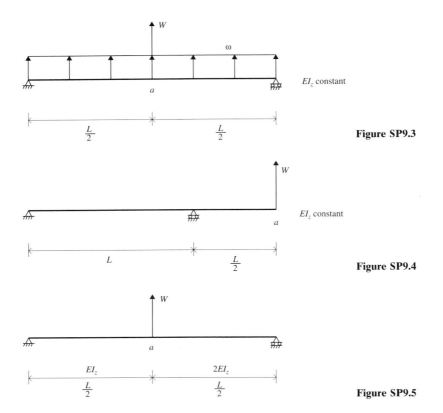

Figure SP9.3

Figure SP9.4

Figure SP9.5

SP9.6 and **SP9.7** Compute the horizontal and vertical translations of point a for the plane trusses shown in Figures SP9.6 and SP9.7 using the Principle of Virtual Work.

SP9.8 Compute the vertical and horizontal translations of joint 4 for the plane truss whose mathematical model is shown in Figure SP9.8. Use the Principle of Virtual Work.

(a) Due to a 40-kip downward force at joint 2 and a 60-kip downward force at joint 5.

(b) Due to a temperature change which causes member 2 to elongate 0.5 inches and member 6 to shorten 0.25 inches.

(c) Due to vertical translations of 0.1 inches down at joint 1 and 0.25 inches down at joint 3.

Suggested Problems

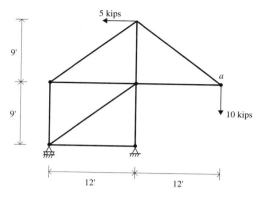

$E = 29{,}000$ kips/in^2
$A_x = 8$ in^2
All Members

Figure SP9.6

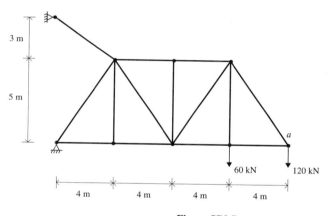

$E = 200{,}000$ MPa
$A_x = 6500$ mm^2
All Members

Figure SP9.7

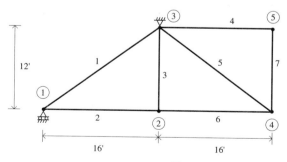

$E = 29{,}000$ kips/in^2
$A_x = 5$ in^2
All Members

Figure SP9.8

SP9.9 Compute the vertical translation of point *a* for the statically indeterminate plane truss shown in Figure SP9.9 using the Principle of Virtual Work. An analysis by another engineer has shown that the vertical reactive force at the right support is 21.861 kips up.

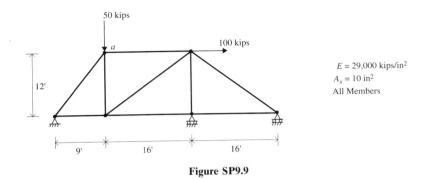

Figure SP9.9

SP9.10 to SP9.12 Compute the vertical translation and the rotation at point *a* for the beams shown in Figures SP9.10 to SP9.12 using the Principle of Virtual Work. Treat the quantity EI_z as a symbol during the computations.

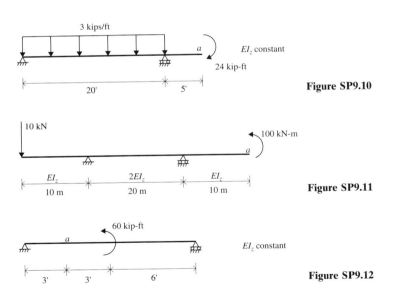

Figure SP9.10

Figure SP9.11

Figure SP9.12

SP9.13 Compute the horizontal translation of point *a* in the plane frame shown in Figure SP9.13 using the Principle of Virtual Work. Consider both axial and bending deformations in the members.

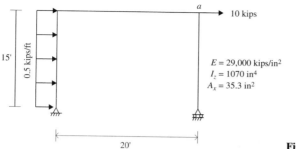

Figure SP9.13

SP9.14 Compute the vertical translation at point a for the beam shown in Figure SP9.14 using the Principle of Virtual Work. Consider the effects of both shear and bending deformations.

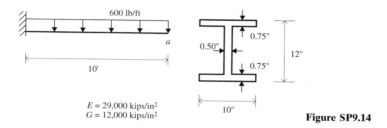

Figure SP9.14

SP9.15 Compute the vertical translation and the rotation at point a in the statically indeterminate beam shown in Figure SP9.15 using the Principle of Virtual Work. Include only bending deformation and treat the quantity EI_z as a symbol during the computations. An analysis by another engineer has shown that the vertical reactive force at the right support is 3.636 kN up.

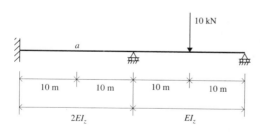

Figure SP9.15

10

The Müller–Breslau Principle

DEVELOPMENT OF THE MÜLLER–BRESLAU PRINCIPLE

In 1886, Henrich Müller-Breslau developed a very powerful and simple procedure for generating influence lines for elastic structures. The procedure is called the *Müller–Breslau Principle*. By using this procedure, a structural engineer can easily sketch the approximate shape of the influence line for any force or moment quantity for essentially any structure by drawing the deformed shape of the structure for a particular imposed deformation. For many structures, the ordinates of the influence line can be easily computed using only simple geometric relationships without performing any type of equilibrium calculations. The basic equation for the Müller-Breslau Principle can be derived from simple work–energy relationships for an elastic structure using Betti's Law, which was developed in the previous chapter.

Consider a linear elastic stable structure which is subjected to two load systems W_A and W_B. For load system W_A, assume that a unit concentrated force acts at point i in an arbitrary direction and that the restraint against deformation at point j is removed and is replaced by a load r_j, which corresponds to the action of the restraint at j, as shown in Figure 10.1a. The structure is shown in generic form in this figure since it may consist of any type of statically determinate or statically indeterminate truss or frame. The magnitude of the load r_j will be equal to the load which would be applied to the structure by the restraint due to the unit force at point i. If the restraint at point j resists translation, then r_j will be a force, and if the restraint resists rotation, then r_j will be a moment. The displacement corresponding to the restraint at point j will be zero for this load system since the load

Development of the Müller–Breslau Principle

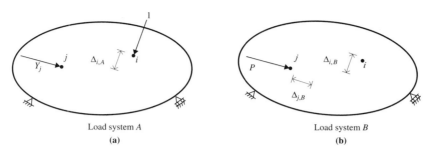

Figure 10.1

r_j has the same effect on the structure displacements as the restraint. The translation at point i, in the direction of the unit force, due to this load system will be designated as $\Delta_{i,A}$.

For the load system $\mathbf{W_B}$, assume that the restraint at point j is removed and is replaced by a load, corresponding to the action of the restraint, of arbitrary magnitude P. This load will be either a force or a moment to be consistent with the restraint. There is no load at point i for this load system. The translation at point i, in the direction of the unit force at that point in load system $\mathbf{W_A}$, due to load system $\mathbf{W_B}$ will be designated as $\Delta_{i,B}$ and the displacement at point j, which will be either a translation or a rotation to correspond to the action of the removed restraint, will be $\Delta_{j,B}$.

The work performed by the loads in load system $\mathbf{W_A}$ moving through the displacements due to load system $\mathbf{W_B}$ will be

$$\text{Work}_{A,B} = (1)(\Delta_{i,B}) + (r_j)(\Delta_{j,B}) \tag{10.1}$$

while the work performed by the loads in load system $\mathbf{W_B}$ moving through the displacements due to load system $\mathbf{W_A}$ will be zero since the load at point i in load system $\mathbf{W_B}$ is zero and the displacement at point j due to load system $\mathbf{W_A}$ is zero

$$\text{Work}_{B,A} = (0)(\Delta_{i,A}) + (P)(0) \tag{10.2}$$

According to Betti's Law, $\text{Work}_{A,B}$ must be equal to $\text{Work}_{B,A}$ Therefore,

$$(1)(\Delta_{i,B}) + (r_j)(\Delta_{j,B}) = 0 \tag{10.3}$$

from which

$$r_j = -\frac{\Delta_{i,B}}{\Delta_{j,B}} \tag{10.4}$$

Since the magnitude of the load P in load system $\mathbf{W_B}$ is arbitrary, it will be chosen such that the displacement $\Delta_{j,B}$ has a unit magnitude, for which r_j will become

$$r_j = -\Delta_{i,B} \tag{10.5}$$

The physical interpretation of this relationship is that the load corresponding to the restraint at point j in the structure, due to a unit force at point i, is equal to the translation at point *i*, in the opposite direction of the unit force, which would be produced if the restraint were removed at point *j* and a corresponding unit deformation was introduced into the structure. The translation at point *i* is measured in the opposite direction of the unit force at point *i* due to the negative sign on the right side of Eq. (10.5). Since the influence line ordinate at point *i* for r_j is equal to the value of r_j due to a unit force at point *i*, the previous procedure represents a process for determining the influence line ordinates at any point on a structure for any force or moment quantity. The specific procedure is known as the Müller–Breslau Principle and can be stated as

> The ordinate of the influence line at any point on a structure, for any force or moment quantity, is equal to the translation of that point, opposite to the direction of a positive concentrated force at the point, which is obtained by removing the restraint in the structure corresponding to the action of the force or moment quantity and introducing a corresponding unit deformation.

The structure with the removed restraint will be known as the *Müller–Breslau structure* throughout the remainder of this chapter. The beam shown in Figure 10.2a can be used to demonstrate the use of this procedure by considering the following three cases.

Influence Line for Reactive Force for a Beam

The influence line ordinate at any point *i* on the beam, for the vertical reaction force at the right support, will be equal to the vertical translation Δ_i which results from removing the vertical translation restraint at the right support and introduc-

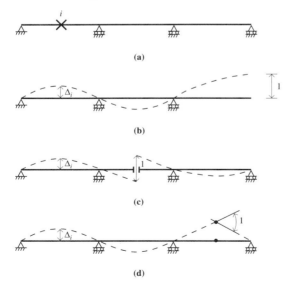

Figure 10.2

Development of the Müller–Breslau Principle

ing a vertical unit translation, as shown in Figure 10.2b. The Müller–Breslau structure for this case will only have three supports since the single vertical restraint for the right support has been removed. The unit translation is up at the right end of the beam to correspond to the positive direction of the vertical reactive force. The point where the vertical translation restraint has been removed is free to rotate as it moves upward since there is no moment restraint at that point.

Influence Line for Shear Force for a Beam

The influence line ordinate at any point i on the beam, for the shear force at a point in the center span, will be equal to the vertical translation Δ_i which results from removing the vertical shear deformation restraint at the point in the center span and introducing a unit vertical translation between the two ends of the beam on either side of the point, as shown in Figure 10.2c. The Müller–Breslau structure for this case will have a vertical slide at the point where the shear restraint was removed, which permits relative vertical translation of the two ends on either side of the slide but prevents relative rotation.. The left end moves down and the right end moves up to correspond to the action of a positive set of shear forces at the point. The slopes of the two ends on either side of the vertical slide will be equal in the deformed beam since relative rotation is still restrained .

Influence Line for Bending Moment for a Beam

The influence line ordinate at any point i on the beam, for the bending moment at a point in the right span, will be equal to the vertical translation Δ_i which results from removing the rotation restraint at the point in the right span and introducing a unit kink angle in the beam, as shown in Figure 10.2d. The Müller–Breslau structure for this case will have a pin at the point where the rotation restraint was removed which permits relative rotation of the two ends on either side of the pin but prevents relative translation. The end to the left of the pin rotates counterclockwise and the end to the right rotates clockwise to correspond to the action of a set of positive bending moments at the point. The point is free to translate in the vertical direction, but both ends must translate together.

For each of the cases shown in Figures 10.2b through 10.2d the translation at point i is up. This corresponds to a positive influence line ordinate according to Eq. (10.5) since the standard practice is to consider a downward force to be positive for an influence line, as described in previous chapters. The deflected shape of the beams, as represented by the dashed line in each of these figures, corresponds to the shape of the influence lines for the specified quantities.

Influence Line for Bending Moment in a Frame

It is usually possible to draw a fairly accurate shape for an influence line using the Müller-Breslau Principle, even though the exact values for the ordinates cannot be

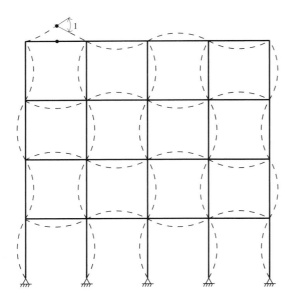

Figure 10.3

determined for some structures without a considerable amount of computation. In many cases, the shape of the deformed Müller–Breslau structure is all that is needed to give a structural engineer the information which is needed to perform a specific type of analysis. For example, consider the four story by four bay plane frame shown by the solid lines in Figure 10.3. The problem facing the design engineer is how to locate the uniformly distributed downward live load forces on the beams in the frame to produce the maximum positive bending moment at the center of the top left beam in the frame. Should all of the beams be loaded over their entire lengths with the live load, or would a larger moment be produced if only selected portions of the frame were loaded? This problem can be easily answered by using the Müller–Breslau Principle. The dashed lines in Figure 10.3 correspond to the deformed shape of the frame if the moment restraint is removed and a positive kink angle is introduced at the center of the top left beam. According to the Müller–Breslau Principle, the translation of any point in the structure from its original position corresponds to the influence of a concentrated force at that point on the bending moment at the point where the kink angle is introduced.

Even though exact numerical values for the bending moment, due to the loads on the structure, cannot be determined from this simple sketch, the deformed shape does give the required information to determine how the structure should be loaded. The vertical translation of any location along any horizontal line of beams represents the contribution of a downward force at that location to the desired bending moment. Therefore, only those points on the beams with upward translations should be loaded with the downward live load to obtain the maximum positive bending moment. If the maximum negative bending moment were desired, then only those points on the beams with downward translations should be loaded.

Influence Lines for Statically Determinate Beams

This type of loading is usually called a *checkerboard loading* due to its similarity to the alternating colored squares on a checkerboard. This statically indeterminate frame now could be analyzed for this loading by the program F2DII in FATPAK II - Student Edition to determine the desired bending moment.

The horizontal translations at any locations along the vertical column lines of the deformed structure correspond to the contribution of a horizontal concentrated force at that location to the bending moment at the center of the top left beam. These values could be used to determine the effect of horizontal loads, such as wind or seismic loads, on the structure.

INFLUENCE LINES FOR STATICALLY DETERMINATE BEAMS

The Müller–Breslau Principle can greatly simplify the generation of influence lines for reactions, shear forces and bending moments in statically determinate beams, compared to the equilibrium analysis procedures discussed previously in Chapters 2 and 5. Since a statically determinate beam has exactly the number of internal and external restraints which are required for stability, the removal of any restraint will result in an unstable Müller–Breslau structure. The result of this is that the deformations which are introduced into the Müller–Breslau structure will not produce any stresses and the individual members in the deformed mathematical model of the beam will remain straight. The vertical translations of all points along the deformed beam can be computed by simple geometric relationships using the known unit deformation at the point where the restraint was removed.

Example Problem 10.1

Generate the influence lines for the two vertical reactive forces and for the shear force and the bending moment at point i by the Müller–Breslau Principle for the beam whose mathematical model is shown in Figure 10.4a. The deformed shapes of the beam, due to unit vertical translations at the points where the restraints corresponding to the reactive forces R_{2Y} and R_{3Y} were removed, are shown in Figures 10.4b and 10.4c. These deformed shapes should be exactly the same shapes as the influence lines for R_{2Y} and R_{3Y} and the vertical translations at any points should be equal to the influence line ordinates. This can be verified by comparing these deformed shapes to the influence lines in Figure 2.36, which were developed for this same beam in Example Problem 2.6 in Chapter 2. Since the beam remains straight while it deforms, the vertical translations of any points along the beam can be easily determined by the geometric relationships for similar triangles. The units of the vertical translations correspond to meters of vertical translation per meter of imposed support deformation (i.e.; m/m). This agrees with the units assigned previously in Chapter 2, where the reactive force influence lines were considered to be dimensionless.

Figures 10.5a and 10.5b show the deformed shapes of the beam for a unit shear deformation and a unit moment deformation at point i. It is left as an exercise for the reader to verify that the vertical translations at any point on these deformed beams are equal to the influence line ordinates for the vertical shear force V_{iy} and the bending moment M_{iz}. The units of the ordinates of the influence line for V_{iy} will be the

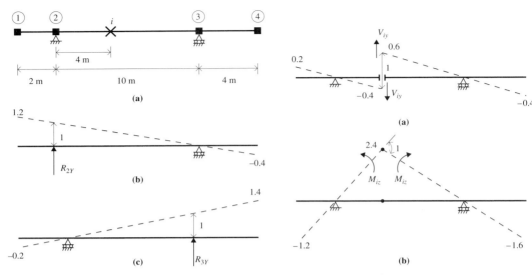

Figure 10.4 Example Problem 10.1.

Figure 10.5 Example Problem 10.1.

same as those for the influence lines for R_{2Y} and R_{3Y}. The units of the translations of the deformed beam due to the kink angle at point i are meters per radian, which, since radians are considered to be dimensionless, translates to meters. This agrees with the units of the ordinates of the influence line for bending moment of kN-m/kN, since the kN unit can be canceled out.

Example Problem 10.2

Generate the influence lines for the three vertical reactive forces and the shear forces to the left and right of the interior support by the Müller–Breslau Principle for the beam whose mathematical model is shown in Figure 10.6a. Figures 10.6b, 10.6c and 10.6d show the deformed shapes of the beam for unit vertical translations corresponding to the actions of each of the vertical reactive forces. These deformed shapes agree with the influence lines for R_{1Y}, R_{2Y} and R_{4Y} shown in Figure 2.37, which were computed for this same beam in Example Problem 2.7 in Chapter 2.

Figures 10.7a and 10.7b show the deformed shapes of the beam for unit vertical shear deformations to the left and right, respectively, of the center support. Note that the slopes of the segments of the deformed beams are equal on either side of the vertical step as required. These deformed shapes agree with the influence lines for these shear forces shown in Figures 5.25 and 5.26 in Chapter 5, which were generated for this beam by the program BEAMIL.

Example Problem 10.3

Generate the influence lines for the shear force and the bending moment at point i by the Müller–Breslau Principle for the bridge girder shown in Figure 10.8. The loads for this bridge run across a system of simply supported stringers which are supported on the girder at their ends.

Influence Lines for Statically Determinate Beams

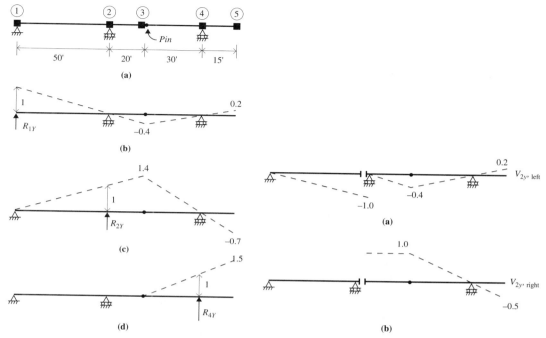

Figure 10.6 Example Problem 10.2.

Figure 10.7 Example Problem 10.2.

The path which the loads take as they move along a structure is known as the *load line* for the structure. When the Müller–Breslau Principle is used to generate the influence line, the translations which correspond to the ordinates of the influence line will be the translations of the various points on the load line. For this particular structure, these ordinates will correspond to the vertical translations of the stringer system as the girder is deformed rather than the vertical translations of the girder itself. Therefore, the generation of the influence lines for the shear force V_{iy} and bending moment M_{iz} in the girder must be performed in two steps:

Step 1. The first step is to generate the deformed shapes of the girder due to a unit vertical shear deformation and a unit kink angle at point *i*, as shown by the dashed lines in Figures 10.9a and 10.9b. These shapes correspond to the shapes of the influence lines if the loads were acting directly on the girder.

Step 2. The second step is to generate the shapes of the stringer system as it deforms due to the displacements of the points where it is attached to the gird-

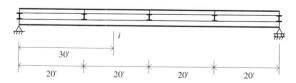

Figure 10.8 Example Problem 10.3.

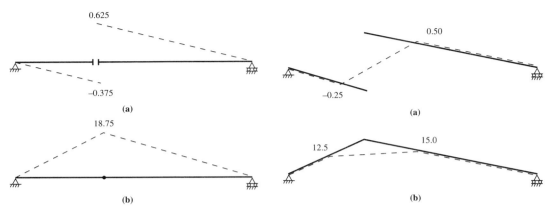

Figure 10.9 Example Problem 10.3.

Figure 10.10 Example Problem 10.3.

er, as shown by the dashed lines in Figures 10.10a and 10.10b. The vertical translations at any point along the stringer system are equal to the ordinates of the desired influence lines.

We can see from this structure that the shapes of the influence lines are highly dependent upon how the loads are transferred from the points where they are applied to the structure to the point where the force or moment quantity corresponding to the influence line occurs. The influence line for the shear force in the girder at point i has a linear transition from a negative ordinate of -0.25 kip/kip to a positive ordinate of 0.50 kip/kip as the unit vertical force moves along the second stringer, rather than the abrupt change which would occur if the force was applied directly to the girder. A similar linear change occurs in the bending moment influence line from an ordinate of 12.5-kip ft/kip to an ordinate of 15.0 kip-ft/kip. The maximum magnitudes of both the shear force and bending moment at point i are reduced compared to the magnitudes which would exist without the stringers.

This same process can be used to generate the influence lines for any force or moment quantity in a statically determinate beam. The only restriction on how fast the influence lines can be drawn depends upon the drafting skills of the engineer. The influence line ordinates always can be computed using simple high school level geometry principles which will usually result in fewer calculations than would be required if an equilibrium analysis were performed.

INFLUENCE LINES FOR STATICALLY INDETERMINATE BEAMS

The Müller–Breslau Principle also can be used to develop influence lines for statically indeterminate beams. The primary difference between the application of this procedure to this type of beam compared to a statically determinate beam is that the Müller–Breslau structure will still be stable after the restraint correspond-

Influence Lines for Statically Indeterminate Beams

ing to the quantity for which the influence line is being developed is removed. Therefore, the Müller–Breslau structure will resist the imposed deformations, which will cause the members in the mathematical model of the beam to deform. The influence lines will no longer be composed of straight line segments, as illustrated previously by the sketches for the influence lines for the statically indeterminate beam in Figure 10.2.

It is a fairly simple task to sketch the shape of the influence line for any force or moment quantity for a statically indeterminate beam. However, since the influence lines will no longer be composed of straight line segments, the computation of the numerical values for the ordinates will be much more difficult than for a statically determinate beam. If the beam is only one degree statically indeterminate, the Müller–Breslau structure will be statically determinate. However, if the beam is two or more degrees statically indeterminate, the Müller–Breslau structure will also be statically indeterminate, which will greatly complicate the analysis process.

Example Problem 10.4

Compute the ordinate of the influence line at point i for the vertical reactive force at the right support by the Müller–Breslau Principle for the prismatic beam whose mathematical model is shown in Figure 10.11a. According to the Müller–Breslau Principle, the ordinate of the influence line for the vertical reactive force R_{2Y}, at any point along the beam, is equal to the vertical translation at that point due to a unit vertical translation at the right end of the beam. Therefore, we must develop a procedure for computing the vertical translation at any point along the beam due to a unit vertical translation at the right end. The simplest process to use is an indirect approach rather than using a direct approach in which a known displacement is introduced into the beam.

For this indirect approach, we will consider the beam, with the vertical translation restraint removed at the right support, subjected to a vertical force P, as shown in Figure 10.11b. The vertical translations at point i and at the point of application of the force P are Δ_i and Δ_2, respectively. The translation Δ_i would be exactly equal to the ordinate of the influence line for R_{2Y} if the translation Δ_2 had a unit value. However, since the magnitude of the force P which is required to produce a unit trans-

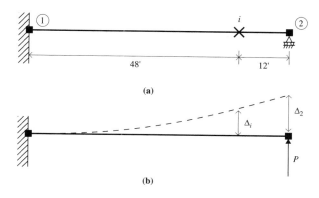

Figure 10.11 Example Problem 10.4.

lation is not readily known, the effect of a unit translation can be achieved by dividing all translations along the beam by the magnitude of Δ_2, which is caused by an arbitrary magnitude of the force P. Since the translations are directly proportional to the force P, the force P will cancel out during the division. The result of this calculation at any point i along the beam will be the desired influence line ordinate at that point.

Since the original beam is one degree statically indeterminate, the beam with the removed restraint is statically determinate. The bending moments along the beam due to the force P can be easily determined and the translations Δ_i and Δ_2 can be computed by any of the procedures discussed previously in Chapters 8 and 9. Fortunately, the computations for Δ_i and Δ_2 can be greatly simplified for this particular beam since an algebraic expression was derived in Example Problem 8.3 in Chapter 8 for the displacement Y at any point on a prismatic cantilever beam subjected to a positive concentrated force at the free end. Substituting the appropriate values for X and L for this beam in feet units into the expression gives

$$\Delta_i = \frac{50{,}688P}{EI_z}$$

and

$$\Delta_2 = \frac{72{,}000P}{EI_z}$$

from which the ordinate of the influence line is

$$\frac{\Delta_i}{\Delta_2} = 0.704$$

This solution can be verified by the program BEAMIL, which can generate influence lines for both statically determinate and statically indeterminate beams. Figure 10.12 shows a listing of the input data in kip and feet units for the program while Figure 10.13 shows the plotted influence line for the vertical reactive force at the right support. Nine interior calculation points were specified for the analysis in order to obtain a realistic plot of the curved influence line since the program plots straight line segments between these interior points. This point spacing also placed an interior calculation point at point i, so that the influence line ordinate would be listed in the output for that point. The listed influence line ordinate at the point 48 feet from the left end of the beam agrees with the manual solution.

This same procedure can be used for any statically indeterminate beam. Of course, if the beam is two degrees or more statically indeterminate, the Müller-Bres-

```
Example Problem 10.4 - Analysis by Program BEAMIL
2,1,1,0,2
Joint Coordinates
1,0.0
2,60.0
Support Restraints
1,1,1,1
2,0,1,0
```

Figure 10.12 Example Problem 10.4, BEAMIL input.

Influence Lines for Statically Indeterminate Beams

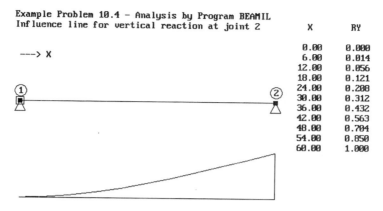

Figure 10.13 Example Problem 10.4, BEAMIL output.

lau beam will be statically indeterminate and the computation of the displacements will become much more difficult.

Example Problem 10.5

Compute the influence line ordinates at point i for the vertical reactive force and the reactive moment at the left support by the Müller–Breslau Principle for the prismatic beam shown in Figure 10.14a. As in the previous example problem, the shape of the influence line for either reactive quantity will be exactly equal to the deformed shape of the beam which would result from removing the restraint corresponding to the reactive quantity and introducing a corresponding unit deformation. This deformation will consist of an upward translation at the left end for the influence line for the reactive force, while it will be a counterclockwise rotation at the left end for the reactive moment, as shown in Figures 10.14b and 10.14c, respectively. Unfortunately, we now have the problem that we cannot compute the displacement at any point

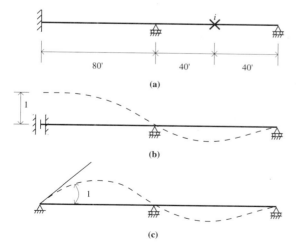

Figure 10.14 Example Problem 10.5.

```
Example Problem 10.5 - Influence Line for RLY - Analysis by FATPAK II
9,8,1,1,0,3,0,0,0,0,1
Joint Coordinates
1,0.0,0.0
2,240.0,0.0
3,480.0,0.0
4,720.0,0.0
5,960.0,0.0
6,1200.0,0.0
7,1440.0,0.0
8,1680.0,0.0
9,1920.0,0.0
Material Data
1,29000.0
Cross Section Data
1,10.0,0.0,1000.0
Member Data
1,1,2
2,2,3
3,3,4
4,4,5
5,5,6
6,6,7
7,7,8
8,8,9
Support Restraints
1,1,1,1
5,0,1,0
9,0,1,0
Displaced Support Joints
1,0.0,1.0,0.0
```

Figure 10.15 Example Problem 10.5, F2DII input.

along this statically indeterminate beam since we cannot compute the bending moments due to the support displacements for use in any of the previously described procedures for computing beam displacements. The analysis of statically indeterminate beams will not be discussed until Chapter 11. However, we do have a tool available to us, in the form of the structural analysis package FATPAK II, which can be used to perform a direct analysis for the influence line ordinates. The programs in FATPAK II have the capability to analyze a statically determinate or statically indeterminate structure for any combination of specified displacements at the supports. The output from the program will consist of the displacements at the joints of the mathematical model of the structure, the member end loads and the reactive loads. For the particular analysis which we will be performing here we will only be interested in the joint translations.

Figure 10.15 shows a listing of the input data in kip and inch units for the plane frame analysis program F2DII in FATPAK II for the generation of the influence line for the vertical reactive force by introducing a unit vertical translation at the left support of the beam. Figure 10.16 shows a plot of the mathematical model of the beam

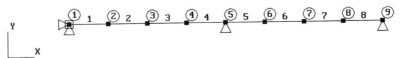

Figure 10.16 Example Problem 10.5, geometry plot from F2DII.

Influence Lines for Statically Indeterminate Beams

which consists of nine joints and eight members. The intermediate joints and members have been located between the supports since the program only computes the displacements at the joints in the mathematical model. A good definition of the influence line should be obtained with this number of computed points. Note that joint 7 in the mathematical model is located at the point i for the beam. Arbitrary values of 29,000 kip/inch^2 for E and 1000 inch^4 for I_z were specified. Figure 10.17 shows a listing of the results of the analysis and Figure 10.18 shows a plot of the deformed shape of the mathematical model with a displacement multiplication factor of 250. The output shows a vertical translation of -0.161 for joint 7, which corresponds to an influence line ordinate at point i for the left reactive force of -0.161 kip/kip. The negative sign means that the reactive force will be down for a downward concentrated force at point i.

This solution can be verified by performing a separate analysis of the type performed in Example Problem 10.4. The Müller-Breslau structure for this analysis will be a beam with the vertical translation restraint removed at the left support. Figure 10.19 shows this Müller–Breslau structure, with a vertical slide at the left end, subjected to a vertical force corresponding to the action of the left reactive force. Figure 10.20 shows a listing of the input data in kip and inch units for F2DII for computing the vertical translations in this beam. The magnitude of the vertical force of 10 kips was chosen in order to obtain displacements with a sufficient number of significant figures of accuracy when printed to the three decimal place output used in F2DII. Of course, if desired, any magnitude of the load could be used by merely printing the output in exponential format, which is one of the program options. The output from F2DII for this analysis is shown in Figure 10.21. By using the procedure in the previous example problem, the influence line ordinate at joint 7 in the mathematical model, for the vertical reactive force at joint 1, can be obtained by dividing the vertical translation at joint 7 by the vertical translation at joint 1:

$$\frac{\Delta_7}{\Delta_1} = \frac{-7.150}{44.491} = -0.161$$

which agrees with the solution obtained in the previous analysis.

The influence line for the left reactive moment can be generated directly by F2DII by using the input data listed in Figure 10.22 in which a unit rotation is introduced at the left support of the beam. The mathematical model for the beam is identical to that shown in Figure 10.16 for the analysis for the vertical reactive force. Figure 10.23 shows the output from F2DII while Figure 10.24 shows a plot of the deformed shape of the mathematical model. This deformed shape represents the desired influence line. The vertical translation of -51.429 inches at joint 7 corresponds to an influence line ordinate of -51.429 kip inch/kip at point i in the beam. The verification of this solution by the procedure used to verify the influence line ordinate for the vertical reactive force is left as an exercise for the reader.

This same process can be used to develop influence lines for the reactive forces and moments for any statically indeterminate beam. A similar process can be used to develop influence lines for shear forces and bending moments by intro-

```
FATPAK II -- Frame and Truss Package -- Student Edition
(C) Copyright 1995 by Structural Software Systems
Program F2DII -- Plane Frame Analysis


Example Problem 10.5 - Influence Line for RLY - Analysis by FATPA


Joint Displacements

Joint       X-Tran          Y-Tran          Z-Rot

  1         0.000           1.000           0.00000
  2         0.000           0.884          -0.00089
  3         0.000           0.607          -0.00134
  4         0.000           0.277          -0.00134
  5         0.000           0.000          -0.00089
  6         0.000          -0.141          -0.00031
  7         0.000          -0.161           0.00011
  8         0.000          -0.100           0.00036
  9         0.000           0.000           0.00045


Member End Loads

Member  Joint        Sx              Vy              Mz

  1       1         0.000           0.225           134.86
          2         0.000          -0.225           -80.92
  2       2         0.000           0.225            80.92
          3         0.000          -0.225           -26.97
  3       3         0.000           0.225            26.97
          4         0.000          -0.225            26.97
  4       4         0.000           0.225           -26.97
          5         0.000          -0.225            80.92
  5       5         0.000          -0.084           -80.92
          6         0.000           0.084            60.69
  6       6         0.000          -0.084           -60.69
          7         0.000           0.084            40.46
  7       7         0.000          -0.084           -40.46
          8         0.000           0.084            20.23
  8       8         0.000          -0.084           -20.23
          9         0.000           0.084             0.00


Reactions

Joint            RX              RY              MZ

  1             0.000           0.225           134.86
  5             0.000          -0.309             0.00
  9             0.000           0.084             0.00
```

Figure 10.17 Example Problem 10.5, F2DII output.

Influence Lines for Statically Indeterminate Beams

```
Example Problem 10.5 - Influence Line for RLY - Analysis by FATPAK II
Displacement Multiplication Factor = 250
```

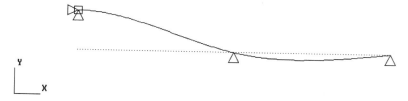

Figure 10.18 Example Problem 10.5, deformed geometry plot from F2DII.

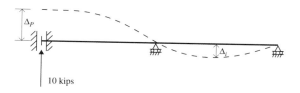

Figure 10.19 Example Problem 10.5.

```
Example Problem 10.5 - Deformed MB Beam for RLY - Analysis by FATPAK II
9,8,1,1,0,3,0,1,0,0,0
Joint Coordinates
1,0.0,0.0
2,240.0,0.0
3,480.0,0.0
4,720.0,0.0
5,960.0,0.0
6,1200.0,0.0
7,1440.0,0.0
8,1680.0,0.0
9,1920.0,0.0
Material Data
1,29000.0
Cross Section Data
1,10.0,0.0,1000.0
Member Data
1,1,2
2,2,3
3,3,4
4,4,5
5,5,6
6,6,7
7,7,8
8,8,9
Support Restraints
1,1,0,1
5,0,1,0
9,0,1,0
Joint Loads
1,0.0,10.0,0.0
```

Figure 10.20 Example Problem 10.5, F2DII input.

```
FATPAK II -- Frame and Truss Package -- Student Edition
(C) Copyright 1995 by Structural Software Systems
Program F2DII -- Plane Frame Analysis


Example Problem 10.5 - Deformed MB Beam for RLY - Analysis by FAT

Joint Displacements

Joint       X-Tran          Y-Tran          Z-Rot

  1          0.000          44.491          0.00000
  2          0.000          39.327         -0.03972
  3          0.000          27.013         -0.05959
  4          0.000          12.315         -0.05959
  5          0.000           0.000         -0.03972
  6          0.000          -6.257         -0.01366
  7          0.000          -7.150          0.00497
  8          0.000          -4.469          0.01614
  9          0.000           0.000          0.01986


Member End Loads

Member   Joint           Sx              Vy              Mz

  1        1             0.000          10.000         6000.01
           2             0.000         -10.000        -3600.02
  2        2             0.000          10.000         3600.02
           3             0.000         -10.000        -1200.00
  3        3             0.000          10.000         1200.00
           4             0.000         -10.000         1200.00
  4        4             0.000          10.000        -1200.00
           5             0.000         -10.000         3600.01
  5        5             0.000          -3.750        -3600.01
           6             0.000           3.750         2700.01
  6        6             0.000          -3.750        -2700.01
           7             0.000           3.750         1800.01
  7        7             0.000          -3.750        -1800.01
           8             0.000           3.750          900.00
  8        8             0.000          -3.750         -900.01
           9             0.000           3.750            0.00


Reactions

Joint            RX              RY              MZ

  1             0.000           0.000         6000.01
  5             0.000         -13.750            0.00
  9             0.000           3.750            0.00
```

Figure 10.21 Example Problem 10.5, F2DII output.

Influence Lines for Trusses

```
Example Problem 10.5 - Influence Line for MLZ - Analysis by FATPAK II
9,8,1,1,0,3,0,0,0,0,1
Joint Coordinates
1,0.0,0.0
2,240.0,0.0
3,480.0,0.0
4,720.0,0.0
5,960.0,0.0
6,1200.0,0.0
7,1440.0,0.0
8,1680.0,0.0
9,1920.0,0.0
Material Data
1,29000.0
Cross Section Data
1,10.0,0.0,1000.0
Member Data
1,1,2
2,2,3
3,3,4
4,4,5
5,5,6
6,6,7
7,7,8
8,8,9
Support Restraints
1,1,1,1
5,0,1,0
9,0,1,0
Displaced Support Joints
1,0.0,0.0,1.0
```

Figure 10.22 Example Problem 10.5, F2DII input.

ducing either a vertical shear deformation or a kink angle in the beam at the desired point. Unfortunately, the program F2DII in FATPAK II does not have the capability to directly introduce these types of deformations in the interior of a beam. It can only impose support displacements and axial deformations in a member. There might be other commercial programs which have this capability.

INFLUENCE LINES FOR TRUSSES

The Müller–Breslau Principle also can be used to develop influence lines for both the reactive forces at the support joints and the axial forces in the members for trusses. The influence lines for the reactive forces can be generated by the same procedures demonstrated in Example Problems 10.4 and 10.5. The program T2DII in FATPAK II can be used to perform the computations for both statically determinate and statically indeterminate plane trusses. The determination of the influence line ordinates for the axial force in any member in a truss requires a somewhat different approach in which an axial deformation is introduced into a member.

The restraint which corresponds to the axial force in a member in a truss prevents the two cross sections on either side of any point on the member from translating with respect to each other along the longitudinal axis of the member. Therefore, if this restraint were removed and a corresponding deformation were

```
FATPAK II -- Frame and Truss Package -- Student Edition
(C) Copyright 1995 by Structural Software Systems
Program F2DII -- Plane Frame Analysis


Example Problem 10.5 - Influence Line for MLZ - Analysis by FATPA


Joint Displacements

Joint       X-Tran          Y-Tran          Z-Rot

  1         0.000           0.000           1.00000
  2         0.000         147.857           0.27679
  3         0.000         154.286          -0.17857
  4         0.000          83.572          -0.36607
  5         0.000           0.000          -0.28571
  6         0.000         -45.000          -0.09821
  7         0.000         -51.429           0.03571
  8         0.000         -32.143           0.11607
  9         0.000           0.000           0.14286


Member End Loads

Member  Joint        Sx              Vy              Mz

  1       1         0.000         134.859         103571.34
          2         0.000        -134.859         -71205.38
  2       2         0.000         134.859          71205.38
          3         0.000        -134.859         -38839.33
  3       3         0.000         134.859          38839.34
          4         0.000        -134.859          -6473.24
  4       4         0.000         134.859           6473.24
          5         0.000        -134.859          25892.85
  5       5         0.000         -26.972         -25892.86
          6         0.000          26.972          19419.67
  6       6         0.000         -26.972         -19419.68
          7         0.000          26.972          12946.46
  7       7         0.000         -26.972         -12946.47
          8         0.000          26.972           6473.23
  8       8         0.000         -26.972          -6473.23
          9         0.000          26.972              0.00


Reactions

Joint              RX              RY              MZ

  1               0.000         134.859         103571.34
  5               0.000        -161.830              0.00
  9               0.000          26.972              0.00
```

Figure 10.23 Example Problem 10.5, F2DII output.

Influence Lines for Trusses

Figure 10.24 Example Problem 10.5, deformed geometry plot from F2DII.

introduced, the two cross sections would either move apart or overlap each other, depending upon whether the deformation were negative or positive. Since a positive deformation corresponds to the case where positive work is performed by a set of positive axial forces S_n on either side of the point while moving through the deformation, a positive deformation, for any member n in the truss, must correspond to an overlapping of the two cross sections, as shown in Figure 10.25. Therefore, a positive unit axial deformation in a truss member, for use in the Müller–Breslau Principle, must correspond to introducing a unit negative change in the length of the member. The application of this process is demonstrated in Figure 10.26. The original geometry of a plane truss is shown in Figure 10.26a and the deformed shape of the truss, which results from removing the axial restraint in a member and introducing a positive unit axial deformation, is shown in Figure 10.26b. According to the Müller–Breslau Principle, the vertical translation of any joint is equal to the axial force which would be produced in the member due to a downward unit force at that joint. Therefore, the deformed shape of the load line for the truss will correspond to the influence line for the axial force in the member.

It now appears that we have a very simple process for generating the influence line for the axial force in any member in a truss without performing any type of equilibrium calculations. Unfortunately, the process is not as easy as it sounds. It probably will be found that, except for trusses with very simple geometries, that the geometric calculations which are required to compute the displaced position of each joint due to a unit axial deformation in a member could be far more difficult than performing the calculations to directly compute the influence line ordinates for a unit force at each joint on the load line. The result is that this process

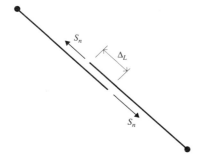

Figure 10.25 Positive member axial deformation.

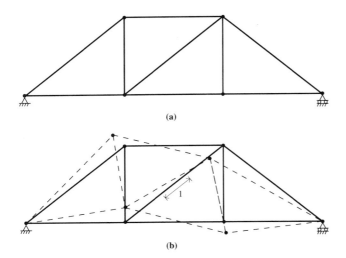

Figure 10.26 Deformed truss for Müller–Breslau principle.

is really not practical for manual calculations. However, it is ideally suited for a computer analysis using a program such as T2DII in FATPAK II, since this program has the capability to compute the joint translations in a plane truss for an initial axial deformation in one or more members. All of the influence line ordinates can be determined by a single computer analysis, rather than performing multiple analyses corresponding to a unit force at each individual joint on the load line. This analysis process can be used in T2DII for both statically determinate and statically indeterminate trusses.

Example Problem 10.6

Compute the influence line ordinates for the force in member 5 by the Müller–Breslau Principle for the plane truss whose mathematical model is shown in Figure 10.27. The load line joints are the joints along the bottom of the truss. Figure 10.28 shows a listing of the input data for the program T2DII for this truss in kip and inch units for an initial axial deformation of -1.0 inches in member 5. The output from the program is shown in Figure 10.29 and a plot of the deformed shape of the truss, with a

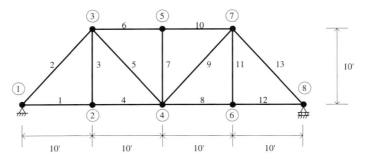

Figure 10.27 Example Problem 10.6.

Influence Lines for Trusses

```
Example Problem 10.6 - Influence Line Member 5 - Analysis by FATPAK II
8,13,1,1,2,0,0,1,0
Joint Coordinates
1,0.0,0.0
2,120.0,0.0
3,120.0,120.0
4,240.0,0.0
5,240.0,120.0
6,360.0,0.0
7,360.0,120.0
8,480.0,0.0
Material Data
1,29000.0
Cross Section Data
1,10.0
Member Data
1,1,2
2,1,3
3,2,3
4,2,4
5,3,4
6,3,5
7,4,5
8,4,6
9,4,7
10,5,7
11,6,7
12,6,8
13,7,8
Support Restraints
1,1,1
8,0,1
Initial Member Axial Deformations
5,-1.0
```

Figure 10.28 Example Problem 10.6, T2DII input.

displacement multiplication factor of 25, is shown in Figure 10.30. Note that the vertical translations listed in Figure 10.29 for joints 1, 2, 4, 6 and 8 agree with the ordinates of the influence line for this member, which were shown in Figure 3.21 in Example Problem 3.5, which were computed by an equilibrium analysis of this same truss. The deformed shape of the lower chord of the truss is the same as the shape of the influence line.

The listed values in the output for all of the member forces and reactive forces are zero to the three decimal place accuracy of the output for this particular truss since it is statically determinate, which results in a Müller–Breslau structure which is unstable. A small deformation can be introduced into any member in a statically determinate truss without causing deformations in any other members.

An alternate analysis procedure which can be used to compute the influence line ordinates for any truss is to apply a set of loads on the truss which have the same effect as introducing a unit axial deformation into a particular member. From our sophomore mechanics course, we know that the change in length of a prismatic elastic member subjected to an axial force is

```
FATPAK II -- Frame and Truss Package -- Student  Edition
(C) Copyright 1995 by Structural Software Systems
Program T2DII - Plane Truss Analysis


Example Problem 10.6 - Influence Line Member 5 - Analysis by FATPAK II


Joint Displacements

Joint         X-Tran           Y-Tran

  1           0.000             0.000
  2          -0.000            -0.354
  3           0.354            -0.354
  4          -0.000             0.707
  5           0.354             0.707
  6          -0.000             0.354
  7           0.354             0.354
  8          -0.000             0.000


Member Forces (Tension Positive)

Member         Force

  1           -0.000
  2            0.000
  3            0.000
  4           -0.000
  5            0.000
  6            0.000
  7            0.000
  8            0.000
  9            0.000
 10            0.000
 11            0.000
 12           -0.000
 13           -0.000


Reactions

Joint          RX                RY

  1           0.000             0.000
  8           0.000             0.000
```

Figure 10.29 Example Problem 10.6, T2DII output.

$$\Delta L = \frac{S_x L}{A_x E} \tag{10.6}$$

where S_x is the axial force in the member, L is the length, E is the material modulus of elasticity and A_x is the cross section area. From this equation, it can be de-

Influence Lines for Trusses

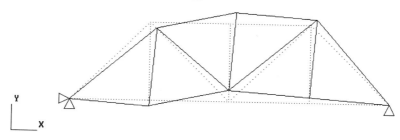

Figure 10.30 Example Problem 10.6, deformed geometry plot from T2DII.

termined that the axial force in the member, which will produce a unit change in length, is

$$S_x = \frac{A_x E}{L} \tag{10.7}$$

Therefore, if a pair of inward forces with magnitudes equal to that expressed by Eq. (10.7) are applied to a truss at the joints at the end of any member, the effect in the truss should be the same as introducing a unit overlap at an interior point in the member. Since this overlap corresponds to the deformation that should be imposed in the truss when applying the Müller–Breslau Principle, this set of forces should produce the same joint translations.

Example Problem 10.7

Compute the joint translations in the truss shown in Figure 10.27 for a pair of equal and opposite inward forces P at joints 3 and 4, acting along the longitudinal axis of member 5, as shown in Figure 10.31. The magnitude of each force will correspond to Eq. (10.7) when evaluated using the properties of member 5.

Substituting the values for A_x, E and L for member 5 in kip and inch units into Eq. (10.7) results in values for the two inclined forces of

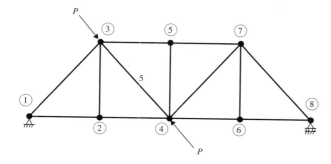

Figure 10.31

$$P = \frac{(10)(29{,}000)}{169.706} = 1708.838 \text{ kips}$$

Since the member is on a one-to-one slope, the components of the forces along the global X and Y axes at joints 3 and 4 will be

$$P_X = P_Y = \frac{P}{\sqrt{2}} = 1208.331 \text{ kips}$$

Figure 10.32 shows a listing of the input data in kip and inch units for the program T2DII for the analysis of the truss for this loading, while Figure 10.33 shows the results of the analysis. The listed values for the translations of the joints are the same as those shown in Figure 10.29 for Example Problem 10.6 by introducing a unit axial deformation into member 5. Therefore, the vertical translation of the load line joints for this loading condition are equal to the ordinates of the influence line for the axial force in member 5.

The listed values for the member forces in the T2DII output show that member 5 is in compression with an axial force of -1708.838 kips, which is exactly equal to the magnitude of the resultant of the inclined joint forces, while the forces in all of

```
Example Problem 10.7 - Influence Line Member 5 - Analysis by FATPAK II
8,13,1,1,2,0,2,0,0
Joint Coordinates
1,0.0,0.0
2,120.0,0.0
3,120.0,120.0
4,240.0,0.0
5,240.0,120.0
6,360.0,0.0
7,360.0,120.0
8,480.0,0.0
Material Data
1,29000.0
Cross Section Data
1,10.0
Member Data
1,1,2
2,1,3
3,2,3
4,2,4
5,3,4
6,3,5
7,4,5
8,4,6
9,4,7
10,5,7
11,6,7
12,6,8
13,7,8
Support Restraints
1,1,1
8,0,1
Joint loads
3,1208.331,-1208.331
4,-1208.331,1208.331
```

Figure 10.32 Example Problem 10.7, T2DII input.

```
FATPAK II -- Frame and Truss Package -- Student Edition
(C) Copyright 1995 by Structural Software Systems
Program T2DII - Plane Truss Analysis


Example Problem 10.7 - Influence Line Member 5 - Analysis by FATPAK II

Joint Displacements

Joint       X-Tran          Y-Tran

  1         0.000           0.000
  2         0.000          -0.354
  3         0.354          -0.354
  4         0.000           0.707
  5         0.354           0.707
  6         0.000           0.354
  7         0.354           0.354
  8         0.000           0.000


Member Forces (Tension Positive)

Member      Force

  1         0.000
  2         0.000
  3         0.000
  4         0.000
  5        -1708.838
  6         0.000
  7         0.000
  8         0.000
  9        -0.000
 10         0.000
 11         0.000
 12         0.000
 13         0.000


Reactions

Joint       RX              RY

  1        -0.000           0.000
  8         0.000          -0.000
```

Figure 10.33 Example Problem 10.7, T2DII output.

the other members are zero. This will always be the case for a statically determinate truss which is loaded in this manner. In addition, the reactive forces are all zero since the active forces on the truss correspond to a load system which is in equilibrium.

The analysis procedures shown in Example Problems 10.6 and 10.7 can be used for a truss with any geometry and any support conditions as long as it is stable. However, the results of the analysis for the member forces and the reactive forces will look somewhat different for a statically indeterminate truss than for a statically determinate truss.

Since the Müller–Breslau structure will be stable for a statically indeterminate truss, the introduction of a unit axial deformation into any member will cause other members to also deform, which in turn will produce stresses in these members. As a result of this, the axial forces in all of the members in the truss will not be zero as they were for the statically determinate truss in Example Problem 10.6. However, this will be considered automatically during the analysis by the program T2DII and the correct joint translations will be computed to correspond to the influence line ordinates.

If a set of forces of the type used in Example Problem 10.7 are applied to a statically indeterminate truss, axial forces will be produced in other members in the truss in addition to the specific member extending between the joints where the loads were applied. However, the deformed shape of the truss will still correspond to the shape that would be produced by a unit overlap at an interior point in that member. The translations of the load line will be equal to the influence line ordinates.

Example Problem 10.8

Compute the influence line ordinates for the axial force in member 3 for the statically indeterminate truss shown in Figure 10.34 by the two procedures used in Example Problems 10.6 and 10.7.

Figure 10.35 shows a listing of the output in kip and inch units from program T2DII for the analysis of the truss for a unit deformation of -1.0 inches in member 3, while Figure 10.36 shows a listing of the output for a pair of inclined forces with magnitudes of 1208.333 kips, as specified by Eq. (10.7), acting inward on joints 2 and 3 along the axis of member 3. These results show that the joint translations are identical for the two types of analyses. Therefore, both procedures are applicable to both statically determinate and statically indeterminate trusses. The only difference in the results of the two analyses is the computed axial force in member 3. In the first analysis, the listed value for this force is 364.798 kips tension while in the second analysis, it is 843.536 kips compression. However, a quick check will show that the difference

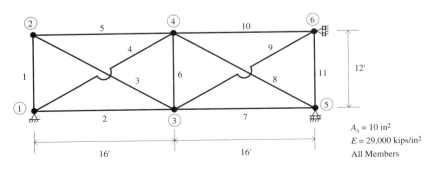

Figure 10.34 Example Problem 10.8.

Influence Lines for Trusses

```
FATPAK II -- Frame and Truss Package -- Student  Edition
(C) Copyright 1995 by Structural Software Systems
Program T2DII - Plane Truss Analysis


Example Problem 10.8 - Unit Deformation in Member 3


Joint Displacements

Joint         X-Tran          Y-Tran

  1           0.000           0.000
  2           0.230          -0.109
  3          -0.232           0.438
  4           0.037           0.352
  5          -0.230           0.000
  6           0.000           0.023


Member Forces (Tension Positive)

Member          Force

  1           -218.879
  2           -350.595
  3            364.798
  4            291.352
  5           -291.838
  6           -172.862
  7              2.599
  8             -3.248
  9            -76.694
 10            -56.158
 11             46.017


Reactions

Joint          RX              RY

  1          117.513          44.068
  5            0.000         -44.068
  6         -117.513           0.000
```

Figure 10.35 Example Problem 10.8, T2DII output.

in these two forces is exactly equal to the magnitude of the inclined joint forces which were applied at the end joints of member 3 in the second analysis. The difference in the two computed axial forces results from these loads, which were not present in the first analysis. The compressive axial force in member 3 in the second analysis is required for the equilibrium of joints 2 and 3. However, the computed force in member 3 is of no real concern to us since we are only interested in the vertical transla-

```
FATPAK II -- Frame and Truss Package -- Student  Edition
(C) Copyright 1995 by Structural Software Systems
Program T2DII - Plane Truss Analysis


Example Problem 10.8 - Inclined Forces at Joints 2 and 3

Joint Displacements

Joint       X-Tran          Y-Tran

  1          0.000           0.000
  2          0.230          -0.109
  3         -0.232           0.438
  4          0.037           0.352
  5         -0.230           0.000
  6          0.000           0.023


Member Forces (Tension Positive)

Member       Force

  1        -218.879
  2        -350.595
  3        -843.536
  4         291.352
  5        -291.838
  6        -172.862
  7           2.599
  8          -3.248
  9         -76.694
 10         -56.158
 11          46.016


Reactions

Joint         RX              RY

  1         117.514          44.067
  5           0.000         -44.067
  6        -117.514           0.000
```

Figure 10.36 Example Problem 10.8, T2DII output.

tions of the load line joints. The remainder of the listed output is unimportant for this analysis.

APPLICATION OF MÜLLER-BRESLAU PRINCIPLE TO 3D STRUCTURES

The Müller–Breslau Principle also can be very useful for three dimensional structures. As an example, Figure 10.37 shows a plot by the space frame analysis program F3DII in FATPAK II of the mathematical model for the floor system in an office building. The floor system consists of a rectangular grid of beams which are connected at the points where they intersect. The grid is simply supported at joints 1, 5, 11, 15, 21 and 25.. Figure 10.38 shows a plot of the deformed shape of the grid due to a unit upward vertical translation of joint 1 in the mathematical model. According to the Müller–Breslau Principle, the vertical translation at any point on the floor system, which is listed in the output by F3DII, is equal to the magnitude of the vertical reactive force at joint 1 which would be produced by a unit downward concentrated force at that point. This deformed shape is known as an *influence surface* since it is in three dimensions. Similar influence surfaces could be developed for the reactive forces or the member forces for three dimensional space trusses of the type used in the roof systems of many auditoriums and sports arenas by using a space truss analysis program such as T3DII in FATPAK II.

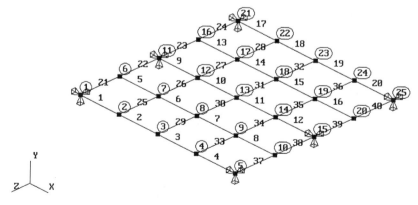

Figure 10.37 Floor grid, original geometry plot from F3DII.

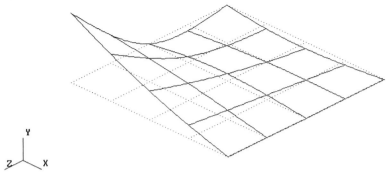

Figure 10.38 Floor grid, deformed geometry plot from F3DII.

EXPERIMENTAL GENERATION OF INFLUENCE LINES

Before computers were available to aid structural engineers in the analysis of complex structures, the computation of the influence line ordinates for statically indeterminate structures was an extremely tedious and time consuming task. A large number of calculations were required to perform the required analyses. To help simplify the analysis process, many engineers resorted to experimental procedures for generating influence lines. The technique which was used was to construct a small scale model of the structure which could then be deformed according to the Müller–Breslau Principle to physically generate the shape of the influence line. By taking careful measurements of the displacements in these models, the values of the influence line ordinates could be determined within the accuracy required for an acceptable analysis of the structure. As an aid in performing these measurements, very accurate optical micrometers were developed which could be used to measure translations in the models to an accuracy of more than 0.001 inches. Specialized devices were also developed for introducing the required deformations into the model. One of these devices was the Beggs' Deformeter (Beggs, 1922, 1927), which was developed by Professor George Beggs of Princeton University in the 1920s. This device had special clamps which could introduce very accurate translations and rotations into the model of the structure. It could be used for the analysis of trusses, beams and frames. Another device which was used was the Moment Deformeter (Norris, 1944), which would deform a model of a beam into the shape of the influence line for the bending moment at any interior point in the beam. Up through the 1960s, a number of structures were built for which essentially the entire analysis was performed by measurements on laboratory models. The use of models for structural analysis quickly declined after the introduction of computers and the development of structural analysis programs. The primary use which is presently made of experimental analysis is to perform measurements on actual structures or large scale models to verify the assumptions for developing the mathematical models of the structures for computer analysis.

It is not necessary to have very sophisticated devices such as the Begg's Deformeter or the Moment Deformeter to perform an experimental analysis for the influence lines for a structure. The experimental model for a statically indeterminate beam can be as simple as a flexible brass rod which is supported on a drawing board by push pins. By introducing a displacement at a support point corresponding to the action of a particular reactive quantity and measuring the translations of the points along the rod with an engineer's scale, the influence line ordinates for the reactive quantity can be determined with surprising accuracy by dividing the measured translations by the imposed displacement at the support, as shown previously in Example Problems 10.4 and 10.5. Performing simple experiments of this type can help the reader to a better understanding of structural behavior.

REFERENCES

BEGGS, GEORGE E. (1922). "An Accurate Mechanical Solution of Statically Indeterminate Structures by Use of Paper Models and Special Gages," *Proceedings of the American Concrete Institute,* Vol. 18.

——— (1927). "The Use of Models in the Solution of Indeterminate Structures," *Journal of the Franklin Institute*, Vol. 203, No. 3.

NORRIS, C. H. (1944). "Model Analysis of Structures," *Experimental Stress Analysis*, Vol. 1, No. 2.

SUGGESTED PROBLEMS

For Suggested Problems SP10.1 to SP10.6, draw the influence lines for the vertical reactive forces for the beams shown previously at the end of Chapter 2 by using the Müller–Breslau Principle.

SP10.1   Problem SP2.11
SP10.2   Problem SP2.12
SP10.3   Problem SP2.13
SP10.4   Problem SP2.14
SP10.5   Problem SP2.15
SP10.6   Problem SP2.16

For Suggested Problems SP10.7 to SP10.10, draw the influence lines for the indicated quantities for the beams shown previously at the end of Chapter 5 by using the Müller–Breslau Principle.

SP10.7   Problem SP5.19
SP10.8   Problem SP5.20

SP10.9 Problem SP5.21

SP10.10 Problem SP5.22

For Suggested Problems SP10.11 to SP10.15, compute the ordinates of the influence lines for the forces in the indicated members for the plane trusses shown previously at the end of Chapter 3 by using the Müller Breslau Principle and the program T2DII in FATPAK II. Plot the final deformed shape of the trusses to verify that the deflected shape of the load lines agree with the previously computed shape of the influence lines. Use arbitrary values of 10,000 for E and 10 for A_x for the members.
(a) By introducing a unit deformation in the member.
(b) By applying a set of inward forces equal to $A_x E/L$ on the joints at the ends of the member.

SP10.11 Problem SP3.21

SP10.12 Problem SP3.22

SP10.13 Problem SP3.23

SP10.14 Problem SP3.24

SP10.15 Problem SP3.25

SP10.16 and **SP10.17** Compute the ordinates of the influence lines for the vertical reactive forces at 2 meter intervals along the statically indeterminate beams shown in Figures SP10.16 and SP10.17. Plot the final deformed shapes of the beams. Use the Müller–Breslau Principle and the program F2DII in FATPAK II. Use any desired arbitrary values for E and I_z. Verify your solutions with the program BEAMIL.
(a) By introducing a unit translation at the support joint.
(b) By removing the support restraint and applying a corresponding joint load as demonstrated in Example Problem 10.5.

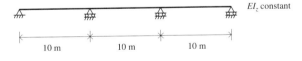

Figure SP10.16 .

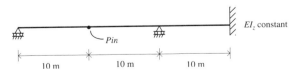

Figure SP10.17 .

Suggested Problems

SP10.18 Compute the ordinates of the influence lines for the forces in the indicated members in the statically indeterminate highway bridge truss shown in Figure SP10.18 using the Müller–Breslau Principle and the program T2DII in FATPAK II. The traffic moves along the bottom chord of the truss. Verify your solutions with the program PTRUSSIL.
(a) By introducing a unit deformation in the member.
(b) By applying a set of inward forces equal to $A_x E/L$ at the ends of the member.

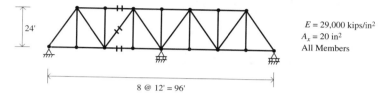

$E = 29{,}000$ kips/in^2
$A_x = 20$ in^2
All Members

8 @ 12' = 96'

Figure SP10.18

11

The Method of Consistent Deformations

STATICALLY DETERMINATE VERSUS STATICALLY INDETERMINATE ANALYSIS

In the previous chapters, the analysis of a statically determinate structure for the reactive loads and the member end loads could be performed using only the equilibrium conditions for the structure. The number of unknown quantities had to be exactly equal to the number of available independent equilibrium equations. However, for a statically indeterminate structure, the number of reactive loads and member end loads will be greater than the number of independent equilibrium equations and a unique solution for these quantities cannot be determined. There will be an infinite number of sets of values which will satisfy the equations of equilibrium. Additional independent equations, which are based on some other condition than equilibrium, are required to perform the analysis.

The three conditions which must be satisfied by the total set of equations for analyzing any statically indeterminate structure are: equilibrium; geometric compatibility; and load-displacement. If it is ensured that these three conditions are satisfied for both the total structure and for each individual joint and member in the mathematical model of the structure, then there will always be exactly as many independent equations as are required to compute the member end loads for each member and the reactive loads at each support joint.

Equilibrium Conditions

The equilibrium conditions for any structural system require that each joint and each member in the mathematical model of the structure is in equilibrium under the combined effects of the active loads and the reactive loads. In the previous chapters, we saw that we could compare the number of independent equilibrium equations for any truss or frame structure to the total number of member end loads and reactive loads to determine whether the structure was unstable, statically determinate or statically indeterminate. The specific relationships obtained were as follows.

for a plane truss,

$\quad$ if $2NJ > NM + NR$, the truss is unstable $\hfill$ (11.1a)

$\quad$ if $2NJ = NM + NR$, the truss is statically determinate $\hfill$ (11.1b)

$\quad$ if $2NJ < NM + NR$, the truss is statically indeterminate $\hfill$ (11.1c)

for a space truss,

$\quad$ if $3NJ > NM + NR$, the truss is unstable $\hfill$ (11.2a)

$\quad$ if $3NJ = NM + NR$, the truss is statically determinate $\hfill$ (11.2b)

$\quad$ if $3NJ < NM + NR$, the truss is statically indeterminate $\hfill$ (11.2c)

for a plane frame,

$\quad$ if $3NJ > 3NM + NR - NC$, the frame is unstable $\hfill$ (11.3a)

$\quad$ if $3NJ = 3NM + NR - NC$, the frame is statically determinate $\hfill$ (11.3b)

$\quad$ if $3NJ < 3NM + NR - NC$, the frame is statically indeterminate $\hfill$ (11.3c)

and for a space frame,

$\quad$ if $6NJ > 6NM + NR - NC$, the frame is unstable $\hfill$ (11.4a)

$\quad$ if $6NJ = 6NM + NR - NC$, the frame is statically determinate $\hfill$ (11.4b)

$\quad$ if $6NJ < 6NM + NR - NC$, the frame is statically indeterminate $\hfill$ (11.4c)

where NJ is the number of joints, NM is the number of members, NR is the number of support restraints and NC is the number of special construction conditions at the joints, such as a pin or a shear slide connection at the end of a member, which result in zero values for specific member end loads. It is also possible for a structure to still be unstable even though the conditions for static determinacy or static indeterminacy are satisfied due to mechanisms which are formed due to the geometry of the structure or the arrangement of the support restraints.

Geometric Compatibility Conditions

The geometric compatibility conditions for any structural system require that the displacement of each point in the structure be compatible with the restraints at that point. These restraints may consist of either external restraints at the support joints or internal restraints in the members. The external restraints at the support joints resist translation or rotation in specific directions at these joints. For example, a fixed support joint will be restrained against translation and rotation in all directions, while a pinned support joint will be restrained against translation but free to rotate. The internal restraints in the members define how the members must deform. The axial restraint at any point in a member resists relative translation of the two cross sections on either side of the point along the longitudinal axis of the member, the shear restraint resists relative translation of the two cross sections perpendicular to the axis of the member and the moment restraint resists relative rotation of the two cross sections. The external support loads and the internal member end loads are dependent upon the type and location of these restraints.

Load-Displacement Relationships

The load-displacement relationships for any structure define how each point in the structure will displace under any set of active and reactive loads. The displacement of any point will depend on: the overall geometry of the structure; the type of restraints at the support joints and at the connections of each member to its end joints; the properties of the material; and the type and location of the active loads. If the structure is either a plane truss or a space truss, then the displacements of any point will be due only to axial deformations in the members, whereas if the structure is a space frame, then axial deformations, shear deformations, bending deformations and torsional deformations in the members can contribute to the displacements. Methods for considering the effects of each of these types of member deformations were discussed in previous chapters.

ANALYSIS OF STATICALLY INDETERMINATE STRUCTURES

Each internal or external restraint in a structure, over and above the number required for stability, will result in an additional internal or external load which must be computed during the analysis of the structure. The internal loads will consist of member end loads due to additional internal restraints in the structure, while the external loads will be reactive loads due to additional support restraints. The additional restraints are known as *redundant restraints* since they are not required for stability of the structure. The internal or external loads corresponding to these redundant restraints are known as the *redundant loads*. The number of redundant restraints will be equal to the number of degrees of static indeterminacy of the structure, which can be expressed as follows:

for a plane truss,

$$NDI = NM + NR - 2NJ \qquad (11.5)$$

Analysis of Statically Indeterminate Structures

for a space truss,
$$NDI = NM + NR - 3NJ \tag{11.6}$$
for a plane frame,
$$NDI = 3NM + NR - NC - 3NJ \tag{11.7}$$
and for a space frame,
$$NDI = 6NM + NR - NC - 6NJ \tag{11.8}$$

where NDI is the number of degrees of static indeterminacy, and NJ, NM, NR and NC are as defined previously.

If the redundant restraints are removed from a statically indeterminate structure, the resulting *reduced structure* will be statically determinate. However, the reduced structure will not behave like the original structure since displacements will occur at the points where the redundant restraints have been removed. These displacements will be referred to as the *redundant displacements*. They are usually zero in the original structure. The following five steps can be performed to compute the loads corresponding to the redundant restraints for any statically indeterminate structure:

Step 1. Remove the redundant restraints from the structure so that the reduced structure is statically determinate and stable. The loads corresponding to the redundant restraints will be designated as R_1 through R_{NDI}, where NDI is the number of degrees of static indeterminacy for the type of structure which is being analyzed, as defined in Eqs. (11.5) through (11.8).

Step 2. Compute the displacements due to the active loads on the reduced structure, at the points where the redundant restraints were removed, corresponding to the action of those restraints. These displacements will be designated as Δ'_1 though Δ'_{NDI}.

Step 3. Apply a set of loads to the reduced structure, at the points where the redundant restraints were removed, which correspond to the action of those restraints, and compute the resulting displacements at each of these redundant loads as a function of all of the redundant loads. These displacements will be designated as Δ''_1 through Δ''_{NDI}. They can be related to the redundant loads by a set of *redundant displacement coefficients* $\delta_{i,j}$ in a set of equations of the form

$$\begin{aligned} \Delta''_1 &= \delta_{1,1} R_1 + \delta_{1,2} R_2 + \cdots + \delta_{1,\,NDI} R_{NDI} \\ \Delta''_2 &= \delta_{2,1} R_1 + \delta_{2,2} R_2 + \cdots + \delta_{2,\,NDI} R_{NDI} \\ &\vdots \\ \Delta''_{NDI} &= \delta_{NDI,1} R_1 + \delta_{NDI,2} R_2 + \cdots + \delta_{NDI,\,NDI} R_{NDI} \end{aligned} \tag{11.9}$$

where the redundant displacement coefficient $\delta_{i,j}$ represents the displacement at redundant load R_i due to a unit positive value of the redundant load R_j.

The coefficient $\delta_{i,j}$ must be equal to the coefficient $\delta_{j,i}$ by Maxwell's Theorem.

Step 4. Write a set of geometric compatibility equations by equating the sum of the displacements Δ'_i and Δ''_i at each redundant load R_i to the actual displacement Δ_i, which will occur at that point in the original structure. The result will be set of linear simultaneous equations, with the redundant loads as the unknowns, in the form

$$\begin{aligned}
\Delta'_1 + \delta_{1,1}R_1 + \delta_{1,2}R_2 + \cdots + \delta_{1,\text{NDI}}R_{\text{NDI}} &= \Delta_1 \\
\Delta'_2 + \delta_{2,1}R_1 + \delta_{2,2}R_2 + \cdots + \delta_{2,\text{NDI}}R_{\text{NDI}} &= \Delta_2 \\
&\vdots \\
\Delta'_{\text{NDI}} + \delta_{\text{NDI},1}R_1 + \delta_{\text{NDI},2}R_2 + \cdots + \delta_{\text{NDI},\text{NDI}}R_{\text{NDI}} &= \Delta_{\text{NDI}}
\end{aligned} \quad (11.10)$$

These equations can then be solved for the redundant loads. Since the displacements Δ_1 through Δ_{NDI} at each of the redundant loads will usually be zero, the right sides of all of the previous equations will usually be zero. However, if one or more redundant displacements are not zero, such as a settlement at a support joint or a temperature deformation in a member, then the magnitude of those displacements should be inserted on the right sides of the appropriate equations.

Step 5. After the redundant loads have been determined, the remaining unknown reactive loads and member end loads can be determined by using the equilibrium equations for the original structure.

This procedure can be used for any statically indeterminate structure as long as the material at all points in the structure remains linear elastic under the active loads. The procedure is known by several different names. The two most common are the *Method of Consistent Displacements* and the *Method of Consistent Deformations*.

One Degree Statically Indeterminate Structures

Equation (11.10) will reduce to one equation with one unknown for a one degree statically indeterminate structure since there will only be one redundant load. It will only be necessary to compute a single Δ' and Δ'' during the analysis.

Example Problem 11.1

Compute the reactions for the long slender prismatic beam shown in Figure 11.1a. The mathematical model for the beam, which consists of two joints and one member, is shown in Figure 11.1b. The beam has four support restraints, which consist of a horizontal translation restraint at joint 1, a vertical translation restraint at joint 1 and at joint 2 and a rotation restraint at joint 1. These are no special construction conditions for this beam. Therefore, substituting into Eq. (11.7) shows that the beam, which is merely a special geometric form of a plane frame, is one degree statically indeterminate.

Analysis of Statically Indeterminate Structures

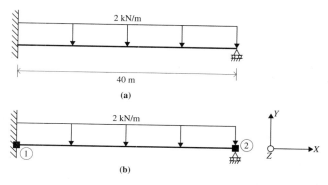

Figure 11.1 Example Problem 11.1.

There are five different choices which can be made for the redundant load for this beam, as shown in Figure 11.2. For Cases I, II and III, the redundant load is considered to be an external reactive load, while for Cases IV and V, it is considered to be a pair of internal member cross section loads.

Case I. The redundant load R_1 is the vertical reactive force at the right support and the redundant displacement Δ'_1 is a vertical translation at that point, as shown in Figure 11.2a. The reduced structure is a cantilever beam with a fixed support at the left end.

Case II. The redundant load R_1 is the vertical reactive force at the left support and the redundant displacement Δ'_1 is a vertical translation at that point, as

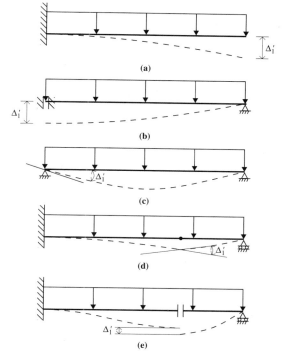

Figure 11.2 Example Problem 11.1.

shown in Figure 11.2b. The reduced structure is a beam with a vertical slide support at the left end and a horizontal roller support at the right end.

Case III. The redundant load R_1 is the reactive moment at the left support and the redundant displacement Δ'_1 is a rotation at that point, as shown in Figure 11.2c. The reduced structure is a beam with a pinned support at the left end and a horizontal roller support at the right end.

Case IV. The redundant load R_1 is the pair of equal and opposite bending moments acting on the two cross sections on either side of an interior point in the beam and the redundant displacement Δ_1 is a kink angle at that point, as shown in Figure 11.2d. The reduced structure is a beam with the original support conditions and a pin at the interior point. This pin can be located at any point along the beam except directly over the right support since the moment at this point in the beam is known to be zero. In addition, if it is located directly over the left support, the redundant load and the redundant displacement will correspond to the same condition shown in Figure 11.2c.

Case V. The redundant load R_1 consists of the pair of equal and opposite vertical shear forces acting on the two cross sections on either side of an interior point in the beam and the redundant displacement Δ'_1 consists of a vertical shear deformation at that point, as shown in Figure 11.2e. The reduced structure is a beam with the original support conditions and a vertical shear slide at the interior point. This shear slide can be located at any point along the beam. If it is located just to the right of the left support, the redundant load and the redundant displacement will correspond to the same condition shown in Figure 11.2b. If it is located just to the left of the right support, the redundant load and the redundant displacement will correspond to the same condition shown in Figure 11.2a.

The analysis for each of these cases will result in the correct solution for the reactive loads for the beam as long as the displacements Δ'_1 and Δ''_1 are computed correctly. These displacements can be computed using any of the procedures discussed previously in Chapters 8 and 9. For demonstration purposes, we will carry through the analyses for Case I and Case III.

Case I Analysis. The translations Δ'_1 and Δ''_1 can be computed using the expressions developed previously in Example Problem 9.4 for a cantilever beam subjected to a uniform distributed force and a concentrated end force. Since this beam is long and slender, only the effects of the bending deformations in the beam will be considered in computing Δ'_1 and Δ''_1.

The displacement Δ'_1 in meters for a downward uniform load of 2 kN/m on the cantilever reduced beam will be

$$\Delta'_1 = \frac{\omega L^4}{8EI_z} = \frac{(-2)(40)^4}{8EI_z} = -\frac{640{,}000}{EI_z}$$

Since this displacement is negative, it is down. The displacement Δ''_1 can be expressed in terms of the redundant load R_1 as

$$\Delta''_1 = \delta_{1,1} R_1$$

Analysis of Statically Indeterminate Structures

where $\delta_{1,1}$ is the displacement in the direction of R_1 due to a unit load at that point in the direction of R_1. The coefficient $\delta_{1,1}$ can be computed by applying an upward unit force at the end of the beam. Using the expression from Example Problem 9.4 gives

$$\delta_{1,1} = \frac{WL^3}{3EI_z} = \frac{(1)(40)^3}{3EI_z} = \frac{21{,}333.333}{EI_z}$$

Substituting these values into Eq. (11.10) and recognizing that the actual vertical translation at the right supprt is zero results in one equation with one unknown

$$-\frac{640{,}000}{EI_z} + \frac{21{,}333.333 R_1}{EI_z} = 0$$

from which we can determine that

$$R_1 = 30 \text{ kN}$$

or, using the notation for the mathematical model of the beam,

$$R_{2Y} = 30 \text{ kN} \uparrow$$

The remaining reactive loads can now be determined by using the equations of equilibrium for the original beam.

$$\Sigma F_Y \rightarrow R_{1Y} = 50 \text{ kN} \uparrow$$

$$\Sigma M_Z \rightarrow M_{1Z} = 400 \text{ kN-m counterclockwise}$$

$$\Sigma F_X \rightarrow R_{1X} = 0$$

These values agree with those which were computed in Example Problem 8.5 in Chapter 8 for this same beam.

Case III Analysis. One of several methods which can be used to compute the rotations Δ'_1 and Δ''_1 for the reduced simply supported beam for this case is the Conjugate Beam Method. Figure 11.3a shows the mathematical model of the reduced beam subjected to the downward uniform load and the corresponding conjugate beam subjected to the conjugate loading. The rotation Δ'_1 in radians will be equal to the vertical shear force V^*_{1Y} at the left end of the conjugate beam, which in turn is equal to the left vertical reactive force R^*_{1Y}. The subscripts which are being used here for these quantities correspond to the joint numbers for the mathematical model of the beam. Summing moments about the right end of the conjugate beam gives

$$R^*_{1Y} = \Delta'_1 = -\frac{5333.333}{EI_z}$$

Figure 11.3b shows the reduced beam subjected to a unit moment in the positive direction of R_1 and the corresponding conjugate beam. Summing moments about the right end of the conjugate beam gives

$$R^*_{1Y} = \delta_{1,1} = \frac{13.333}{EI_z}$$

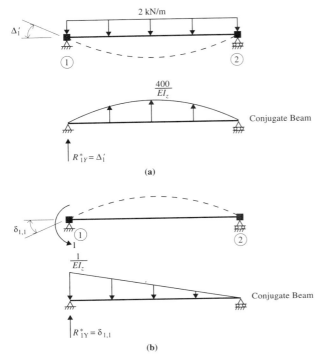

Figure 11.3 Example Problem 11.1.

The units of $\delta_{1,1}$ will be radians per kN-m when E and I_z are expressed in kN and meter units. Substituting these two values into Eq. (11.10) gives

$$-\frac{5333.333}{EI_z} + \frac{13.333 R_1}{EI_z} = 0$$

from which

$$R_1 = 400 \text{ kN-m}$$

or, using the notation for the mathematical model of the beam

$$M_{1Z} = 400 \text{ kN-m counterclockwise}$$

The other reactive loads will be the same as in the previous solution for Case I. Therefore, the solutions using either choice for the redundant load are identical as expected.

The displacements Δ_1' and Δ_1'' also can be computed using the Principle of Virtual Work. Figure 11.4a shows the mathematical model of the reduced beam subjected to the real load system **W**, and Figure 11.4b shows this same mathematical model subjected to the virtual load system **Q**, which can be used to compute Δ_1'. The expressions for the bending moments due to the real load system and the virtual load system, with the origin of the local coordinate system of the single member in the mathematical model at the left end, are

$$M_{z,W} = 40x - x^2$$

Analysis of Statically Indeterminate Structures

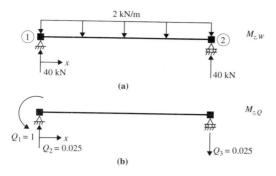

Figure 11.4 Example Problem 11.1.

and

$$M_{z,Q} = -1 + 0.025x$$

The effect of the shear deformations in the beam will be ignored. Substituting these expressions into Eq. (9.42), and recognizing that the displacements at each of the virtual loads except Q_1 are zero, gives

$$\Delta'_1 = \int \frac{M_{z,Q} M_{z,W}}{EI_z} dx = \int_0^{40} \frac{(40x - x^2)(-1 + 0.025x)}{EI_z} dx$$

from which

$$\Delta'_1 = -\frac{5333.333}{EI_z}$$

Since the load Q_1 in the virtual load system in Figure 11.4b corresponds to a unit load at R_1 in the direction of R_1, the quantity $\delta_{1,1}$ can be computed by the expression

$$\delta_{1,1} = \int \frac{M_{z,Q} M_{z,Q}}{EI_z} dx = \int_0^{40} \frac{(-1 + 0.025x)(-1 + 0.025x)}{EI_z} dx$$

which gives

$$\delta_{1,1} = \frac{13.333}{EI_z}$$

The values for Δ'_1 and Δ''_1 agree with those obtained previously.

Example Problem 11.2

Compute the reactions and the member forces for the plane truss whose mathematical model is shown in Figure 11.5. The truss has five joints, seven members and four independent reactive force components. Substituting into Eq. (11.5) shows that it is one degree statically indeterminate.

As with the beam in the previous example, there are several different choices which can be made for the redundant load for this truss. It could be considered to be any of the four global reactive force components, since any one of these components could be removed to leave a statically determinate and stable reduced structure.

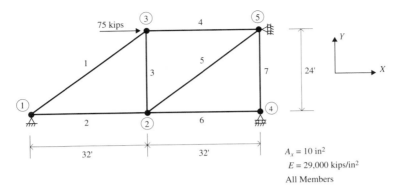

Figure 11.5 Example Problem 11.2.

It could also be considered to be the axial force in one of the members, as long as removing that member force results in a stable reduced structure. For demonstration purposes, we will carry through the analysis for two different cases. All displacements will be computed using the Principle of Virtual Work.

Case I Analysis. The redundant load for this case will be the horizontal reactive force at joint 5. Figure 11.6a shows the reduced structure, without the horizontal translation restraint at joint 5, subjected to the real load system **W**, and Figure 11.6b shows the reduced structure subjected to the virtual load system **Q** which can be used to compute the redundant displacement Δ'_1 by the Principle of Virtual Work. This displacement corresponds to the horizontal translation of joint 5. Applying Eq. (9.43), which was developed previously in Chapter 9, and recognizing that the translations at each of the virtual loads are zero except at joint 5, gives

$$\Delta'_1 = \sum_{n=1}^{7} \left(\frac{S_{x,Q} S_{x,W} L}{A_x E} \right)_n$$

which can be evaluated in the table in Figure 11.6c to give

$$\Delta'_1 = 0.11792 \text{ inches}$$

Since the load Q_1 in the virtual load system in Figure 11.6b corresponds to a unit load at R_1 in the direction of R_1, the quantity $\delta_{1,1}$ can be computed by the expression

$$\delta_{1,1} = \sum_{n=1}^{7} \left(\frac{S_{x,Q} S_{x,Q} L}{A_x E} \right)_n$$

which, upon evaluating the summation in a table similar to that in Figure 11.6c, gives

$$\delta_{1,1} = 0.0022342 \text{ inches/kip}$$

The redundant load now can be computed by substituting the values for Δ'_1 and Δ''_1 into Eq. (11.10)

$$0.11792 + 0.0022342 R_1 = 0$$

Analysis of Statically Indeterminate Structures

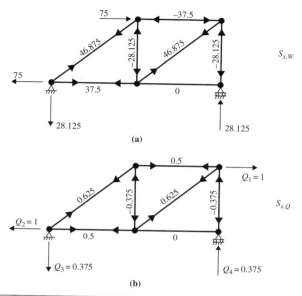

Figure 11.6 Example Problem 11.2.

from which

$$R_1 = -52.779 \text{ kips}$$

Therefore, the horizontal reactive force for the truss at joint 5 is

$$R_{5X} = 52.779 \text{ kips} \leftarrow$$

The remaining reactions now can be computed by using the external equations of equilibrium for the truss and the member forces can be determined by the Method of Joints after the reactive forces have been determined. The final results of the analysis are shown in Figure 11.7. This solution can be verified by using the program T2DII in FATPAK II. Figure 11.8 shows a listing of the input data for T2DII in kip and inch units and Figure 11.9 shows a listing of the output. The two solutions agree

396 The Method of Consistent Deformations   Chap. 11

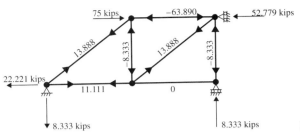

Figure 11.7 Example Problem 11.2.

```
Example Problem 11.2 - Analysis by FATPAK II
5,7,1,1,3,0,1,0,0
Joint Coordinates
1,0.0,0.0
2,384.0,0.0
3,384.0,288.0
4,768.0,0.0
5,768.0,288.0
Material Data
1,29000.0
Cross Section Data
1,10.0
Member Data
1,1,3
2,1,2
3,2,3
4,3,5
5,2,5
6,2,4
7,4,5
Support Restraints
1,1,1
4,0,1
5,1,0
Joint Loads
3,75.0,0.0
```

Figure 11.8 Example Problem 11.2, T2DII input.

within expected roundoff error. Of course, if the calculations for Δ'_1 and $\delta_{1,1}$ were carried out to fewer significant figures, the roundoff error in the manual solution would be greater. For example, if only three significant figure accuracy had been used in evaluating Δ'_1 and $\delta_{1,1}$, the computed value for R_1 would have been -52.326 kips. This number would still be acceptable for essentially any structural engineering application.

Case II Analysis. The redundant load for this case will be the axial force in member 4. Since this force restrains relative axial translation along the axis of the member, for the two cross sections on either side of any point along the member, the removal of the restraint corresponds to cutting the member at some interior point. Figure 11.10a shows the reduced structure, with a cut in member 4, subjected to the real load system **W**, and Figure 11.10b shows the reduced structure subjected to the virtual load system **Q**, which can be used to compute the redundant displacement Δ'_1 by the Principle of Virtual Work. This displacement corresponds to the relative translations of the two ends on either side of the cut in member 4. The table in Figure 11.10c shows the evaluation of the virtual work equation

Analysis of Statically Indeterminate Structures

```
FATPAK II -- Frame and Truss Package -- Student Edition
(C) Copyright 1995 by Structural Software Systems
Program T2DII - Plane Truss Analysis

Example Problem 11.2 - Analysis by FATPAK II

Joint Displacements

Joint          X-Tran          Y-Tran

  1            0.000           0.000
  2            0.015          -0.066
  3            0.085          -0.074
  4            0.015           0.000
  5            0.000          -0.008

Member Forces (Tension Positive)

Member         Force

  1            13.889
  2            11.111
  3            -8.333
  4           -63.889
  5            13.889
  6             0.000
  7            -8.333

Reactions

Joint          RX              RY

  1           -22.222         -8.333
  4             0.000          8.333
  5           -52.778          0.000
```

Figure 11.9 Example Problem 11.2, T2DII output.

$$\Delta'_1 = \sum_{n=1}^{7} \left(\frac{S_{x,Q} S_{x,w} L}{A_x E} \right)_n$$

from which

$$\Delta'_1 = 0.57105 \text{ inches}$$

A similar evaluation of the relationship

$$\delta_{1,1} = \sum_{n=1}^{7} \left(\frac{S_{x,Q} S_{x,Q} L}{A_x E} \right)_n$$

gives

$$\delta_{1,1} = 0.0089378 \text{ inches/kip}$$

Substituting these values into Eq. (11.10) gives

398 The Method of Consistent Deformations Chap. 11

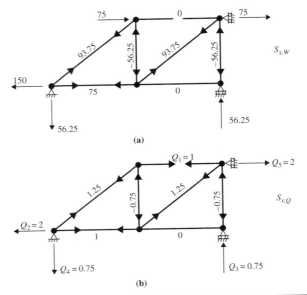

Figure 11.10 Example Problem 11.2.

$$0.57105 + 0.0089378R_1$$

from which

$$R_1 = -63.892 \text{ kips}$$

which corresponds to an axial force in member 4 of 63.892 kips compression. With this value known, the remaining member forces and the reactive forces can be computed by using a combination of the internal and external equations of equilibrium for the truss. This is left as an exercise for the reader. The final solution will agree with the T2DII solution within acceptable roundoff error.

This same approach can be used for any one degree statically indeterminate structure. It will usually be possible to choose more than one of the internal member loads or external reactive loads as the redundant load during the analysis. A considerable amount of time and effort can be saved in the analysis if expressions are already available for the displacements in the reduced structure for the active loads and for the redundant loads, as in the Case I analysis in Example Problem 11.1. Many engineers keep a catalog of solutions for cases, which have been analyzed previously, for future use. Expressions are also available in a number of structural engineering design manuals and handbooks for the displacements for beams under various loadings.

Multi-Degree Statically Indeterminate Structures

The analysis of a multi-degree statically indeterminate structure is more complex than the analysis of a single degree indeterminate structure since more than one equation must be generated and solved to compute the redundant loads. Multiple values of the displacements Δ' and Δ'' must be computed.

Example Problem 11.3

Compute the reactive forces and the member forces for the plane truss whose mathematical model is shown in Figure 11.11. The truss has five joints, eight members and four independent reactive force components. Therefore, according to Eq. (11.5) it is two degrees statically indeterminate.

The horizontal reactive force at joint 4 and the axial force in member 5 will be chosen to be the redundant loads R_1 and R_2, respectively, for this analysis. Figure 11.12a shows the reduced structure subjected to the real load system **W**, and Figures 11.12b and 11.12c show the reduced structure subjected to the virtual load systems **Q1** and **Q2**, which can be used to compute the redundant displacements Δ'_1 and Δ'_2 by the Principle of Virtual Work.

The redundant displacements Δ'_1 and Δ'_2 will be equal to

$$\Delta'_1 = \sum_{n=1}^{8} \left(\frac{S_{x,Q1} S_{x,W} L}{A_x E} \right)_n = 0.060690 \text{ inches}$$

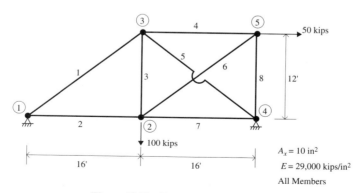

Figure 11.11 Example Problem 11.3.

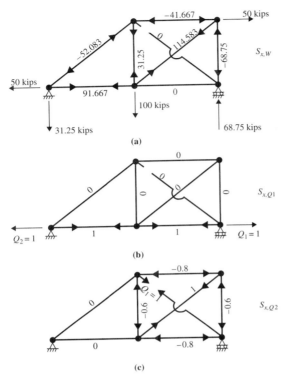

Figure 11.12 Example Problem 11.3.

$$\Delta'_2 = \sum_{n=1}^{8} \left(\frac{S_{x,Q2} S_{x,W} L}{A_x E} \right)_n = 0.12801 \text{ inches}$$

while the displacements Δ''_1 and Δ''_2 will be

$$\Delta''_1 = \delta_{1,1} R_1 + \delta_{1,2} R_2$$

$$\Delta''_2 = \delta_{2,1} R_1 + \delta_{2,2} R_2$$

where

$$\delta_{1,1} = \sum_{n=1}^{8} \left(\frac{S_{x,Q1} S_{x,Q1} L}{A_x E} \right)_n = 0.0013241 \text{ inches/kip}$$

$$\delta_{1,2} = \sum_{n=1}^{8} \left(\frac{S_{x,Q1} S_{x,Q2} L}{A_x E} \right)_n = -0.00052960 \text{ inches/kip}$$

$$\delta_{2,1} = \sum_{n=1}^{8} \left(\frac{S_{x,Q2} S_{x,Q1} L}{A_x E} \right)_n = -0.00052960 \text{ inches/kip}$$

Analysis of Statically Indeterminate Structures

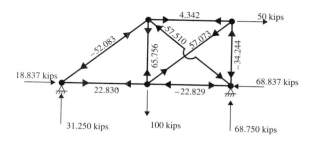

Figure 11.13 Example Problem 11.3.

$$\delta_{2,2} = \sum_{n=1}^{8} \left(\frac{S_{x,Q2} S_{x,Q2} L}{A_x E} \right)_n = 0.0028598 \text{ inches/kip}$$

Note that $\delta_{1,2}$ is equal to $\delta_{2,1}$, as expected from Maxwell's Theorem. Substituting these values into Eq. (11.10) gives the set of equations

$$0.060690 + 0.0013241 R_1 - 0.00052960 R_2 = 0$$

$$0.12801 - 0.00052960 R_1 + 0.0028598 R_2 = 0$$

from which

$$R_1 = -68.837 \text{ kips} \rightarrow R_{4X} = 68.837 \text{ kips} \leftarrow$$

$$R_2 = -57.510 \text{ kips} \rightarrow S_5 = 57.510 \text{ kips compression}$$

The remainder of the reactive forces and the member forces can be determined by an equilibrium analysis of the truss. The final results of this analysis are shown in Figure 11.13. The analysis for this truss can be verified by the program T2DII in FATPAK II. This is left as an exercise for the reader. It will be found that the computed reactive forces and member forces will agree within expected roundoff error.

This same approach can be used for any type of truss or frame. The displacements Δ_i and the coefficients $\delta_{i,j}$ can always be computed by the Principle of Virtual Work for any structure by

$$\Delta_i = \sum_{n=1}^{NM} \left(\int \frac{S_{x,Qi} S_{x,W}}{A_x E} dx + \int \frac{V_{y,Qi} V_{y,W}}{A_y G} dx + \int \frac{V_{z,Qi} V_{z,W}}{A_z G} dx \right.$$

$$\left. + \int \frac{M_{x,Qi} M_{x,W}}{J_x G} dx + \int \frac{M_{y,Qi} M_{y,W}}{EI_y} dx + \int \frac{M_{z,Qi} M_{z,W}}{EI_z} dx \right)_n \quad (11.11)$$

and

$$\delta_{i,j} = \sum_{n=1}^{NM} \left(\int \frac{S_{x,Qi}S_{x,Qj}}{A_x E} dx + \int \frac{V_{y,Qi}V_{y,Qj}}{A_y G} dx + \int \frac{V_{z,Qi}V_{z,Qj}}{A_z G} dx \right.$$
$$\left. + \int \frac{M_{x,Qi}M_{x,Qj}}{J_x G} dx + \int \frac{M_{y,Qi}M_{y,Qj}}{EI_y} dx + \int \frac{M_{z,Qi}M_{z,Qj}}{EI_z} dx \right)_n \quad (11.12)$$

where **W** is the real load system acting on the reduced structure and **Qi** and **Qj** are virtual load systems acting on the reduced structure with unit loads at R_i and R_j, respectively, and virtual reactive loads at the supports. Of course, only those member deformation effects which correspond to the type of structure being analyzed would be included during the computations. Since the number of calculations which are required to perform the analysis increases rapidly with the number of degrees of static indeterminacy for the structure, this procedure is really not practical for manual calculations for any structure which is more than two or three degrees statically indeterminate. In the next chapter, we will see how this analysis procedure can be formulated as a set of simple matrix operations which is valid for any type of truss or frame in two or three dimensions and how the analysis process can be easily implemented in a simple computer program.

SUGGESTED PROBLEMS

Compute the reactions for each of the beams shown in Figures SP11.1 through SP11.4 using the Method of Consistent Deformations. Draw the shear force and bending moment diagrams after the reactions have been determined. Verify your solutions with either the program BEAMVM or the program F2DII in FATPAK II.

SP11.1 For the beam shown in Figure SP11.1:
(a) Treat the reactive force at the right support as the redundant.
(b) treat the moment at the left support as the redundant.

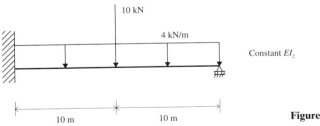

Figure SP11.1.

SP11.2 For the beam shown in Figure SP11.2:
(a) Treat the reactive force at the center support as the redundant.
(b) Treat the reactive force at the right support as the redundant.

Suggested Problems

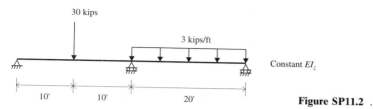

Figure SP11.2 .

SP11.3 For the beam shown in Figure SP11.3:
  (a) Treat the reactive forces at the two interior supports as the redundants.
  (b) Treat the moments over the two interior supports as the redundants.

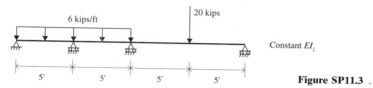

Figure SP11.3 .

SP11.4 For the beam shown in Figure SP11.4:
  (a) Treat the reactive forces at the two roller supports as the redundants.
  (b) Treat the moment at the fixed end and over the interior support as the redundants.

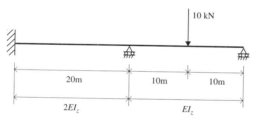

Figure SP11.4

SP11.5 and **SP11.6** Compute the reactions for the plane frames shown in Figures SP11.5 and SP11.6 by the Method of Consistent Deformations. Draw the shear force and bending moments for the frame. Verify your solutions with the program F2DII in FATPAK II.

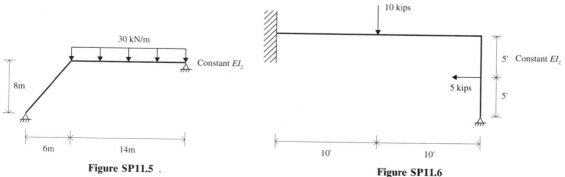

Figure SP11.5 .

Figure SP11.6

SP11.7 to SP11.10 Compute the reactions and the member forces for the plane trusses shown in Figures SP11.7 to SP11.10 using the Method of Consistent Deformations. Treat the most convenient quantities as the redundants. Verify your solutions with the program T2DII in FATPAK II.

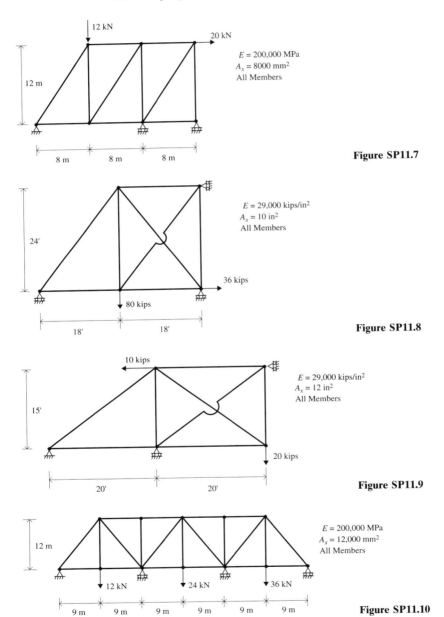

Figure SP11.7

Figure SP11.8

Figure SP11.9

Figure SP11.10

12

The Flexibility Method

BASIC STRUCTURAL MATRICES

The Method of Consistent Deformations can be formulated as a set of simple matrix operations which are ideally suited for implementation in a computer program. The resulting procedure is usually known as the *Flexibility Method*, although it is sometimes called the *Force Method*. Although the Flexibility Method can be used for the analysis of essentially any type of structural system in two or three dimensions, the present discussion will be limited to the analysis of plane trusses and plane frames. A complete discussion of the application of this procedure to the analysis of space trusses and space frames can be found in Fleming, 1989.

The variables which are used to develop the basic equations for the Flexibility Method will be represented in matrix form, with a capital letter enclosed in brackets representing the total matrix and a corresponding lowercase letter representing any individual element in the matrix, with one or two position variables enclosed in parentheses. Curved brackets of the form $\{A\}$ will be used to represent a column matrix and rectangular brackets of the form $[B]$ will be used to represent a rectangular matrix. For any element in a column matrix, $a(i)$, the position variable i represents its row in the matrix, while, for any element in a rectangular matrix, $b(i, j)$, the position variables i and j represent its row and column, respectively. The notation of representing the position of an element in a matrix by enclosing the position variables in parentheses, rather than as subscripts of the form a_i or $b_{i,j}$, is being used here since either parentheses or rectangular brackets are used in essentially all computer programming languages to define the position variables of an element in a matrix.

Global Joint Load Matrix
and Global Joint Displacement Matrix

The active and reactive loads for any structure which is analyzed by the Flexibility Method must be restricted to concentrated loads acting on the joints. For a plane truss, these loads will consist of concentrated forces acting in the plane of the truss, while, for a plane frame, the loads may consist of concentrated forces acting in the plane of the frame and moments about an axis perpendicular to that plane. Any distributed forces which are acting on the members in a frame must be converted to equivalent concentrated joint loads before the analysis can be performed. These concentrated joint loads will consist of the fixed end forces and moments at the end joints of any member which is subjected to a distributed force, as defined previously in Eq. (6.10) through Eq. (6.24) in Chapter 6 during the discussion of the statically determinate plane frame analysis program SDPFRAME.

The joint loads for any structure, such as the plane truss and plane frame whose mathematical models are shown in Figures 12.1a and 12.1b, will be expressed as components in the global right hand orthogonal coordinate system for the structure, as defined previously in Figure 1.3 in Chapter 1, and will be represented by a matrix $\{W\}$, which is known as the *global joint load matrix*, of the form

$$\{W\} = \begin{Bmatrix} w(1) \\ w(2) \\ w(3) \\ w(4) \end{Bmatrix} \tag{12.1}$$

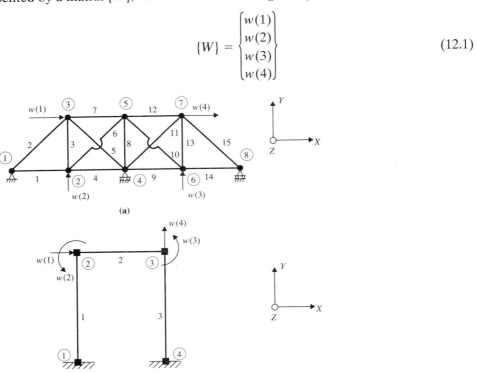

Figure 12.1 Joint loads.

Basic Structural Matrices

in which any element $w(i)$ represents a particular joint load component. The load $w(i)$ may be either a force or a moment as long as it is consistent with the type of loads which are permitted on the structure. The number of elements in the matrix will be equal to the number of joint load components which are considered to act on the structure during the analysis. Force components are positive if they act along one of the coordinate axes and moments are positive about an axis by the right hand rotation rule.

The displacements of the joints will be represented by a matrix $\{D\}$, which is known as the *global joint displacement matrix*, of the form

$$\{D\} = \begin{Bmatrix} d(1) \\ d(2) \\ d(3) \\ d(4) \end{Bmatrix} \tag{12.2}$$

in which any element $d(i)$ represents the displacement at load $w(i)$ in the direction of that load. If $w(i)$ is a force, then $d(i)$ will be a translation, while, if $w(i)$ is a moment, then $d(i)$ will be a rotation. The displacement $d(i)$ is positive if it is in the positive direction of the joint load $w(i)$.

Only the specific displacements which are included in $\{D\}$ will be computed during the analysis. Therefore, if it is desired to compute a particular displacement, a place must be reserved for that displacement in $\{D\}$ by defining a load in $\{W\}$ at that point on the structure even though the magnitude of that load is zero. If all of the displacements of each of the joints in the structure are required, then it will be necessary to include a load in $\{W\}$ corresponding to every displacement degree of freedom in the structure. The number of displacement degrees of freedom NDOF will vary depending upon the type of structure. For a plane truss,

$$\text{NDOF} = 2\text{NJ} - \text{NR} \tag{12.3}$$

for a space truss or a plane frame,

$$\text{NDOF} = 3\text{NJ} - \text{NR} \tag{12.4}$$

and for a space frame,

$$\text{NDOF} = 6\text{NJ} - \text{NR} \tag{12.5}$$

where NJ is the number of joints in the mathematical model of the structure and NR is the number of support restraints. It is usually not necessary to include this many elements in $\{W\}$ and $\{D\}$ to gain the information needed to determine how a structure behaves under a set of loads. It is prudent to plan ahead to determine which joint displacements are actually needed since the effort to perform the analysis increases with the number of elements in $\{W\}$ and $\{D\}$.

Global Flexibility Matrix

The N joint loads $w(1)$ through $w(N)$ acting on any structure can be related to the N joint displacements $d(1)$ through $d(N)$ by the *global flexibility coefficients* in a set of equations of the form

$$\begin{aligned} f(1,1)w(1) + f(1,2)w(2) + \cdots + f(1,N)w(N) &= d(1) \\ f(2,1)w(1) + f(2,2)w(2) + \cdots + f(2,N)w(N) &= d(2) \\ &\vdots \\ f(N,1)w(1) + f(N,2)w(2) + \cdots + f(N,N)w(N) &= d(N) \end{aligned} \quad (12.6)$$

where any global flexibility coefficient $f(i, j)$ is equal to the magnitude of the displacement $d(i)$ due to a unit magnitude of the load $w(j)$ with all other elements in $\{W\}$ being zero. This set of equations can be expressed in matrix form as

$$\begin{bmatrix} f(1,1) & f(1,2) & \cdots & f(1,N) \\ f(2,1) & f(2,2) & \cdots & f(2,N) \\ \vdots & \vdots & \ddots & \vdots \\ f(N,1) & f(N,2) & \cdots & f(N,N) \end{bmatrix} \begin{Bmatrix} w(1) \\ w(2) \\ \vdots \\ w(N) \end{Bmatrix} = \begin{Bmatrix} d(1) \\ d(2) \\ \vdots \\ d(N) \end{Bmatrix} \quad (12.7)$$

or

$$[F]\{W\} = \{D\} \quad (12.8)$$

where $[F]$ is the *global flexibility matrix* for the structure. This matrix has two important properties: first, it is always square since the number of elements in $\{W\}$ will always be equal to the number of elements in $\{D\}$; and second, it is always symmetric as a result of Maxwell's Theorem since $f(i, j)$ must be equal to $f(j, i)$ for a linear elastic structure.

The matrix $[F]$ can be considered to be a general solution for the joint displacements $\{D\}$ for the structure for a set of joint loads $\{W\}$. A particular solution can be obtained by substituting numerical values for the elements in $\{W\}$ and performing the matrix multiplication represented by Eq. (12.8) to obtain a specific set of values for the joint displacements. As we will see later in this chapter, the Flexibility Method will provide a very simple and convenient procedure for generating $[F]$ for any type of statically determinate or statically indeterminate structure.

Local Member End Load Matrix and Local Member Deformation Matrix

As the external loads are applied to the joints of a structure, the joints apply internal loads to the ends of the members and the members apply equal and opposite loads to the joints to hold the joints in equilibrium. The loads applied to the ends of the members cause the members to deform. These member end loads and the corresponding member deformations will be represented by separate column

Basic Structural Matrices

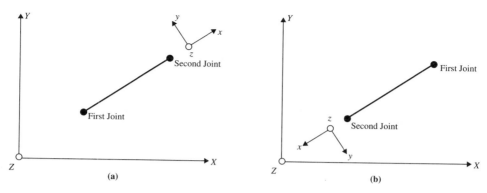

Figure 12.2

matrices for each individual member in the structure. For any individual member n in the structure, the member end loads will be represented by the *local member end load matrix* $\{S\}_n$ and the corresponding member deformations will be represented by the *local member deformation matrix* $\{U\}_n$.

The elements in the matrix $\{S\}_n$, for any member n, will correspond to the statically independent components of the end loads for the member in the local xyz coordinate system of the member. There will be a unique right hand orthogonal local coordinate system for each member in the structure, which is oriented with the positive direction of the local x axis extending along the member from the first joint toward the second joint. For a plane truss or a plane frame member, the local x and y axes lie in the plane of the structure, with the positive local z axis extending out of the plane in the same direction as the global Z axis. Since either end joint may be chosen as the first joint during the analysis, there are two possible directions for the x and y axes for each member in a plane frame, as shown in Figure 12.2. The y and z axes must correspond to the principal axes of the cross section for each member in a plane frame; otherwise, the structure must be analyzed as a space frame. The orientation of the principal axes is not important for a truss member since bending deformations will not be considered during the analysis.

Plane Truss Member. The end loads for a plane truss member consist of a pair of axial forces acting at each end of the member, with the positive directions corresponding to the positive direction of the local x axis, as shown in Figure 12.3. In order to satisfy equilibrium for the member, these two forces must be equal in magnitude and opposite in direction. Therefore, only one of the end loads is statically independent and the local member end load matrix $\{S\}_n$ will only contain one element.

$$\{S\}_n = \{s(1)\}_n \tag{12.9}$$

Since either end load for the member can be considered to be the statically independent quantity, there are two possible definitions for the single element $s(1)_n$ in

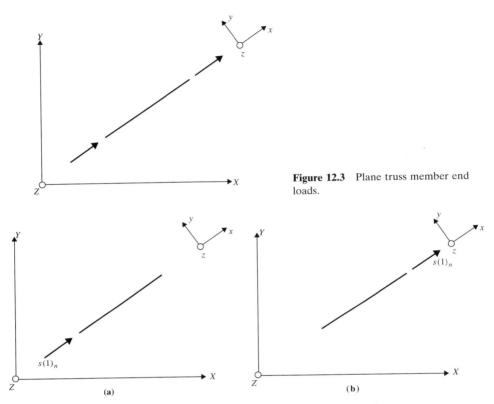

Figure 12.3 Plane truss member end loads.

Figure 12.4 Plane truss member independent end load.

$\{S\}_n$, as shown in Figure 12.4. For convenience in the analysis, the definition shown in Figure 12.4b will be used, since a positive value for this end load corresponds to tension in the member. If the definition shown in Figure 12.4a were used, then a positive value for the end load would correspond to compression in the member, which could be confusing since tension is normally considered to be positive for the axial force in a truss member.

The local member deformation matrix $\{U\}_n$ will contain one element, which corresponds to the deformation produced in the member by $s(1)_n$:

$$\{U\}_n = \{u(1)\}_n \qquad (12.10)$$

Figure 12.5 shows the definition of the deformation $u(1)_n$ for a plane truss member. This deformation is equal to the relative axial displacement along the member of the second end with respect to the first end. A physical interpretation is to consider the first end of the member to be restrained against translation by the statically dependent end load which balances $s(1)_n$. This axial deformation will be equal to the change in length of the member. Using this definition for $u(1)_n$ permits the total internal work which is performed as the truss is loaded to be com-

Basic Structural Matrices

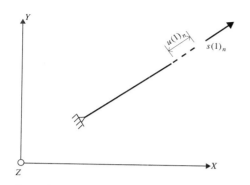

Figure 12.5 Plane truss member deformation.

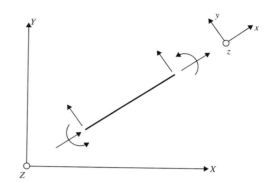

Figure 12.6 Plane frame member end loads.

puted by summing the work performed as each member deforms, by the simple expression

$$\text{Work}_{\text{int}} = \sum_{n=1}^{NM} \frac{\{S\}_n^T \{U\}_n}{2} \qquad (12.11)$$

As we will see later, this is one of the expressions which will be used in developing the basic equations for the Flexibility Method.

Plane Frame Member. The member end loads for a plane frame member consist of a force component along the local x axis, a force component along the local y axis and a moment about the local z axis at each end of the member, as shown in Figure 12.6. Since there are three independent equations of equilibrium for a rigid body in a plane, only three of these end loads will be statically independent. Therefore, the matrix $\{S\}_n$ for a plane frame member will contain three elements.

$$\{S\}_n = \begin{Bmatrix} s(1) \\ s(2) \\ s(3) \end{Bmatrix}_n \qquad (12.12)$$

There are several different choices which can be made for which three of the six end loads are considered to be statically independent. Figures 12.7a and 12.7b show two sets of end loads which are often used for analysis of a plane frame in the Flexibility Method. However, any set of independent end loads may be used if desired. After the three independent end loads have been computed for any member, the three dependent end loads can be determined by an equilibrium analysis of the member.

The member local deformation matrix $\{U\}_n$ will also contain three elements for a plane frame member:

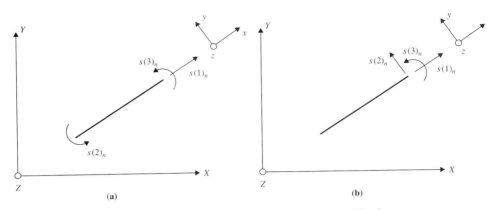

Figure 12.7 Plane frame member independent end loads.

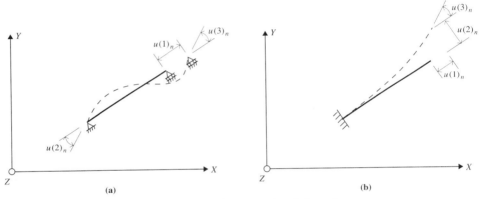

Figure 12.8 Plane frame member deformations.

$$\{U\}_n = \begin{Bmatrix} u(1) \\ u(2) \\ u(3) \end{Bmatrix}_n \tag{12.13}$$

Each element $u(i)_n$ corresponds to the action of the end load $s(i)_n$ in deforming the member. Figures 12.8a and 12.8b show the definitions of these deformations for the two sets of end loads shown in Figures 12.7a and 12.7b. These deformations are the relative displacements which would occur in the member if the member were restrained by the action of the statically dependent end loads. Rigid body displacements of the member are not included in these deformations. Therefore, a physical interpretation for these deformations is to consider the member to have fictitious supports which consist of a pin at the first joint and a longitudinal roller at the second joint for the set of member deformations shown in Figure 12.8a and a fixed support at the first joint with the second joint free to translate and rotate for the set of member deformations shown in Figure 12.8b.. These definitions

Basic Structural Matrices

are consistent with the requirements of Eq. (12.11) for computing the internal work to deform a plane frame, as described previously for a plane truss.

Local Member Flexibility Matrix

The member deformations can be expressed in terms of the member end loads, for any member n in the structure, by the *local member flexibility matrix* $[F_m]_n$, as

$$\{U\}_n = [F_m]_n \{S\}_n \qquad (12.14)$$

where any element $f_m(i, j)_n$ in $[F_m]_n$ is equal to the magnitude of $u(i)_n$ due to a unit $s(j)_n$ with all other elements in $\{S\}_n$ being zero. It now becomes obvious from this definition for the elements in $[F_m]_n$ why the member end loads in $\{S\}_n$ can only consist of the statically independent end loads for the member. The statically dependent end loads must assume values to balance the unit end loads to ensure that the member is in equilibrium. The magnitudes of these dependent end loads will be equal to the reactive loads at the fictitious restraints for the member corresponding to the dependent end loads.

Plane Truss Member. The matrix $[F_m]_n$ will contain only one element for a plane truss member since the matrices $\{S\}_n$ and $\{U\}_n$ only contain single elements. From our sophomore mechanics course, we know that the change in length of an elastic prismatic member subjected to an axial force of magnitude $s(1)_n$ will be

$$\Delta L_x = \frac{s(1)_n L}{A_x E} \qquad (12.15)$$

where L is the length of the member, A_x is the cross section area and E is the modulus of elasticity of the material. Therefore, since $u(1)_n$ corresponds to the change in length of the member due to $s(1)_n$, it follows that the matrix $[F_m]_n$ for a prismatic plane truss member is

$$[F_m]_n = \left[\frac{L}{A_x E} \right]_n \qquad (12.16)$$

Plane Frame Member. The member deformations produced by the end loads $s(2)_n$ and $s(3)_n$, for either of the sets of independent end loads shown in Figures 12.7a and 12.7b, can be determined using the procedures described in Chapter 8 for computing the displacements for a beam in bending.

For the set of independent end loads in Figure 12.7a, the three member deformations for the member, with the fictitious simple supports corresponding to the dependent end loads, can be expressed in terms of the member end loads, if shear deformations in the member are ignored, as

$$u(1)_n = \frac{s(1)_n L}{A_x E} \tag{12.17}$$

$$u(2)_n = \frac{s(2)_n L}{3EI_z} - \frac{s(3)_n L}{6EI_z} \tag{12.18}$$

$$u(3)_n = -\frac{s(2)_n L}{6EI_z} + \frac{s(3)_n L}{3EI_z} \tag{12.19}$$

where I_z is the moment of inertia of the cross section about the local z axis. Therefore, the local member flexibility matrix is

$$[F_m]_n = \begin{bmatrix} \dfrac{L}{A_x E} & 0 & 0 \\ 0 & \dfrac{L}{3EI_z} & -\dfrac{L}{6EI_z} \\ 0 & -\dfrac{L}{6EI_z} & \dfrac{L}{3EI_z} \end{bmatrix}_n \tag{12.20}$$

If shear deformations are included, the matrix $[F_m]_n$ must be modified to include the contribution of these additional deformations to the quantities $u(2)_n$ and $u(3)_n$. The resulting matrix is

$$[F_m]_n = \begin{bmatrix} \dfrac{L}{A_x E} & 0 & 0 \\ 0 & \dfrac{L}{3EI_z} + \dfrac{1}{LA_y G} & -\dfrac{L}{6EI_z} + \dfrac{1}{LA_y G} \\ 0 & -\dfrac{L}{6EI_z} + \dfrac{1}{LA_y G} & \dfrac{L}{3EI_z} + \dfrac{1}{LA_y G} \end{bmatrix}_n \tag{12.21}$$

where A_y is the effective shear area of the cross section as described previously in Chapter 9.

For the set of independent end loads in Figure 12.7b the member deformations for the cantilever member with the fictitious fixed support can be expressed in terms of the member end loads, if shear deformations in the member are ignored, as

$$u(1)_n = \frac{s(1)_n L}{A_x E} \tag{12.22}$$

$$u(2)_n = \frac{s(2)_n L^3}{3EI_z} + \frac{s(3)_n L^2}{2EI_z} \tag{12.23}$$

Basic Structural Matrices

$$u(3)_n = \frac{s(2)_n L^2}{2EI_z} + \frac{s(3)_n L}{EI_z} \qquad (12.24)$$

from which the member flexibility matrix is

$$[F_m]_n = \begin{bmatrix} \dfrac{L}{A_x E} & 0 & 0 \\ 0 & \dfrac{L^3}{3EI_z} & \dfrac{L^2}{2EI_z} \\ 0 & \dfrac{L^2}{2EI_z} & \dfrac{L}{EI_z} \end{bmatrix}_n \qquad (12.25)$$

If shear deformations are included, the local member flexibility matrix for this set of independent end loads will become

$$[F_m]_n = \begin{bmatrix} \dfrac{L}{A_x E} & 0 & 0 \\ 0 & \dfrac{L^3}{3EI_z} + \dfrac{L}{A_y G} & \dfrac{L^2}{2EI_z} \\ 0 & \dfrac{L^2}{2EI_z} & \dfrac{L}{EI_z} \end{bmatrix}_n \qquad (12.26)$$

The effect of shear deformations usually can be ignored for long slender plane frame members, as demonstrated previously for a beam in Chapter 9. However, for short deep members, the shear deformations can have a significant effect upon the overall response of a frame.

Total Structure Member End Load Matrix and Total Structure Member Deformation Matrix

The total set of statically independent member end loads for the structure and the corresponding member deformations will be defined by the *total structure member end load matrix* $\{S\}$ and the *total structure member deformation matrix* $\{U\}$. For a plane truss, the matrix $\{S\}$ and the matrix $\{U\}$ will contain one element for each member, while for a plane frame, they will contain three elements for each member. For example, Figure 12.9 shows one of the permissible definitions for the nine elements for $\{S\}$ for the plane frame shown in Figure 12.1b. These particular end loads correspond to the statically independent end loads for each member shown in Figure 12.7a. However, any set of independent loads may be used for any member in the frame. The matrix $\{U\}$ for this frame will also have nine elements to correspond to the size of $\{S\}$, where each element $u(i)$ represents the deformation in the member corresponding to the action of $s(i)$ for the particular set of inde-

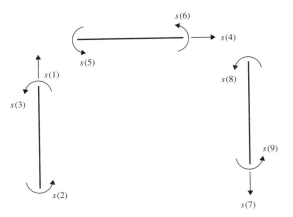

Figure 12.9

pendent end loads which are being used for that member. Since the end loads in Figure 12.9 have been numbered consecutively for each member in the frame in turn, the matrices $\{S\}$ and $\{U\}$ will be in the form

$$\{S\} = \begin{Bmatrix} s(1) \\ s(2) \\ s(3) \\ \vdots \\ s(9) \end{Bmatrix} = \begin{Bmatrix} \{S\}_1 \\ \{S\}_2 \\ \{S\}_3 \end{Bmatrix} \qquad (12.27)$$

and

$$\{U\} = \begin{Bmatrix} u(1) \\ u(2) \\ u(3) \\ \vdots \\ u(9) \end{Bmatrix} = \begin{Bmatrix} \{U\}_1 \\ \{U\}_2 \\ \{U\}_3 \end{Bmatrix} \qquad (12.28)$$

where $\{S\}_1$, $\{S\}_2$ and $\{S\}_3$ are the local member end load matrices, and $\{U\}_1$, $\{U\}_2$ and $\{U\}_3$, are the corresponding local member deformation matrices for members 1, 2 and 3, respectively.

If some other numbering scheme had been used for the elements in $\{S\}$ and $\{U\}$, the matrices would still contain the same number of elements, but the elements would be in other positions in the matrices. The numbering scheme used here is the most convenient for performing an analysis by the Flexibility Method. For a plane truss, the member end loads in $\{S\}$ will automatically be numbered consecutively for each member in turn since there is only one independent end load for each member.

Basic Structural Matrices

Total Structure Member Flexibility Matrix

The member deformations $\{U\}_n$, for any member n in the structure can be related to the member end loads $\{S\}_n$ by the local member flexibility matrix $[F_m]_n$ as shown previously in Eq. (12.14). Therefore, for the plane frame in Figure 12.1a the relationships for the three members are

$$\{U\}_1 = [F_m]_1 \{S\}_1$$
$$\{U\}_2 = [F_m]_2 \{S\}_2 \qquad (12.29)$$
$$\{U\}_3 = [F_m]_3 \{S\}_3$$

which can be combined into one matrix equation of the form

$$\begin{Bmatrix} \{U\}_1 \\ \{U\}_2 \\ \{U\}_3 \end{Bmatrix} = \begin{bmatrix} [F_m]_1 & & \\ & [F_m]_2 & \\ & & [F_m]_3 \end{bmatrix} \begin{Bmatrix} \{S\}_1 \\ \{S\}_2 \\ \{S\}_3 \end{Bmatrix} \qquad (12.30)$$

If we now compare the two column matrices in this expression to the forms of $\{S\}$ and $\{U\}$ shown in Eqs. (12.27) and (12.28), for the numbering scheme for the end loads in Figure 12.9, we see that it can be written as

$$\{U\} = [FM]\{S\} \qquad (12.31)$$

where matrix $[FM]$ is known as the *total structure member flexibility matrix*. It contains the individual local member flexibility matrices situated in turn on the main diagonal. If the member end loads had been numbered in some other order in $\{S\}$, the matrix $[FM]$ would still contain the same elements, but it would not be in this diagonal form. The elements would be scattered throughout the matrix. The simplest form for $[FM]$ is obtained if the elements in $\{S\}$ are numbered consecutively for each member in turn.

Global Equilibrium Matrix

The member end loads $\{S\}$ can be expressed in terms of the joint loads $\{W\}$ by the *global equilibrium matrix* $[B]$ as

$$\{S\} = [B]\{W\} \qquad (12.32)$$

where any element $b(i, j)$ in $[B]$ is equal to the magnitude of the member end load $s(i)$ due to a unit magnitude of $w(j)$ with all other elements in $\{W\}$ being zero. It is a trivial task to perform this simple matrix multiplication to determine the magnitude of the member end loads, for any magnitudes of the joint loads, after the matrix $[B]$ has been determined for any structure. The elements in this matrix can be easily computed for a statically determinate structure by an equilibrium analysis, since each column j in the matrix corresponds to the magnitudes of the elements in $\{S\}$ due to a unit magnitude of $w(j)$. However, it is not possible to compute the elements in [B] for a statically indeterminate structure by an equilibrium analysis alone. The Flexibility Method provides a simple procedure for generating this matrix for a statically indeterminate structure.

MATRIX FORMULATION OF THE PRINCIPLE OF VIRTUAL WORK

One of the basic tools which will be used in the derivation of the basic equations for the Flexibility Method is the Principle of Virtual Work. However, before we can proceed, we must redevelop this principle in matrix form. The procedure will be very similar to that used in the original development of the Principle of Virtual Work in Chapter 9, except that the expressions for the work performed as the structure deforms will be written as a set of matrix operations.

Consider a stable linear elastic structure subjected to three different loading conditions. The structure may be either statically determinate or statically indeterminate. The mathematical model of the structure will be assumed to consist of a set of joints which are connected by elastic members.

Case I. Assume that the structure is subjected to the load system **Q**, which consists of a set of concentrated loads $\{Q\}$ at the joints, as shown in Figure 12.10a. A generic representation of the structure is shown here since it may be of any type and have any geometry. These loads will consist of both the active and the reactive loads on the structure, which will result in a load system which is in equilibrium. The load system **Q** will produce joint displacements $\{D_Q\}$, member end loads $\{S_Q\}$ and member deformations $\{U_Q\}$. The member end loads $\{S_Q\}$ will be in equilibrium with the joint loads $\{Q\}$ at each joint and the member deformations $\{U_Q\}$ will be geometrically compatible with the joint displacements $\{D_Q\}$. The external work performed by these joint loads moving through the joint displacements will be

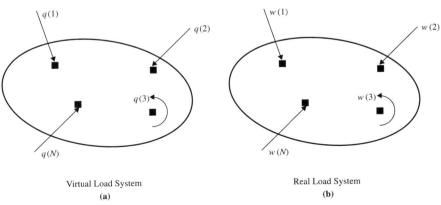

Virtual Load System
(a)

Real Load System
(b)

Figure 12.10

$$\text{Work}_{\text{ext}, Q} = \frac{\{Q\}^T \{D_Q\}}{2} \tag{12.33}$$

and the internal work performed by the member end loads moving through the member deformations will be

$$\text{Work}_{\text{int}, Q} = \sum_{n=1}^{NM} \left(\frac{\{S_Q\}_n^T \{U_Q\}_n}{2} \right)_n = \frac{\{S_Q\}^T \{U_Q\}}{2} \tag{12.34}$$

Case II. Assume that the structure is subjected to the load system **W**, which consists of a set of concentrated loads $\{W\}$ at the joints as shown in Figure 12.10b. These loads will also consist of both the active and the reactive loads on the structure, which will result in a load system which is in equilibrium. The joint loads $\{W\}$ must be of the same type and in the same locations as the joint loads $\{Q\}$ in the load system **Q**. Therefore, the number of elements in $\{W\}$ will be equal to the number of elements in $\{Q\}$. The load system **W** will produce joint displacements $\{D\}$, member end loads $\{S\}$ and member deformations $\{U\}$. The member end loads $\{S\}$ will be in equilibrium with the joint loads $\{W\}$ at each joint and the member deformations $\{U\}$ will be geometrically compatible with the joint displacements $\{D\}$. The external work performed by these joint loads moving through the joint displacements will be

$$\text{Work}_{\text{ext}, W} = \frac{\{W\}^T \{D\}}{2} \tag{12.35}$$

and the internal work performed by the member end loads moving through the member deformations will be

$$\text{Work}_{\text{int}, W} = \sum_{n=1}^{NM} \left(\frac{\{S\}_n^T \{U\}_n}{2} \right)_n = \frac{\{S\}^T \{U\}}{2} \tag{12.36}$$

Case III. Assume that the structure is subjected to the load system **Q** and then while this load system is still on the structure, it is subjected to the load system **W**. Since the structure will be in equilibrium under each load system separately, it will be in equilibrium under the combined load systems. The total external work for this case will be

$$\text{Work}_{\text{ext}, QW} = \text{Work}_{\text{ext}, Q} + \text{Work}_{\text{ext}, W} + \{Q\}^T \{D\} \tag{12.37}$$

where the last term on the right side is the work performed by the joint loads $\{Q\}$ moving through the joint displacements $\{D\}$ which are produced by the joint loads $\{W\}$. The factor of one half is not present since the loads $\{Q\}$ are acting during the entire time that the displacements $\{D\}$ are produced. The total internal work for this case will be

$$\text{Work}_{\text{int}, QW} = \text{Work}_{\text{int}, Q} + \text{Work}_{\text{int}, W} + \{S_Q\}^T \{U\} \tag{12.38}$$

where the last term on the right side is the work performed by the member end loads $\{S_Q\}$ which are produced by the joint loads $\{Q\}$ moving through the member deformations $\{U\}$ which are produced by the joint loads $\{W\}$.

The external work and the internal work which is performed during each load case must be equal. Therefore, equating Eqs. (12.37) and (12.38) and eliminating equal terms on each side corresponding to load Cases I and II gives

$$\{Q\}^T\{D\} = \{S_Q\}^T\{U\} \tag{12.39}$$

which is the basic equation for the Principle of Virtual Work in matrix form.

If a structure, which is subjected to a set of joint loads $\{Q\}$, which produces a set of member end loads $\{S_Q\}$, undergoes a set of external joint displacements $\{D\}$ and geometrically compatible internal member deformations $\{U\}$, the external work performed by the loads $\{Q\}$ moving through the displacements $\{D\}$ will be equal to the internal work performed by the member end loads $\{S_Q\}$ moving through the member deformations $\{U\}$.

The only requirements are: $\{Q\}$ must be a set of joint loads which are in equilibrium; the member end loads $\{S_Q\}$ must be in equilibrium with the loads $\{Q\}$ at all joints in the structure; and the joint displacements $\{D\}$ and the member deformations $\{U\}$ must be geometrically compatible at all points in the structure. The joint displacements $\{D\}$ and the member deformations $\{U\}$ may be a result of a load system **W** or they may be the result of any other type of external or internal action on the structure. There are no requirements between $\{Q\}$ and $\{W\}$, $\{S_Q\}$ and $\{S\}$, $\{D_Q\}$ and $\{D\}$ or $\{U_Q\}$ and $\{U\}$ except that the pairs of matrices must be the same size and the elements must be of the same type and occur at the same points in the structure. The load system **Q** corresponds to the virtual load system and the load system **W** corresponds to the real load system described previously in Chapter 9 during the development of the Principle of Virtual Work.

FORMULATION OF THE FLEXIBILITY METHOD

The basic analysis procedure which is used in the Flexibility Method is identical to the Method of Consistent Deformations, which was described in Chapter 11, except that the equations are expressed as a set of matrix operations. In the following sections in this chapter, expressions will be developed which can be used to compute the global flexibility matrix $[F]$ and the global equilibrium matrix $[B]$ for any structure. The joint displacements $\{D\}$ and the member end loads $\{S\}$ then can be easily computed for any magnitudes for the joint loads $\{W\}$ by Eqs. (12.8) and (12.32).

Formulation of the Flexibility Method

Computation of Redundant Loads

The first step in the analysis is to compute the redundant loads for the structure. These redundant loads are due to the redundant internal and/or external restraints in the statically indeterminate structure, as defined previously in the development of the Method of Consistent Deformations. For example, for both the plane truss and the plane frame shown in Figure 12.1, there are three redundant restraints since Eqs. (11.5) and (11.7) in Chapter 11 show that both structures are three degrees statically indeterminate. There are a number of choices which can be made for the redundant restraints. One of the many possible choices for the reduced structures for the truss and the frame and the corresponding redundant loads are shown in Figures 12.11a and 12.11b, respectively. For the truss, the redundant loads correspond to the vertical reactive force at joint 4 and the axial forces in members 6 and 10. For the frame, the redundant loads correspond to the horizontal and vertical reactive forces and the reactive moment at joint 4.

The redundant loads for any statically indeterminate structure can be represented by a matrix $\{R\}$ of the form

$$\{R\} = \begin{Bmatrix} r(1) \\ r(2) \\ \vdots \\ r(\text{NDI}) \end{Bmatrix} \tag{12.40}$$

where NDI is the number of degrees of static indeterminacy for the structure. These redundant loads can be computed by the following steps:

Step 1. Remove the redundant restraints from the structure to leave a reduced structure which is statically determinate and stable and compute the member end loads $\{S_1\}$ and the displacements $\{D_1\}$, which correspond to the action of the redundant loads, due to the joint loads $\{W\}$ acting on the reduced structure.

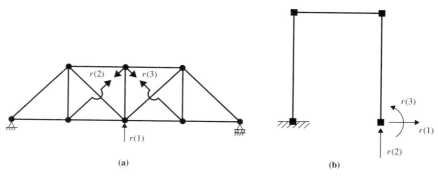

Figure 12.11

The member end loads $\{S_1\}$ can be related to the joint loads $\{W\}$ by using an equilibrium matrix, which we will call $[B_1]$, by the relationship

$$\{S_1\} = [B_1]\{W\} \tag{12.41}$$

where any element $b_1(i, j)$ in $[B_1]$ is equal to the magnitude of $s_1(i)$ in the reduced structure due to a unit magnitude of $w(j)$ with all other elements in $\{W\}$ being zero. The number of rows in $[B_1]$ will be equal to the number of member end loads, and the number of columns will be equal to the number of joint loads. Since the reduced structure is statically determinate, the elements in $[B_1]$ can be computed by performing an equilibrium analysis of the reduced structure. The elements in any column j in $[B_1]$ will be equal to the magnitudes of the member end loads in $\{S_1\}$ due to a unit magnitude of $w(j)$ with all other elements in $\{W\}$ being zero.

The redundant displacements $\{D_1\}$ can be related to the joint loads by using a load-displacement matrix, which we will call $[G_1]$, by the relationship

$$\{D_1\} = [G_1]\{W\} \tag{12.42}$$

where any element $g_1(i, j)$ in $[G_1]$ is equal to the magnitude of $d_1(i)$ due to a unit magnitude of $w(j)$ with all other elements in $\{W\}$ being zero. Any element $d_1(i)$ in $\{D_1\}$ will be equal to the displacement at $r(i)$ corresponding to the action of $r(i)$. If $r(i)$ is a force, then $d_1(i)$ will be a translation, while, if $r(i)$ is a moment, then $d_1(i)$ will be a rotation. The number of rows in $[G_1]$ will be equal to the number of degrees of static indeterminacy of the structure, and the number of columns will be equal to the number of joint loads. We will develop an expression for computing the elements in $[G_1]$ later. Note that any redundant displacement $d_1(i)$ is equivalent to the displacement Δ'_i, which was used in Chapter 11 in the Method of Consistent Deformations.

For the reduced structure shown in Figure 12.11a, the displacement $d_1(1)$ corresponds to the vertical translation of joint 4 and the displacements $d_1(2)$ and $d_1(3)$ correspond to the relative translations of the two sections on either side of the interior cuts in members 6 and 10. For the reduced structure in Figure 12.11b the displacements $d_1(1)$ and $d_1(2)$ correspond to the horizontal and vertical translations of joint 4 and the displacement $d_1(3)$ corresponds to the rotation of joint 4.

Step 2. Place the redundant loads $\{R\}$ on the reduced structure and compute the member end loads $\{S_2\}$ and the displacements $\{D_2\}$ which correspond to the action of the redundant loads. Each individual displacement $d_2(i)$ in $\{D_2\}$ is at the same point and in the same direction as the displacement $d_1(i)$ in $\{D_1\}$.

The member end loads can be related to the redundant loads by using an equilibrium matrix, which we will call $[B_2]$, by

$$\{S_2\} = [B_2]\{R\} \tag{12.43}$$

Formulation of the Flexibility Method

where any element $b_2(i, j)$ in $[B_2]$ is equal to the magnitude of $s_2(i)$ due to a unit magnitude of $r(j)$ with all other elements in $\{R\}$ being zero. The number of rows in $[B_2]$ will be equal to the number of member end loads, and the number of columns will be equal to the number of degrees of static indeterminacy of the structure. The elements in $[B_2]$ can be computed by an equilibrium analysis of the reduced structure. The elements in any column j will be equal to the magnitudes of the member end loads in $\{S_2\}$ due to a unit magnitude of $r(j)$ with all other elements in $\{R\}$ being zero.

The redundant displacements $\{D_2\}$ can be related to the redundant loads using a load-displacement matrix which we will call $[G_2]$ by the relationship

$$\{D_2\} = [G_2]\{R\} \qquad (12.44)$$

where any element $g_2(i, j)$ in $[G_2]$ is equal to the magnitude of $d_2(i)$ due to a unit magnitude of $r(j)$ with all other elements in $\{R\}$ being zero. The matrix $[G_2]$ will always be square since the number of rows and the number of columns in the matrix are equal to the number of degrees of static indeterminacy of the structure. The matrix will also always be symmetric as a result of Maxwell's Theorem. We will develop an expression for computing the elements in $[G_2]$ later. Note that any displacement $d_2(i)$ is equivalent to the displacement Δ_i'' and any matrix element $g_2(i, j)$ is equivalent to the redundant displacement coefficient $\delta_{i,j}$ which were used in Chapter 11 in the Method of Consistent Deformations.

Step 3. Since the redundant displacements must be zero in the original structure due to the restraints corresponding to the redundant loads, we can add the expressions for the displacements $\{D_1\}$ and $\{D_2\}$ in Eqs. (12.42) and (12.44) and set the result to zero to obtain

$$[G_1]\{W\} + [G_2]\{R\} = 0 \qquad (12.45)$$

which can be solved for the redundant loads:

$$\{R\} = -[G_2]^{-1}[G_1]\{W\} \qquad (12.46)$$

Therefore, if we can develop expressions for computing the elements in the matrices $[G_1]$ and $[G_2]$, the redundant loads $\{R\}$ can be easily computed. Expressions for computing the elements in $[G_1]$ and $[G_2]$ can be developed by using the Principle of Virtual Work.

Matrix $[G_1]$. To generate an expression for the matrix $[G_1]$ let the joint loads $\{Q\}$ in the virtual work relationship in Eq. (12.39) correspond to the redundant loads $\{R\}$ and the associated reactions in the reduced structure. The member end loads $\{S_Q\}$ will then correspond to the end loads $\{S_2\}$, where the relationship in Eq. (12.43) ensures that $\{R\}$ and $\{S_2\}$ are in equilibrium at each joint as required by the Principle of Virtual Work. In addition, let the joint displacements $\{D\}$ correspond to the redundant displacements $\{D_1\}$ in the reduced structure due to the joint loads $\{W\}$. The member deformations $\{U\}$ will then correspond to the member deformations $\{U_1\}$ in the reduced structure due to $\{W\}$. Equation (12.42) and the relationship

$$\{U_1\} = [FM]\{S_1\} = [FM][B_1]\{W\} \quad (12.47)$$

will ensure that $\{D_1\}$ and $\{U_1\}$ are geometrically compatible as required for the virtual work relationship. We can now substitute these quantities into Eq. (12.39) to obtain

$$\{R\}^T[G_1]\{W\} = \{[B_2]\{R\}\}^T[FM][B_1]\{W\} \quad (12.48)$$

which can be written as

$$\{R\}^T[G_1]\{W\} = \{R\}^T[[B_2]^T[FM][B_1]]\{W\} \quad (12.49)$$

If this expression is to be satisfied for any $\{W\}$ and the corresponding $\{R\}$, then

$$[G_1] = [B_2]^T[FM]\{B_1\} \quad (12.50)$$

Therefore, the elements in the matrix $[G_1]$ can be computed by a simple set of matrix operations using the equilibrium matrices $[B_1]$ and $[B_2]$ for the reduced structure and the total structure member flexibility matrix $[FM]$. A check of the size of the individual matrices on the right side of Eq. (12.50) will show that these matrix operations will result in the correct size for $[G_1]$.

Matrix $[G_2]$. To generate an expression for the matrix $[G_2]$ let the joint loads $\{Q\}$ in the virtual work relationship in Eq. (12.39) correspond to the redundant loads $\{R\}$ and the associated reactions and let the member end loads $\{S_Q\}$ correspond to the end loads $\{S_2\}$ in the reduced structure, as in the previous derivation of the expression for $[G_1]$. However, for this case, let the joint displacements $\{D\}$ correspond to the displacements $\{D_2\}$ in the reduced structure due to $\{R\}$. The member deformations $\{U\}$ will then correspond to the member deformations $\{U_2\}$ in the reduced structure due to $\{R\}$. Equation (12.44) and the relationship

$$\{U_2\} = [FM]\{S_2\} = [FM][B_2]\{R\} \quad (12.51)$$

will ensure that $\{D_2\}$ and $\{U_2\}$ are geometrically compatible as required for the virtual work relationship. We can now substitute these quantities into Eq. (12.39) to obtain

$$\{R\}^T[G_2]\{R\} = \{[B_2]\{R\}\}^T[FM][B_2]\{R\} \quad (12.52)$$

which can be written as

$$\{R\}^T[G_2]\{R\} = \{R\}^T[[B_2]^T[FM][B_2]]\{R\} \quad (12.53)$$

If this expression is to be satisfied by any $\{R\}$, then

$$[G_2] = [B_2]^T[FM][B_2] \quad (12.54)$$

Note that this expression is similar to the expression for $[G_1]$ in Eq. (12.50). The only difference is the last matrix on the right side. This set of matrix operations

Formulation of the Flexibility Method

will result in a square matrix with the number of rows and columns being equal to the number of redundant loads in the structure, as required for the matrix $[G_2]$.

The numerical values for the elements in $[G_1]$ and $[G_2]$ now can be computed for any statically indeterminate structure by the operations shown in Eqs. (12.50) and (12.54). After these two matrices have been determined, the redundant loads for the structure can be computed by Eq. (12.46).

Computation of Member End Loads

The member end loads for the structure can be computed by adding the end loads $\{S_1\}$ in the reduced structure due to $\{W\}$ and the end loads $\{S_2\}$ in the reduced structure due to $\{R\}$:

$$\{S\} = \{S_1\} + \{S_2\} \tag{12.55}$$

Substituting the expressions for $\{S_1\}$ and $\{S_2\}$ in Eqs. (12.41) and (12.43) gives

$$\{S\} = [B_1]\{W\} + [B_2]\{R\} \tag{12.56}$$

which, upon substituting the expression for $\{R\}$ in Eq. (12.46), becomes

$$\{S\} = [[B_1] - [B_2][G_2]^{-1}[G_1]]\{W\} \tag{12.57}$$

Comparing this expression to Eq. (12.32) shows that the elements in the global equilibrium matrix $[B]$ can be computed by

$$[B] = [B_1] - [B_2][G_2]^{-1}[G_1] \tag{12.58}$$

We will not bother substituting the expressions for $[G_1]$ and $[G_2]$ into this expression to find a final expression for $[B]$ in terms of only $[B_1]$, $[B_2]$ and $[FM]$. It is more convenient during the analysis to compute the elements in $[G_1]$ and $[G_2]$ by Eqs. (12.50) and (12.54) and then to use these values directly in Eq. (12.58) to compute the elements in $[B]$. After these computations have been completed, the member end loads can be computed by Eq. (12.32)

Computation of Joint Displacements

An expression for computing the elements in the global flexibility matrix $[F]$ can be developed by using the Principle of Virtual Work. Let the joint loads $\{Q\}$ in the virtual work relationship in Eq. (12.39) correspond to the global joint loads $\{W\}$ and the associated reactions in the original structure. The member end loads $\{S_Q\}$ will then correspond to the end loads $\{S\}$, where the relationship in Eq. (12.32) ensures that $\{W\}$ and $\{S\}$ are in equilibrium at each joint as required. Let the joint displacements $\{D\}$ in Eq. (12.39) correspond to the global joint displacements in the original structure due to the global joint loads $\{W\}$. The member deformations $\{U\}$ in Eq. (12.39) will then correspond to the member deformations $\{U\}$ in the original structure due to the global joint loads $\{W\}$. Equation (12.8) and the relationship

$$\{U\} = [FM]\{S\} = [FM][B]\{W\} \tag{12.59}$$

will ensure that $\{D\}$ and $\{U\}$ are geometrically compatible as required. Substituting these quantities into Eq. (12.39) gives

$$\{W\}^T[F]\{W\} = \{[B]\{W\}\}^T[FM][B]\{W\} \tag{12.60}$$

which can be written as

$$\{W\}^T[F]\{W\} = \{W\}^T[[B]^T[FM][B]]\{W\} \tag{12.61}$$

If this expression is to be satisfied by any $\{W\}$, then

$$[F] = [B]^T[FM][B] \tag{12.62}$$

After the elements in $[F]$ have been computed by this expression, the joint displacements can be computed by Eq. (12.8).

Geometric Compatibility Check

One of the major problems in any manual analysis by the Flexibility Method is the high probability of arithmetic errors due to the large number of calculations in performing the matrix operations. Although the probability of arithmetic errors is small in a computer analysis, after the input data has been properly entered into a computer program, there is still the problem of roundoff errors in the final results. Therefore, it would be beneficial if a simple procedure were available for checking the accuracy of the final results of the analysis. A procedure can be developed to check the results of the analysis for any structure using the Principle of Virtual Work.

Let the matrix $\{D\}$ in the virtual work relationship in Eq. (12.39) correspond to the displacements $\{D_R\}$ at the redundant loads in the reduced structure due to the combination of the joint loads $\{W\}$ and the redundant loads $\{R\}$ which are produced by $\{W\}$. The matrix $\{U\}$ will then correspond to the member deformations in the original structure produced by $\{W\}$. The requirement that $\{D_R\}$ and $\{U\}$ are geometrically compatible can be ensured by defining a geometric matrix $[C]$ for the original structure which relates $\{D_R\}$ to $\{U\}$ by the relationship

$$\{D_R\} = [C]\{U\} \tag{12.63}$$

where any element $c(i, j)$ in $[C]$ is equal to the magnitude of the displacement $d_R(i)$ at the redundant restraint corresponding to $r(i)$ due to a unit magnitude of $u(j)$ with all other elements in $\{U\}$ being zero. In addition, let the matrix $\{Q\}$ correspond to the redundant loads $\{R\}$ in the reduced structure and the associated reactions. The matrix $\{S_Q\}$ will then correspond to the member end loads $\{S_2\}$ in the reduced structure due to $\{R\}$. Equation (12.43) ensures that the loads $\{R\}$ and the member end loads $\{S_2\}$ will be in equilibrium. We can now substitute these quantities into Eq. (12.39) to obtain

$$\{R\}^T[C]\{U\} = \{[B_2]\{R\}\}^T\{U\} \tag{12.64}$$

which can be written as

$$\{R\}^T[C]\{U\} = \{R\}^T[B_2]^T\{U\} \qquad (12.65)$$

from which we can determine that

$$[C] = [B_2]^T \qquad (12.66)$$

Substituting this relationship into Eq. (12.63) gives

$$\{D_R\} = [B_2]^T\{U\} \qquad (12.67)$$

which, upon substituting the relationships in Eqs. (12.31) and (12.32) in turn, results in

$$\{D_R\} = [B_2]^T[FM][B]\{W\} \qquad (12.68)$$

However, the displacements $\{D_R\}$ at the redundant restraints due to the action of $\{W\}$ and $\{R\}$ combined must be zero. The only way that this can be ensured in Eq. (12.68) is if

$$[B_2]^T[FM][B] = [0] \qquad (12.69)$$

where [0] is a null matrix with as many rows as the number of degrees of static indeterminacy of the structure and as many columns as the number of joint loads in $\{W\}$. The matrix which results from the operations on the left side of Eq. (12.69) is known as the *compatibility check matrix*. Therefore, computing the elements in this matrix will give an indication of the accuracy of the computed elements in the global equilibrium matrix $[B]$, which in turn will be an indication of the accuracy of the computed values for the member end loads in $\{S\}$ and the joint displacements in $\{D\}$. If the values of the elements in the matrix are significantly larger than zero, then this would indicate a problem in the computations for $[B]$. However, this check will only give an indication of the accuracy of the calculations to obtain $[B]$ starting with a specific set of initial matrices $[B_1]$, $[B_2]$ and $[FM]$. It does not verify that the elements in these initial matrices were correct.

SUMMARY OF FLEXIBILITY METHOD FOR STATICALLY INDETERMINATE STRUCTURES

The steps which are required to perform the analysis of any statically indeterminate structure by the Flexibility Method can be summarized as follows:

Step 1. Decide on how many joint loads are to be included in the analysis and number them to define the elements in the global joint load matrix $\{W\}$.

Step 2. Remove the redundant restraints from the statically indeterminate structure to obtain a stable statically determinate reduced structure and number the corresponding redundant loads to define the elements in the redundant load matrix $\{R\}$.

Step 3. Choose the statically independent end loads which will be used in the analysis for each member and number the end loads consecutively by member to define the elements in the total structure member end load matrix $\{S\}$.

Step 4. Compute the elements in the matrices $[B_1]$, $[B_2]$ and $[FM]$.

Each column j in $[B_1]$ is equal to the magnitude of the end loads in $\{S\}$ due to a unit magnitude of the joint load $w(j)$ on the reduced structure.

Each column j in $[B_2]$ is equal to the magnitude of the end loads in $\{S\}$ due to a unit magnitude of the redundant load $r(j)$ on the reduced structure.

The matrix $[FM]$ consists of the individual local member flexibility matrices situated consecutively along the main diagonal.

Step 5. Perform the matrix operations corresponding to the following equations:

$$[G_1] = [B_2]^T[FM][B_1] \qquad (12.50)$$

$$[G_2] = [B_2]^T[FM][B_2] \qquad (12.54)$$

$$[B] = [B_1] - [B_2][G_2]^{-1}[G_1] \qquad (12.58)$$

$$[F] = [B]^T[FM][B] \qquad (12.62)$$

$$\{S\} = [B]\{W\} \qquad (12.32)$$

$$\{D\} = [F]\{W\} \qquad (12.8)$$

$$[CHECK] = [B_2]^T[FM][B] \quad \text{(Should be a null matrix.)} \qquad (12.68)$$

If desired, the redundant loads also can be computed by the following operations, but these values are often not needed to supply the information which is actually required by the engineer performing the analysis.

$$\{R\} = -[G_2]^{-1}[G_1]\{W\} \qquad (12.46)$$

Step 6. Compute the statically dependent end loads for each member by performing an equilibrium analysis of the member using the values for the statically independent end loads in $\{S\}$.

Step 7. Compute the reactions by performing an equilibrium analysis of each support joint using the end loads for the members attached to the joints.

Steps 6 and 7 are optional, depending upon the final information which is required for the structure. For example, for some structures, the only information which might be needed is the joint displacements.

The matrices $\{S\}$ and $\{D\}$ represent the member end loads and the joint displacements for the statically indeterminate structure due to the specific set of values which were assigned to the joint loads in $\{W\}$. One of the major advantages of this analysis procedure is that it is not necessary to repeat the entire set of calculations if the magnitudes of the joint loads are changed. The $[B]$ and $[F]$ ma-

Computer Program FLEX

trices can be used to compute the new values for $\{S\}$ and $\{D\}$ by the simple matrix operations in Eqs. (12.32) and (12.8) for any new set of joint loads.

ANALYSIS OF STATICALLY DETERMINATE STRUCTURES BY THE FLEXIBILITY METHOD

The Flexibility Method also can be used to analyze a statically determinate structure. The difference between this type of analysis and the analysis of a statically indeterminate structure is that after the matrices $\{W\}$ and $\{S\}$ have been defined, the elements in the global equilibrium matrix $[B]$ can be computed directly by an equilibrium analysis of the structure. Therefore, the only matrix operations which must be performed are those corresponding to the following equations:

$$[F] = [B]^T [FM] [B] \tag{12.62}$$

$$\{S\} = [B]\{W\} \tag{12.32}$$

$$\{D\} = [F]\{W\} \tag{12.8}$$

There is no compatibility check matrix for a statically determinate structure since there are no redundant loads or redundant displacements.

COMPUTER PROGRAM FLEX

Although it is possible to perform all of the matrix operations manually for the analysis of a structure by the Flexibility Method, it will be found that it is a very tedious and error prone set of calculations. The advantage of this analysis procedure is that it is ideally suited for implementation in a computer program. The program FLEX, whose QBasic source code is given in the ASCII disk file FLEX.BAS, demonstrates how a simple computer program can be developed which will analyze any type of statically determinate or statically indeterminate structure. The program consists of three primary parts. The first part of the program reads the input data, which consists of either the matrices $[B]$, $[FM]$ and $\{W\}$ for a statically determinate structure or the matrices $[B_1]$, $[B_2]$, $[FM]$ and $\{W\}$ for a statically indeterminate structure. The second part of the program computes the member end loads $\{S\}$ and the joint displacements $\{D\}$. The matrix operations are performed by four subroutines for matrix subtraction, matrix transposition, matrix multiplication and matrix inversion. Finally, the third part of the program writes out the results of the analysis. This output consists of a listing of the input matrices, the intermediate matrices which are computed during the analysis and the matrices $\{S\}$ and $\{D\}$. The input matrices are listed so that the input can be checked for errors, while the intermediate matrices are listed to give the reader an idea of the form of these matrices since the program was developed for educational purposes. In addition, the compatible check matrix is included in the output for a

430 The Flexibility Method Chap. 12

statically indeterminate structure. A description of the required format for the input data and the instructions for running the program are contained in the ASCII disk file FLEX.DOC.

The following example problems demonstrate the application of the program FLEX to the analysis of statically determinate and statically indeterminate structures.

Example Problem 12.1

Analyze the statically determinate plane truss whose mathematical model is shown in Figure 12.12a using the program FLEX. The output should include both the horizontal and vertical translations of joint 3.

Since the output is to include the horizontal and the vertical translations of joint 3, the matrix $\{W\}$ must contain both a horizontal and a vertical force component at that point. Figure 12.12b shows the definitions of the elements which will be includ-

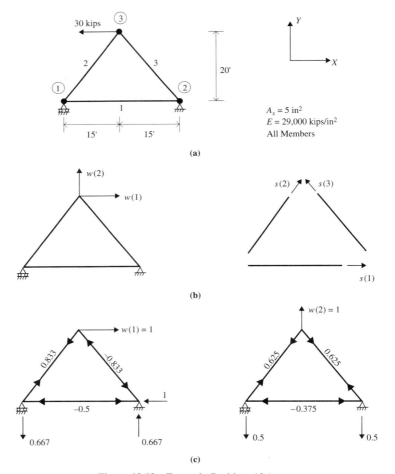

Figure 12.12 Example Problem 12.1.

ed in $\{W\}$ and $\{S\}$. The matrix $\{S\}$ contains three elements since there is one statically independent end load for each truss member.

The input to the program for this statically determinate structure consists of the matrices $[B]$, $[FM]$ and $\{W\}$. The elements in the two columns of the matrix $[B]$ can be computed by an equilibrium analysis of the statically determinate truss by applying unit values of the two joint loads as shown in Figure 12.12c, while the elements in $[FM]$ for each of the members can be computed by Eq. (12.16). The input matrices for the program, in kip and inch units, are

$$[B] = \begin{bmatrix} -0.5 & -0.375 \\ 0.83333 & 0.625 \\ -0.83333 & 0.625 \end{bmatrix}$$

$$[FM] = \begin{bmatrix} 0.0024827 & 0 & 0 \\ 0 & 0.0020689 & 0 \\ 0 & 0 & 0.0020689 \end{bmatrix}$$

$$\{W\} = \begin{Bmatrix} -30.0 \\ 0 \end{Bmatrix}$$

The zero element in $\{W\}$ reserves a space in $\{D\}$ for the vertical translation of joint 3.

Figure 12.13 shows a listing of the input data for FLEX. The values for the elements in the matrices have been entered to five significant figure accuracy to help to eliminate roundoff error in the calculations. Note that the third number in the second line of the input data is zero. This number corresponds to the number of degrees of static indeterminacy of the structure. The zero value informs FLEX that the structure is statically determinate so that it will expect to only read the elements of the matrices $[B]$, $[FM]$ and $\{W\}$ in the input data.

Figure 12.14 shows the output from FLEX. The headings identify each output matrix. Therefore, the numbers should be self explanatory. The elements in $[FM]$ and $[F]$ are listed in exponential format since these numbers are usually very small. The computed values for the member forces in $\{S\}$ and the joint displacements in $\{D\}$ agree with the values obtained for this same truss in Example Problem 9.5 in Chapter 9, in which the joint displacements were computed using the Principle of Virtual Work. In fact, if a careful comparison is made between the individual computations which were performed to compute these displacements by the Flexibility Method and by the Principle of Virtual Work, it will be found that they are identical.

```
Example Problem 12.1 - Analysis by Program FLEX
3,2,0
Matrix [B]
-0.5,-0.375
0.83333,0.625
-0.83333,0.625
Matrix [FM]
1,1,0.0024827,0
2,2,0.0020689,0
3,3,0.0020689,1
Matrix {W}
-30.0
0.0
```

Figure 12.13 Example Problem 12.1, FLEX input.

Example Problem 12.1 - Analysis by Program FLEX

Flexibility Analysis - Statically Determinate

Number of Member End Loads = 3
Number of Joint Loads = 2

Matrix [B]
```
   -0.50000       -0.37500
    0.83333        0.62500
   -0.83333        0.62500
```

Matrix [FM]
```
  0.24827E-02   0.00000E+00   0.00000E+00
  0.00000E+00   0.20689E-02   0.00000E+00
  0.00000E+00   0.00000E+00   0.20689E-02
```

Matrix {W}
```
   -30.000
     0.000
```

Matrix [F]
```
  0.34941E-02   0.46551E-03
  0.46551E-03   0.19655E-02
```

Matrix {S}
```
    15.000
   -25.000
    25.000
```

Matrix {D}
```
    -0.1048
    -0.0140
```

Figure 12.14 Example Problem 12.1, FLEX output.

Example Problem 12.2

Analyze the statically determinate plane frame whose mathematical model is shown in Figure 12.15a using the program FLEX. The output should include both the vertical translation and the rotation of joint 3. Ignore the effect of shear deformations in the members.

Since the output is to include the vertical translation and the rotation of joint 3, the matrix {W} must contain both a vertical force and a moment at that point. Figure 12.15b shows the definitions of the elements which will be included in {W} and {S}. The matrix {S} contains six elements since there are three statically independent end loads for each frame member. The independent end loads, which have been defined for each member, correspond to those shown in Figure 12.7a.

As with the previous plane truss, the input to the program for this statically determinate frame will consist of the matrices [B], [FM] and {W}. The elements in the two columns of the matrix [B] can be computed by an equilibrium analysis of the

Computer Program FLEX

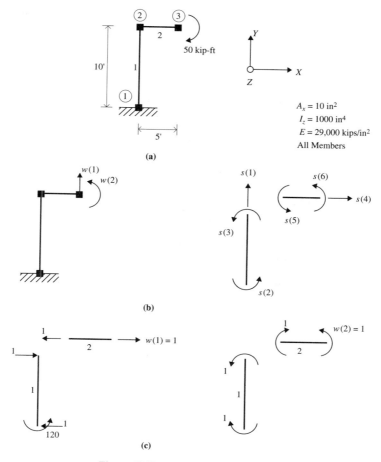

Figure 12.15 Example Problem 12.2.

frame by applying unit values of the two joint loads, as shown in Figure 12.15c, while the elements in $[FM]$ for each of the members can be computed by Eq. (12.20). The resulting input matrices $[B]$ and $[FM]$ for the program, in kip and inch units, are

$$[B] = \begin{bmatrix} 0 & 0 \\ 120.0 & -1.0 \\ 0 & 1.0 \\ 1.0 & 0 \\ 0 & -1.0 \\ 0 & 1.0 \end{bmatrix}$$

$$[FM] = \begin{bmatrix} [F_m]_1 & \\ & [F_m]_2 \end{bmatrix}$$

```
Example Problem 12.2 - Analysis by Program FLEX
6,2,0
Matrix [B]
0.0,0.0
120.0,-1.0
0.0,1.0
1.0,0.0
0.0,-1.0
0.0,1.0
Matrix [FM]
1,1,0.00041379,0
2,2,0.0000013793,0
2,3,-0.00000068966,0
3,2,-0.00000068966,0
3,3,0.0000013793,0
4,4,0.00020690,0
5,5,0.00000068966,0
5,6,-0.00000034483,0
6,5,-0.00000034483,0
6,6,0.00000068966,1
Matrix {W}
0.0
-600.0
```

Figure 12.16 Example Problem 12.2, FLEX input.

where

$$[F_m]_1 = \begin{bmatrix} 0.00041379 & 0 & 0 \\ 0 & 0.0000013793 & -0.00000068966 \\ 0 & -0.00000068966 & 0.0000013793 \end{bmatrix}$$

$$[F_m]_2 = \begin{bmatrix} 0.00020690 & 0 & 0 \\ 0 & 0.00000068966 & -0.00000034483 \\ 0 & -0.00000034483 & 0.00000068966 \end{bmatrix}$$

and the matrix $\{W\}$ is

$$[W] = \begin{Bmatrix} 0 \\ -600.0 \end{Bmatrix}$$

Figure 12.16 above shows a listing of the input data for FLEX and Figure 12.17 on page 434 shows the output. The value of $d(2)$, which corresponds to the rotation of joint 3 in radians, agrees with the value computed in Example Problem 9.2 in Chapter 9 for this same plane frame. The same results for the displacements also would be obtained if the statically independent member end loads in Figure 12.7b were used in the matrix $\{S\}$. This is left as an exercise for the reader.

Example Problem 12.3

Compute the member forces for the two degree statically indeterminate truss whose mathematical model is shown in Figure 12.18a on page 436 using the program FLEX.

Computer Program FLEX

```
Example Problem 12.2 - Analysis by Program FLEX

Flexibility Analysis - Statically Determinate

Number of Member End Loads = 6
Number of Joint Loads = 2

Matrix [B]
     0.00000        0.00000
   120.00000       -1.00000
     0.00000        1.00000
     1.00000        0.00000
     0.00000       -1.00000
     0.00000        1.00000

Matrix [FM]
  0.41379E-03    0.00000E+00    0.00000E+00    0.00000E+00    0.00000E+00
  0.00000E+00
  0.00000E+00    0.13793E-05   -.68966E-06    0.00000E+00    0.00000E+00
  0.00000E+00
  0.00000E+00   -.68966E-06    0.13793E-05    0.00000E+00    0.00000E+00
  0.00000E+00
  0.00000E+00    0.00000E+00    0.00000E+00    0.20690E-03    0.00000E+00
  0.00000E+00
  0.00000E+00    0.00000E+00    0.00000E+00    0.00000E+00    0.68966E-06
 -.34483E-06
  0.00000E+00    0.00000E+00    0.00000E+00    0.00000E+00   -.34483E-06
  0.68966E-06

Matrix {W}
       0.000
    -600.000

Matrix [F]
  0.20069E-01   -.24828E-03
 -.24828E-03    0.62069E-05

Matrix {S}
       0.000
     600.000
    -600.000
       0.000
     600.000
    -600.000

Matrix {D}
       0.1490
      -0.0037
```

Figure 12.17 Example Problem 12.2, FLEX output.

Let the redundant forces correspond to the horizontal reactive force component at joint 4 and the axial force in member 5.

The definitions of the elements in $\{W\}$ and $\{S\}$ are shown in Figure 12.18b. Note that the end load in member 5 is included in $\{S\}$ since the member is still present in the reduced structure. It is merely considered to be cut so that the redundant axial force is zero. The elements in the matrices $[B_1]$ and $[B_2]$ can be determined by

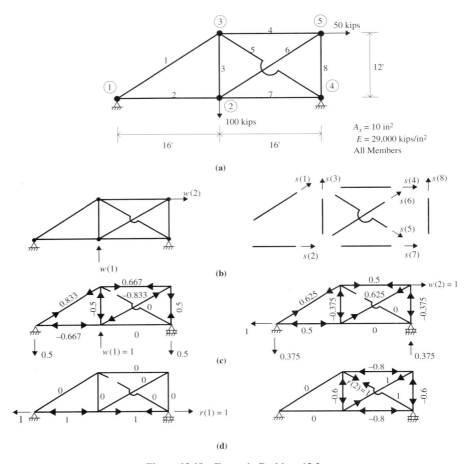

Figure 12.18 Example Problem 12.3.

computing the member forces in the reduced structure due to unit values of the joint loads and redundant loads, as shown in Figures 12.18c and 12.18d. The elements in $[FM]$ for each member can be computed by Eq. (12.16). The resulting input matrices $[B_1]$, $[B_2]$ and $[FM]$ for the program, in kip and inch units, are

$$[B_1] = \begin{bmatrix} 0.83333 & 0.625 \\ -0.66667 & 0.5 \\ -0.5 & -0.375 \\ 0.66667 & 0.5 \\ 0 & 0 \\ -0.83333 & 0.625 \\ 0 & 0 \\ 0.5 & -0.375 \end{bmatrix}$$

$$[B_2] = \begin{bmatrix} 0 & 0 \\ 1.0 & 0 \\ 0 & -0.6 \\ 0 & -0.8 \\ 0 & 1.0 \\ 0 & 1.0 \\ 1.0 & -0.8 \\ 0 & -0.6 \end{bmatrix}$$

$$[FM] = \begin{bmatrix} [F_m]_1 & & \\ & \ddots & \\ & & [F_m]_8 \end{bmatrix}$$

where the one by one matruces $[F_m]_1$ through $[F_m]_8$ are

$$[F_m]_1 = [F_m]_5 = [F_m]_6 = [0.00082759]$$

$$[F_m]_2 = [F_m]_4 = [F_m]_7 = [0.00066207]$$

$$[F_m]_3 = [F_m]_8 = [0.00049655]$$

and the matrix $\{W\}$ is

$$\{W\} = \begin{Bmatrix} -100.0 \\ 50.0 \end{Bmatrix}$$

Figure 12.19 shows a listing of the input data for the program. The nonzero value for the third number in the second row of the input data informs the program that the structure is statically indeterminate and that it should read the elements of the matrices $[B_1]$, $[B_2]$, $[FM]$ and $\{W\}$ as input.

Figure 12.20 shows the output from FLEX. The zero values for all of the elements in the compatibility check matrix indicate that the computed values for the elements in $[B]$ should be accurate to at least the listed precision. Therefore, we should be able to accept the computed elements in $\{S\}$ and $\{D\}$ with confidence. Any differences between the FLEX solution and the manual solution for this truss in Example Problem 11.3 in Chapter 11 is probably due to roundoff errors in the manual solution.

Example Problem 12.4

Analyze the three degree statically indeterminate plane frame whose mathematical model is shown in Figure 12.21a using the program FLEX. Let the redundant loads correspond to the horizontal and vertical reactive force components and the reactive moment at joint 4.

```
Example Problem 12.3 - Analysis by Program FLEX
8,2,2
Matrix [B1]
0.83333,0.625
-0.66667,0.5
-0.5,-0.375
0.66667,0.5
0.0,0.0
-0.83333,0.625
0.0,0.0
0.5,-0.375
Matrix [B2]
0.0,0.0
1.0,0.0
0.0,-0.6
0.0,-0.8
0.0,1.0
0.0,1.0
1.0,-0.8
0.0,-0.6
Matrix [FM]
1,1,0.00082759,0
2,2,0.00066207,0
3,3,0.00049655,0
4,4,0.00066207,0
5,5,0.00082759,0
6,6,0.00082759,0
7,7,0.00066207,0
8,8,0.00049655,1
Matrix {W}
-100.0
50.0
```

Figure 12.19 Example Problem 12.3, FLEX input.

The first step in the analysis of this frame must be to convert the distributed force on member 2 into an equivalent set of concentrated joint loads. These joint loads consist of the vertical forces and moments which the member would apply to its end joints if the ends were fixed against translation and rotation. They can be computed by Eqs. (6.11) and (6.12), which were presented previously in Chapter 6 and verified in Example Problem 8.6 in Chapter 8. Therefore, the equivalent vertical forces on joints 2 and 3 will be

$$W_{2Y} = W_{3Y} = \frac{\omega L}{2} = \frac{(-2)(15)}{2} = -22.5 \text{ kips}$$

and the moments will be

$$M_{2Z} = -M_{3Z} = \frac{\omega L^2}{12} = \frac{(-3)(15)^2}{12} = -56.25 \text{ kip-feet} = -675 \text{ kip-inches}$$

The signs correspond to downward forces on both joints 2 and 3, and clockwise and counterclockwise moments on joints 2 and 3, respectively. Figure 12.21(b) shows the loads which the member applies to the joints and the equal and opposite loads that the joints apply to the ends of the member. The units are kips and inches to be consistent with the units which will be used in the input data for FLEX.

Computer Program FLEX

```
Example Problem 12.3 - Analysis by Program FLEX

Flexibility Analysis - Statically Indeterminate

Number of Member End Loads = 8
Number of Joint Loads = 2
Number of Redundants = 2

Matrix [B1]
     0.83333         0.62500
    -0.66667         0.50000
    -0.50000        -0.37500
     0.66667         0.50000
     0.00000         0.00000
    -0.83333         0.62500
     0.00000         0.00000
     0.50000        -0.37500


Matrix [B2]
     0.00000         0.00000
     1.00000         0.00000
     0.00000        -0.60000
     0.00000        -0.80000
     0.00000         1.00000
     0.00000         1.00000
     1.00000        -0.80000
     0.00000        -0.60000


Matrix [FM]
  0.82759E-03    0.00000E+00    0.00000E+00    0.00000E+00    0.00000E+00
  0.00000E+00    0.00000E+00    0.00000E+00
  0.00000E+00    0.66207E-03    0.00000E+00    0.00000E+00    0.00000E+00
  0.00000E+00    0.00000E+00    0.00000E+00
  0.00000E+00    0.00000E+00    0.49655E-03    0.00000E+00    0.00000E+00
  0.00000E+00    0.00000E+00    0.00000E+00
  0.00000E+00    0.00000E+00    0.00000E+00    0.66207E-03    0.00000E+00
  0.00000E+00    0.00000E+00    0.00000E+00
  0.00000E+00    0.00000E+00    0.00000E+00    0.00000E+00    0.82759E-03
  0.00000E+00    0.00000E+00    0.00000E+00
  0.00000E+00    0.00000E+00    0.00000E+00    0.00000E+00    0.00000E+00
  0.82759E-03    0.00000E+00    0.00000E+00
  0.00000E+00    0.00000E+00    0.00000E+00    0.00000E+00    0.00000E+00
  0.00000E+00    0.66207E-03    0.00000E+00
  0.00000E+00    0.00000E+00    0.00000E+00    0.00000E+00    0.00000E+00
  0.00000E+00    0.00000E+00    0.49655E-03


Matrix {W}
   -100.000
     50.000

Matrix [G1]
   -.44138E-03     0.33104E-03
   -.10428E-02     0.47586E-03


Matrix [G2]
   0.13241E-02    -.52966E-03
   -.52966E-03     0.28601E-02
```

Figure 12.20 Example Problem 12.3, FLEX output.

```
Matrix {R}
    -68.844
    -57.526

Matrix [B]
     0.83333      0.62500
    -0.14917      0.15813
    -0.77625     -0.23719
     0.29834      0.68375
     0.46042     -0.22969
    -0.37291      0.39531
     0.14917     -0.15813
     0.22375     -0.23719

Matrix [F]
     0.12777E-02   0.39041E-03
     0.39041E-03   0.89477E-03

Matrix {S}
    -52.083
     22.823
     65.766
      4.354
    -57.526
     57.057
    -22.823
    -34.234

Matrix {D}
    -0.1082
     0.0057

Compatibility Check Matrix
     0.00000     -0.00000
     0.00000     -0.00000
```

Figure 12.20 *(cont.)*

Figure 12.21c shows the definitions for the elements in the matrices $\{W\}$ and $\{S\}$ which will be used in this analyses. The matrix $\{W\}$ has five elements to correspond to the horizontal force on joint 3 and the four concentrated joint loads corresponding to the effect of the distributed force on member 2. The matrix $\{S\}$ has nine elements to correspond to the three statically independent end loads for each member in the frame. Figure 12.21d shows the definitions of the three elements which are being used to represent the redundant loads in the matrix $\{R\}$.

The elements in the matrices $[B_1]$ and $[B_2]$ can be computed for the statically determinate reduced structure by computing the member end loads for unit values for the joint loads and the redundant loads, respectively. This is a very simple set of calculations due to the cantilever form of the reduced structure. The elements in $[FM]$ for each member in the frame can be computed by Eq. (12.20). The resulting input matrices $[B_1]$, $[B_2]$ and $[FM]$ for the program, in kip and inch units, are

Computer Program FLEX

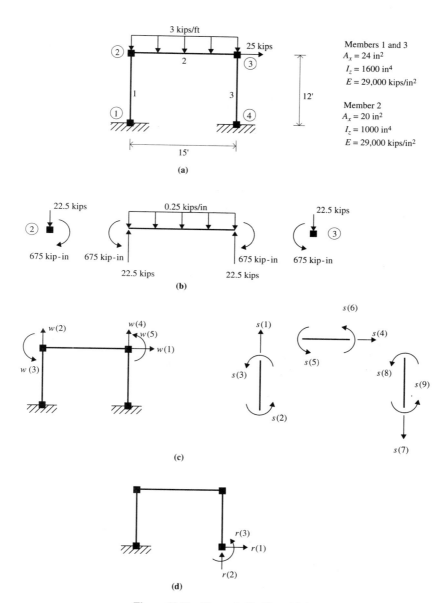

Figure 12.21 Example Problem 12.4.

$$[B_1] = \begin{bmatrix} 0 & 1.0 & 0 & 1.0 & 0 \\ 144.0 & 0 & -1.0 & -180.0 & -1.0 \\ 0 & 0 & 1.0 & 180.0 & 1.0 \\ 1.0 & 0 & 0 & 0 & 0 \\ 0 & 0 & 0 & -180.0 & -1.0 \\ 0 & 0 & 0 & 0 & 1.0 \\ 0 & 0 & 0 & 0 & 0 \\ 0 & 0 & 0 & 0 & 0 \\ 0 & 0 & 0 & 0 & 0 \end{bmatrix}$$

$$[B_2] = \begin{bmatrix} 0 & 1.0 & 0 \\ 0 & -180.0 & -1.0 \\ 144.0 & 180.0 & 1.0 \\ 1.0 & 0 & 0 \\ -144.0 & -180.0 & -1.0 \\ 144.0 & 0 & 1.0 \\ 0 & -1.0 & 0 \\ -144.0 & 0 & -1.0 \\ 0 & 0 & 1.0 \end{bmatrix}$$

$$[FM] = \begin{bmatrix} [F_m]_1 & & \\ & [F_m]_2 & \\ & & [F_m]_3 \end{bmatrix}$$

where

$$[F_m]_1 = [F_m]_3 = \begin{bmatrix} 0.00020690 & 0 & 0 \\ 0 & 0.0000010345 & -0.00000051724 \\ 0 & -0.00000051724 & 0.0000010345 \end{bmatrix}$$

$$[F_m]_2 = \begin{bmatrix} 0.00031034 & 0 & 0 \\ 0 & 0.0000020690 & -0.0000010345 \\ 0 & -0.0000010345 & 0.0000020690 \end{bmatrix}$$

and the matrix $\{W\}$ is

$$\{W\} = \begin{Bmatrix} 25.0 \\ -22.5 \\ -675.0 \\ -22.5 \\ 675.0 \end{Bmatrix}$$

Computer Program FLEX

```
Example Problem 12.4 - Analysis by Program FLEX
9,5,3
Matrix [B1]
0.0,1.0,0.0,1.0,0.0
144.0,0.0,-1.0,-180.0,-1.0
0.0,0.0,1.0,180.0,1.0
1.0,0.0,0.0,0.0,0.0
0.0,0.0,0.0,-180.0,-1.0
0.0,0.0,0.0,0.0,1.0
0.0,0.0,0.0,0.0,0.0
0.0,0.0,0.0,0.0,0.0
0.0,0.0,0.0,0.0,0.0
Matrix [B2]
0.0,1.0,0.0
0.0,-180.0,-1.0
144.0,180.0,1.0
1.0,0.0,0.0
-144.0,-180.0,-1.0
144.0,0.0,1.0
0.0,-1.0,0.0
-144.0,0.0,-1.0
0.0,0.0,1.0
Matrix [FM]
1,1,0.00020690,0
2,2,0.0000010345,0
2,3,-0.00000051724,0
3,2,-0.00000051724,0
3,3,0.0000010345,0
4,4,0.00031034,0
5,5,0.0000020690,0
5,6,-0.0000010345,0
6,5,-0.0000010345,0
6,6,0.0000020690,0
7,7,0.00020690,0
8,8,0.0000010345,0
8,9,-0.00000051724,0
9,8,-0.00000051724,0
9,9,0.0000010345,1
Matrix {W}
25.0
-22.5
-675.0
-22.5
675.0
```

Figure 12.22 Example Problem 12.4, FLEX input.

Figure 12.22 shows a listing of the input data for the program FLEX and Figure 12.23 shows the output. The member end loads listed in $\{S\}$ are correct for members 1 and 3, but they have to be modified for member 2 to consider the effect of the distributed load on the member. Figure 12.24a shows the values of the statically independent end loads from $\{S\}$ for member 2 in inch and kip units. The dependent end loads can be computed from these values by summing forces along the member longitudinal axis and summing moments about each end to give the complete set of end loads shown in Figure 12.24b. The member is in equilibrium under these end loads. However, if we now consider the downward distributed force of 0.25 kips per inch to be acting on the member, as shown in Figure 12.24c, we have a problem since ΣF_Y and ΣM_Z will

```
Example Problem 12.4 - Analysis by Program FLEX

Flexibility Analysis - Statically Indeterminate

Number of Member End Loads = 9
Number of Joint Loads = 5
Number of Redundants = 3

Matrix [B1]
     0.00000         1.00000         0.00000         1.00000         0.00000
   144.00000         0.00000        -1.00000      -180.00000        -1.00000
     0.00000         0.00000         1.00000       180.00000         1.00000
     1.00000         0.00000         0.00000         0.00000         0.00000
     0.00000         0.00000         0.00000      -180.00000        -1.00000
     0.00000         0.00000         0.00000         0.00000         1.00000
     0.00000         0.00000         0.00000         0.00000         0.00000
     0.00000         0.00000         0.00000         0.00000         0.00000
     0.00000         0.00000         0.00000         0.00000         0.00000

Matrix [B2]
     0.00000         1.00000         0.00000
     0.00000      -180.00000        -1.00000
   144.00000       180.00000         1.00000
     1.00000         0.00000         0.00000
  -144.00000      -180.00000        -1.00000
   144.00000         0.00000         1.00000
     0.00000        -1.00000         0.00000
  -144.00000         0.00000        -1.00000
     0.00000         0.00000         1.00000

Matrix [FM]
  0.20690E-03    0.00000E+00    0.00000E+00    0.00000E+00    0.00000E+00
  0.00000E+00    0.00000E+00    0.00000E+00    0.00000E+00
  0.00000E+00    0.10345E-05   -.51724E-06    0.00000E+00    0.00000E+00
  0.00000E+00    0.00000E+00    0.00000E+00    0.00000E+00
  0.00000E+00   -.51724E-06    0.10345E-05    0.00000E+00    0.00000E+00
  0.00000E+00    0.00000E+00    0.00000E+00    0.00000E+00
  0.00000E+00    0.00000E+00    0.00000E+00    0.31034E-03    0.00000E+00
  0.00000E+00    0.00000E+00    0.00000E+00    0.00000E+00
  0.00000E+00    0.00000E+00    0.00000E+00    0.00000E+00    0.20690E-05
 -.10345E-05    0.00000E+00    0.00000E+00    0.00000E+00
  0.00000E+00    0.00000E+00    0.00000E+00    0.00000E+00   -.10345E-05
  0.20690E-05    0.00000E+00    0.00000E+00    0.00000E+00
  0.00000E+00    0.00000E+00    0.00000E+00    0.00000E+00    0.00000E+00
  0.00000E+00    0.20690E-03    0.00000E+00    0.00000E+00
  0.00000E+00    0.00000E+00    0.00000E+00    0.00000E+00    0.00000E+00
  0.00000E+00    0.00000E+00    0.10345E-05   -.51724E-06
  0.00000E+00    0.00000E+00    0.00000E+00    0.00000E+00    0.00000E+00
  0.00000E+00    0.00000E+00   -.51724E-06    0.10345E-05

Matrix {W}
       25.000
      -22.500
     -675.000
      -22.500
      675.000

Matrix [G1]
  -.10415E-01    0.00000E+00    0.22345E-03    0.12066E+00    0.11173E-02
  -.40221E-01    0.20690E-03    0.55863E-03    0.16780E+00    0.11173E-02
  -.22345E-03    0.00000E+00    0.31035E-05    0.11173E-02    0.93105E-05
```

Figure 12.23 Example Problem 12.4, FLEX output.

Computer Program FLEX

```
Matrix [G2]
    0.17192E+00    0.12066E+00    0.13407E-02
    0.12066E+00    0.16800E+00    0.11173E-02
    0.13407E-02    0.11173E-02    0.12414E-04

Matrix {R}
      -18.204
       29.954
     1407.633

Matrix [B]
     0.29816    0.99693   -0.00414    0.00307   -0.00414
    44.54787    0.27608    0.06755   -0.27606   -0.32215
    26.62837   -0.27607    0.52558    0.27606   -0.27098
     0.49428    0.00000    0.00412   -0.00000   -0.00412
   -26.62837    0.27607    0.47442   -0.27606    0.27098
   -27.04024    0.27607    0.27098   -0.27612    0.47442
    -0.29816    0.00307    0.00414    0.99693    0.00414
    27.04024   -0.27607   -0.27098    0.27612    0.52558
    45.78352    0.27607   -0.32215   -0.27611    0.06755

Matrix [F]
    0.48063E-02    0.61689E-04   -.27806E-04   -.61689E-04   -.29085E-04
    0.61689E-04    0.20627E-03   -.85679E-06    0.63466E-06   -.85679E-06
   -.27806E-04   -.85679E-06    0.71076E-06    0.85679E-06    0.79397E-07
   -.61689E-04    0.63466E-06    0.85679E-06    0.20627E-03    0.85679E-06
   -.29085E-04   -.85679E-06    0.79397E-07    0.85679E-06    0.71076E-06

Matrix {S}
      -15.046
      850.653
      128.028
        6.796
     -803.028
     -538.687
      -29.954
     1213.687
     1407.633

Matrix {D}
       0.1193
      -0.0031
      -0.0011
      -0.0062
      -0.0003

Compatibility Check Matrix
    0.00000   -0.00000   -0.00000   -0.00000    0.00000
    0.00000   -0.00000    0.00000   -0.00000    0.00000
    0.00000   -0.00000   -0.00000   -0.00000    0.00000
```

Figure 12.23 *(cont.)*

not be satisfied for this loading. This condition can be corrected by superimposing the fixed end loads which are shown on the ends of the member in Figure 12.21b to obtain the final values for the end loads shown in Figure 12.24d. A quick check will show that the member is now in equilibrium. This solution can be verified by the program F2DII in FATPAK II. A similar procedure can be used for any frame with distributed or any other type of intermediate member loads.

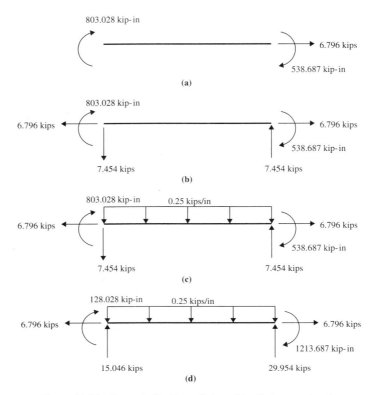

Figure 12.24 Example Problem 12.4, end loads for member 2.

DEVELOPMENT OF INFLUENCE LINES USING FLEX

According to our previous definition, any element $b(i, j)$ in the global equilibrium matrix $[B]$ is equal to the magnitude of the member end load $s(i)$ due to a unit magnitude of the joint load $w(j)$. If we now compare this definition to the previous definition for an influence line ordinate, we see that they are very similar. This suggests a convenient procedure for using the program FLEX to compute the influence line ordinates for the shear force and bending moment at any point in a statically indeterminate beam. By defining the joint load matrix $\{W\}$ to consist of a set of downward forces spaced along the beam, the elements in each row i of the matrix $[B]$ will be equal to a set of influence line ordinates for the member end load $s(i)$. By plotting these values at the locations of the loads in $\{W\}$, and connecting them with a smooth curve, the influence lines can be developed. The following example problem will demonstrate this procedure.

Example Problem 12.5

Generate the influence lines for the shear force and bending moment at point a in the prismatic beam shown in Figure 12.25a. The mathematical model which will be

Development of Influence Lines Using FLEX 447

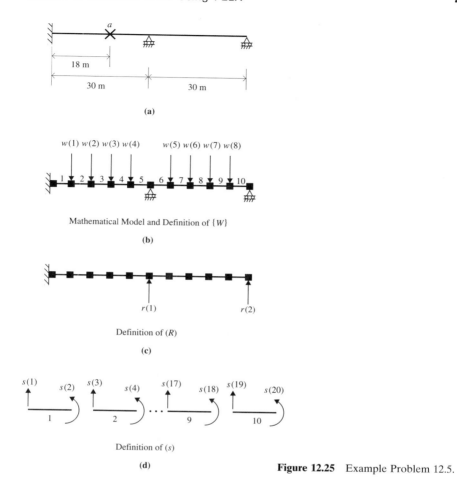

Figure 12.25 Example Problem 12.5.

used to represent the beam in the analysis and the definition of the elements in the joint load matrix $\{W\}$ are shown in Figure 12.25b. The joints have been equally spaced in each span of the beam. These are the points where the influence line ordinates will be computed. Joint loads have only been defined at these eight interior points, since it is known beforehand that the influence line ordinates for the shear force and bending moment, for any point in the beam, will be zero at the supports. The joint loads consist of downward vertical forces to correspond to the definition of an influence line ordinate. The reduced structure and the definition of the elements in the redundant load matrix $\{R\}$ are shown in Figure 12.25c. This choice for the reduced structure will greatly simplify the computation of the elements in $[B_1]$ and $[B_2]$ compared to some of the other choices which could have been made.

Since only vertical forces are being considered in the analysis of this beam, the axial force in each of the members in the mathematical model will be zero. Therefore, since there is no need to consider the effects of axial deformation in the mem-

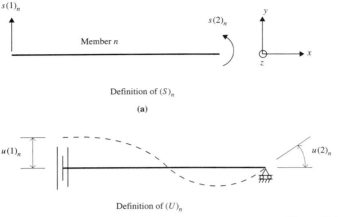

Figure 12.26 Example Problem 12.5.

bers, the definition for the elements in $\{S\}$ can be simplified by eliminating the axial force end loads. The definition of the elements in $\{S\}$ which will be used in this analysis are shown in Figure 12.25d. This corresponds to the definitions shown in Figure 12.26 for $\{S\}_n$ and $\{U\}_n$ for each member. These specific statically independent end loads have been chosen for each member since the upward force on the left end and the counterclockwise moment on the right end correspond to the positive directions in the designer sign convention for shear force and bending moment, as defined previously in Figure 5.2 in Chapter 5. The corresponding local member flexibility matrix is

$$[F_m]_n = \begin{bmatrix} \dfrac{L^3}{3EI_z} & -\dfrac{L^2}{2EI_z} \\ -\dfrac{L^2}{2EI_z} & \dfrac{L}{EI_z} \end{bmatrix}_n \tag{12.70}$$

The elements in $[B_1]$ and $[B_2]$ now can be computed by applying unit forces to the reduced structure corresponding to the elements in $\{W\}$ and $\{R\}$, while the elements in $[FM]$ for each member can be computed by Eq. (12.70). Since the influence line ordinates for shear force and bending moment in a prismatic beam of this type are independent of the specific values for E and I_z, any arbitrary values may be used for these quantities in the computations. In order to simplify the calculation of the numbers in $[FM]$, unit values will be used for E and I_z in the analysis. These values will not affect the numbers in $[B_1]$ and $[B_2]$. The resulting input matrices $[B_1]$, $[B_2]$ and $[FM]$ for the program, in kilonewton and meter units, are

Development of Influence Lines Using FLEX

$$[B_1] = \begin{bmatrix} 1.0 & 1.0 & 1.0 & 1.0 & 1.0 & 1.0 & 1.0 & 1.0 \\ 0 & -6.0 & -12.0 & -18.0 & -30.0 & -36.0 & -42.0 & -48.0 \\ 0 & 1.0 & 1.0 & 1.0 & 1.0 & 1.0 & 1.0 & 1.0 \\ 0 & 0 & -6.0 & -12.0 & -24.0 & -30.0 & -36.0 & -42.0 \\ 0 & 0 & 1.0 & 1.0 & 1.0 & 1.0 & 1.0 & 1.0 \\ 0 & 0 & 0 & -6.0 & -18.0 & -24.0 & -30.0 & -36.0 \\ 0 & 0 & 0 & 1.0 & 1.0 & 1.0 & 1.0 & 1.0 \\ 0 & 0 & 0 & 0 & -12.0 & -18.0 & -24.0 & -30.0 \\ 0 & 0 & 0 & 0 & 1.0 & 1.0 & 1.0 & 1.0 \\ 0 & 0 & 0 & 0 & -6.0 & -12.0 & -18.0 & -24.0 \\ 0 & 0 & 0 & 0 & 1.0 & 1.0 & 1.0 & 1.0 \\ 0 & 0 & 0 & 0 & 0 & -6.0 & -12.0 & -18.0 \\ 0 & 0 & 0 & 0 & 0 & 1.0 & 1.0 & 1.0 \\ 0 & 0 & 0 & 0 & 0 & 0 & -6.0 & -12.0 \\ 0 & 0 & 0 & 0 & 0 & 0 & 1.0 & 1.0 \\ 0 & 0 & 0 & 0 & 0 & 0 & 0 & -6.0 \\ 0 & 0 & 0 & 0 & 0 & 0 & 0 & 1.0 \\ 0 & 0 & 0 & 0 & 0 & 0 & 0 & 0 \\ 0 & 0 & 0 & 0 & 0 & 0 & 0 & 0 \\ 0 & 0 & 0 & 0 & 0 & 0 & 0 & 0 \end{bmatrix}$$

$$[B_1] = \begin{bmatrix} -1.0 & -1.0 \\ 24.0 & 54.0 \\ -1.0 & -1.0 \\ 18.0 & 48.0 \\ -1.0 & -1.0 \\ 12.0 & 42.0 \\ -1.0 & -1.0 \\ 6.0 & 36.0 \\ -1.0 & -1.0 \\ 0 & 30.0 \\ 0 & -1.0 \\ 0 & 24.0 \\ 0 & -1.0 \\ 0 & 18.0 \\ 0 & -1.0 \\ 0 & 12.0 \\ 0 & -1.0 \\ 0 & 6.0 \\ 0 & -1.0 \end{bmatrix}$$

$$[FM] = \begin{bmatrix} [F_m]_1 & & \\ & \ddots & \\ & & [F_m]_{10} \end{bmatrix}$$

where

$$[F_m]_1 = \cdots = [F_m]_{10} = \begin{bmatrix} 72.0 & -18.0 \\ -18.0 & 6.0 \end{bmatrix}$$

Since we are only interested in the computed elements in the matrix $[B]$ for this analysis, the elements in $\{W\}$ can have any arbitrary values. Therefore, all of the elements in $\{W\}$ will be set to zero in the program input, which will result in zero values for all of the elements in $\{R\}$, $\{S\}$ and $\{D\}$.

Figure 12.27 shows the matrix $[B]$ from the FLEX output. The fifth row in $[B]$ corresponds to the values of the shear force at point a and the sixth row corresponds to the bending moment. The plotted influence lines are shown in Figure 12.28. The eight points which correspond to the elements in the rows in $[B]$ are circled. The elements in $[B]$ fully define the influence line for the bending moment when combined with the known zero ordinates at the support joints. However, the computed infor-

```
Example Problem 12.5 - Analysis by Program FLEX

Matrix [B]
     0.92343      0.73029      0.47543      0.21371     -0.12343
    -0.16457     -0.14400     -0.08228
     1.42629     -0.76114     -1.26172     -0.77487      0.49371
     0.65828      0.57599      0.32913
    -0.07657      0.73029      0.47543      0.21371     -0.12343
    -0.16457     -0.14400     -0.08228
     0.96686      3.62057      1.59085      0.50742     -0.24686
    -0.32915     -0.28801     -0.16458
    -0.07657     -0.26971      0.47543      0.21371     -0.12343
    -0.16457     -0.14400     -0.08228
     0.50743      2.00229      4.44343      1.78971     -0.98743
    -1.31657     -1.15200     -0.65829
    -0.07657     -0.26971     -0.52457      0.21371     -0.12343
    -0.16457     -0.14400     -0.08228
     0.04800      0.38400      1.29600      3.07200     -1.72800
    -2.30400     -2.01600     -1.15200
    -0.07657     -0.26971     -0.52457     -0.78629     -0.12343
    -0.16457     -0.14400     -0.08228
    -0.41143     -1.23428     -1.85143     -1.64571     -2.46857
    -3.29142     -2.87999     -1.64571
     0.01371      0.04114      0.06171      0.05486      0.88229
     0.70971      0.49600      0.25486
    -0.32914     -0.98743     -1.48114     -1.31657      2.82515
     0.96686      0.09600     -0.11657
     0.01371      0.04114      0.06171      0.05486     -0.11771
     0.70971      0.49600      0.25486
    -0.24686     -0.74057     -1.11086     -0.98743      2.11886
     5.22515      3.07200      1.41257
     0.01371      0.04114      0.06171      0.05486     -0.11771
    -0.29029      0.49600      0.25486
    -0.16457     -0.49371     -0.74057     -0.65829      1.41257
     3.48343      6.04800      2.94172
     0.01371      0.04114      0.06171      0.05486     -0.11771
    -0.29029     -0.50400      0.25486
    -0.08229     -0.24686     -0.37029     -0.32914      0.70629
     1.74172      3.02400      4.47086
     0.01371      0.04114      0.06171      0.05486     -0.11771
    -0.29029     -0.50400     -0.74514
     0.00000      0.00000      0.00000      0.00000      0.00000
     0.00000      0.00000      0.00000
```

Figure 12.27 Example Problem 12.5, FLEX output—Matrix $[B]$.

Modifications to Program FLEX

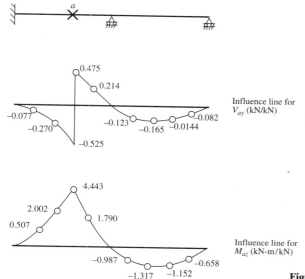

Figure 12.28 Example Problem 12.5.

mation in [B] is incomplete for the shear force influence line since only one of the dual values at point a is provided. Fortunately, the other value can be easily computed since it is known that a unit upward step occurs in the influence line at that point. The units of the shear force influence line ordinates are kN/kN while the units of the bending moment influence line ordinates are kN m/kN. Influence lines for the shear force or bending moment can be plotted in a imilar manner for any points on the beam which correspond to load positions in {W}.

A similar procedure also can be used to compute the influence line ordinates for the member forces and the vertical translations of the load line joints in a statically indeterminate truss. For this case, the loads in {W} will consist of downward vertical forces on the load line joints. Each row in the matrix [B] in the program output will correspond to the influence line ordinates for a particular member force, while the elements in [F] will correspond to the vertical translation influence line ordinates. An example problem which demonstrates this procedure is included in Fleming, 1989.

MODIFICATIONS TO PROGRAM FLEX

The program FLEX is somewhat tedious to use in its present form due to the amount of preprocessing which must be performed by the program user to determine the values for the elements in the matrices $[B_1]$, $[B_2]$ and $[FM]$. However, it is possible to modify the program so that the elements in these matrices are computed automatically from a description of the structure similar to that used for the input data in the programs in FATPAK II. This modification would consist of incorporating SUB Procedures into the program which could perform an analysis of the statically determinate reduced structure to determine the elements in $[B_1]$ and $[B_2]$ and to compute the elements in $[FM]$ for various types of structures. The programs SDPTRUSS and SDPFRAME, which were presented previously in Chapters 3 and 7, could be easily modified to supply the SUB Procedures for performing the static analysis of the reduced structure, while the SUB Procedures for computing the elements in $[FM]$ could be easily written using the previously developed expressions for the elements in the local member flexibility matrices for plane truss and plane frame members. The input to FLEX then could consist of essentially the same input as used in the programs SDPTRUSS and SDPFRAME, with additional information to describe the material properties and the member cross section properties. This would simplify greatly the generation of the input data. The program could be further modified to analyze space trusses and space frames by incorporating the programs SDSTRUSS and SDSFRAME as Sub Procedures. Of course, it then would be necessary to have expressions for the elements in the local member flexibility matrices for these structure types to develop the SUB Procedures for computing the elements in $[FM]$. These expressions are shown in Fleming (1989).

REFERENCE

FLEMING, JOHN F. (1989). *Computer Analysis of Structural Systems.* New York: McGraw-Hill.

SUGGESTED PROBLEMS

Compute the member axial forces for each of the statically indeterminate structures in Suggested Problems SP11.7 through SP11.10 at the end of Chapter 11 with the program FLEX. Use the same redundant quantities for each problem as used in the manual analysis.

SP12.1 Problem SP11.7

SP12.2 Problem SP11.8

SP12.3 Problem SP11.9

SP12.4 Problem SP11.10

SP12.5 to SP12.7 Compute the end loads for each member in the mathematical models of the structures shown in Figures SP12.5 to SP12.7 with the program FLEX. Draw the shear force and bending moment diagram for each member. Use any convenient set of redundants.

SP12.8 Derive expressions for the elements in the local member flexibility matrix for the plane frame member shown in Figure SP12.8. Do not consider the effect of shear deformation. Use the Moment Area Theorems.
(a) Use the definition for $\{S\}_n$ shown in Figure 12.7a.
(b) Use the definition for $\{S\}_n$ shown in Figure 12.7b.

Suggested Problems

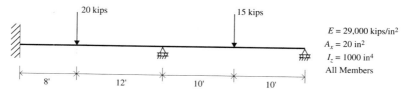

Figure SP12.5

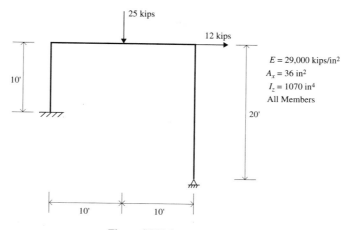

Figure SP12.6

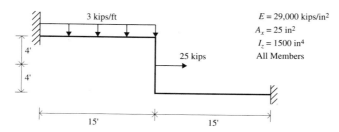

Figure SP12.7

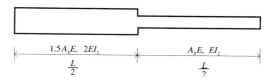

Figure SP12.8

13

The Slope-Deflection Method

BASIC PRINCIPLES OF THE SLOPE-DEFLECTION METHOD

During the analysis of any statically indeterminate structure by the Method of Consistent Deformations, or its counterpart the Flexibility Method, a set of geometric compatibility equations was generated with the redundant loads for the structure as the unknowns. The equations were developed as a function of the flexibility of the structure. Another approach which can be used is to generate a set of equilibrium equations, with the displacements corresponding to the independent displacement degrees of freedom of the structure as the unknowns, as a function of the stiffness of the structure. One such procedure is the classical analysis procedure known as the *Slope-Deflection Method*, which was introduced by George A. Maney in 1915. As originally formulated, this analysis procedure can be used to analyze both statically determinate and statically indeterminate beams and plane frames. The primary limitation of the procedure is that it only includes the effects of bending deformations in the members in the mathematical model of the structure in the computations for the displacements at the joints. The effects of axial deformations and shear deformations are not considered. However, as we saw in Chapter 9, the effects of the shear deformations in any member in the mathematical model of a beam or frame are small compared to the bending deformations as long as the depth to length ratio of the member is small. Therefore, ignoring this effect should not have a significant effect upon the response of most structures. In addition, the axial deformations in the members in the mathematical model of a beam usually can be ignored since the axial forces in the members are either zero or very small. However, the effects of the axial deformations in a

Basic Principles of the Slope-Deflection Method

frame will depend on the geometry of the frame, the ratio of the cross section area to the moment of inertia for the members and the type of loading. Several example problems will be considered later in this chapter where the analysis by the Slope-Deflection Method will be compared to an analysis by the program F2DII in FATPAK II in which axial deformations in the members are included.

The basic concept of the Slope-Deflection Method is to develop a set of relationships for the structure which satisfy the conditions of geometric compatibility, load-deformation and equilibrium. The geometric compatibility relationships will consist of a set of equations which relate the various displacements at the joints in the mathematical model of the structure to the independent displacement degrees of freedom for the structure. The load-deformation relationships will consist of a set of equations which relate the deformations which occur in each member to the active loads on the member and the moments which are applied to the ends of the member by the joints. And, finally, the equilibrium relationships consist of a set of equations, which are written in terms of the end moments for the members, which ensure that each joint and the overall structure are in equilibrium. As we will see later, the number of independent equilibrium equations which will be generated will be equal to the number of independent displacement degrees of freedom. By combining the equations which are developed for each of these conditions, the final equations for computing the joint displacements can be developed.

Geometric Compatibility Relationships

The first step in the analysis of any structure by the Slope-Deflection Method is to define the independent displacement degrees of freedom for the mathematical model of the structure. In Chapter 12, the following relationship was shown in Eq. (12.4) for computing the number of independent degrees of freedom, NDOF, for a plane frame:

$$NDOF = 3NJ - NR$$

where NJ is the number of joints in the mathematical model of the frame and NR is the number of support restraints. However, the number of degrees of freedom will be less than this number, for any frame in which the axial deformations in the members are ignored, due to the additional restrictions which ignoring these deformations place upon the joint translations. For example, if we apply Eq. (12.4) to the plane frame whose mathematical model is shown in Figure 13.1a, we will find that it has six degrees of freedom, which consist of a horizontal and a vertical translation and a rotation at both joint 2 and joint 3. The translations and rotations of both joint 1 and joint 4 will be zero due to the six support restraints. However, if we now assume that the axial deformations of the members in the frame are zero, the vertical translations of joints 2 and 3 must be zero and the horizontal translations of these two joints must be equal. Therefore, the number of independent degrees of freedom for the frame will be reduced to three, consisting of

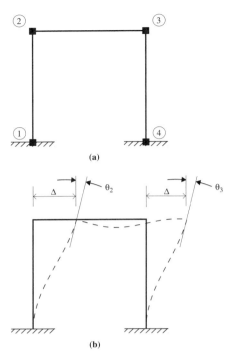

Figure 13.1

the rotations θ_2 and θ_3 of joints 2 and 3 and the common horizontal translation Δ of these same joints, as shown in Figure 13.1b. These are the displacements which will be used in an analysis of this structure by the Slope-Deflection Method.

Structures which are analyzed by the Slope-Deflection Method can be classified into two primary groups: those which have only rotation degrees of freedom at the joints; and those which have both rotation and translation degrees of freedom. As we will see later, it will be somewhat easier to generate the equilibrium equations for those structures which have only rotation degrees of freedom at the joints.

Structures with Only Rotation Degrees of Freedom. Figure 13.2 shows examples of the mathematical models for several structures which will have only rotation degrees of freedom at the joints.

> The beam in Figure 13.2a has four degrees of freedom, consisting of the rotations at each of the four joints. The vertical translations of all four joints and the horizontal translation of joint 1 must be zero due to the support restraints. The horizontal translations of joints 2, 3 and 4 also must be equal to the zero horizontal translation of joint 1 since the axial deformations in the members are considered to be zero.
>
> The frame in Figure 13.2b has three degrees of freedom, consisting of the rotations of joints 1, 2 and 4. The translations of joints 1, 3 and 5 and the ro-

Basic Principles of the Slope-Deflection Method

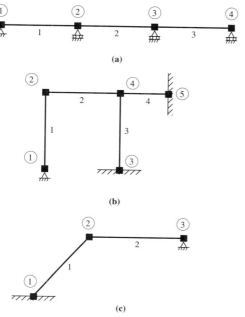

Figure 13.2

tations of joints 3 and 5 must be zero due to the support restraints. The horizontal and the vertical translations of joints 2 and 4 must be zero since the axial deformations in the members are considered to be zero. The horizontal translations of joints 2 and 4 must be equal to the zero horizontal translation of joint 5 and the vertical translations of joints 2 and 4 must be equal to the zero vertical translations of joints 1 and 3.

The frame in Figure 13.2c has two degrees of freedom, consisting of the rotations of joints 2 and 3. The translations of joints 1 and 3 and the rotation of joint 1 must be zero due to the support restraints. The vertical and horizontal translations of joint 2 must be zero since the axial deformations of the two members are considered to be zero. Any translation which could occur for joint 2 would have to be along the arc centered at joint 1, with a radius equal to the length of member 1, and the arc centered at joint 3, with a radius equal to the length of member 2, as shown in Figure 13.3. Therefore, the only two possible positions for the joint are at the intersection points of these arcs. However, at least one of the members would have to deform axially during the translation of the joint to the lower intersection point. Since even a temporary axial deformation in a member is not permitted, the only possible position for the joint is at the upper intersection point, which corresponds to its original location.

Structures with Both Translation and Rotation Degrees of Freedom. Figure 13.4 shows examples of the mathematical models of several structures which have both translation and rotation degrees of freedom at the joints.

458   The Slope-Deflection Method   Chap. 13

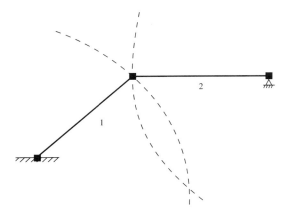

Figure 13.3

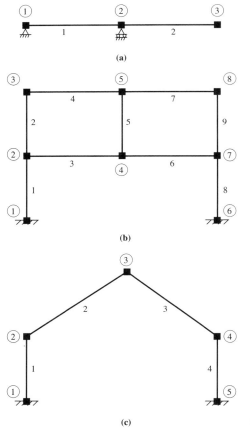

Figure 13.4

Basic Principles of the Slope-Deflection Method

The beam in Figure 13.4a has four degrees of freedom, consisting of the rotations of joints 1, 2 and 3 and the vertical translation of joint 3. The vertical translations of joints 1 and 2 and the horizontal translation of joint 1 must be zero due to the support restraints. The horizontal translations of joints 2 and 3 must be equal to the zero horizontal translation of joint 1 since the axial deformations in the members are considered to be zero.

The frame in Figure 13.4b has nine degrees of freedom, consisting of: the rotations of joints 2, 3, 4, 5, 7 and 8; the equal horizontal translations of joints 2, 4 and 7; the equal horizontal translations of joints 3, 5 and 8; and the equal vertical translations of joints 4 and 5. The translations and rotations of joints 1 and 6 must be zero due to the support restraints, while: the vertical translations of joints 2, 3, 7 and 8 must be zero; the horizontal translations of joints 2, 4 and 7 must be equal; the horizontal translations of joints 3, 5 and 8 must be equal; and the vertical translations of joints 4 and 5 must be equal since the axial deformations in the members are considered to be zero.

The frame shown in Figure 13.4c has five degrees of freedom, consisting of the rotations of joints 2, 3 and 4 and the horizontal translations of joints 2 and 4. The translations and rotations of joints 1 and 5 must be zero due to the support restraints and the vertical translations of joints 2 and 4 must be zero since the axial deformations in the members are considered to be zero. The horizontal and vertical translations of joint 3 are not independent degrees of freedom since they can be expressed in terms of the horizontal translations at joints 2 and 4. The final position of joint 3 must be at the intersection of the arc centered at joint 2, with a radius equal to the length of member 2, and the arc centered at joint 4, with a radius equal to the length of member 3, as shown in Figure 13.5. This location can be computed by simple geometric relationships.

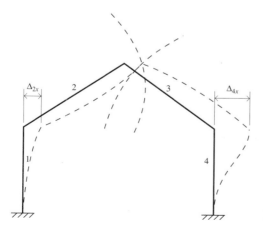

Figure 13.5

For some structures the relationships between the various joint translations are not obvious. Extreme care must be taken to ensure that only independent degrees of freedom are considered in the analysis.

The relationships between the various translations must be compatible with both the geometry of the structure and its support conditions and with the assumption that the axial deformations in all of the members in the mathematical model are zero.

Load-Deformation Relationships

The load-deformation relationships which are used in the Slope-Deflection Method consist of a pair of equations for each member which relate the moments which are applied to the member ends by the joints to the active loads on the member and the rotations and translations of the member ends. This set of equations is usually called the *Slope-Deflection Equations*.

To develop these equations, we will consider a typical member in a plane frame, which is rigidly connected at its ends to joints i and j, as shown in Figure 13.6a. The local right hand orthogonal coordinate axes xyz for the member are oriented with the local x axis extending along the length of the member from joint i to joint j and the local z axis extending outward from the plane of the structure. As the structure is loaded, the joints will translate and rotate, which will cause the member to deform, as shown in Figure 13.6b. The translations of the ends i and j

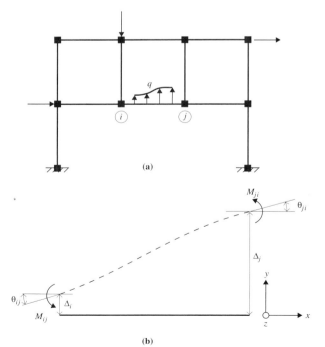

Figure 13.6

Basic Principles of the Slope-Deflection Method

of the member, perpendicular to its original longitudinal axis, will be designated as Δ_i and Δ_j, while the rotations will be designated as θ_{ij} and θ_{ji}. The moments which will be applied to the ends of the member by the joints and the equal and opposite moments which will be applied to the joints by the ends of the member will be designated as M_{ij} and M_{ji} at ends i and j, respectively. The rotations of the ends of the member and the moments applied to the ends of the member by the joints will be positive about the local z axis in the counterclockwise direction (i.e.; by the right hand rotation rule), while the directions of the translations of the member ends and any intermediate active forces on the member will be positive in the positive direction of the local member y axis. The final deformed shape of the member can be obtained by superimposing the three deformed shapes shown in Figure 13.7 below. Figure 13.7a shows the deformed shape due to the positive intermediate active loads, assuming that the ends are restrained against translation and rotation. Figure 13.7b shows the deformed shape due to the positive end rotations θ_{ij} and θ_{ji}, assuming that the ends are restrained against translation. And, finally, Figure 13.7c shows the deformed shape due to the positive end translations Δ_i and Δ_j, assuming that the ends are restrained against rotation. The final end moments will be equal to the sum of the end moments which correspond to these three deformed shapes. The following sections discuss the contributions of the intermediate member forces, the member end rotations and the member end translations to the member end moments.

Contribution of Intermediate Member Forces. The contribution to the end moments of any intermediate active forces on the member will correspond to the fixed end moments for the member, as described in previous chapters. These fixed end moments will be designated as M_{ij}^F and M_{ji}^F. They are equal to the moments which will be applied to the ends of the member by the joints if the ends of the member are totally restrained against rotation and translation while it deforms under the active loads. Several typical cases are shown in Figure 13.8.

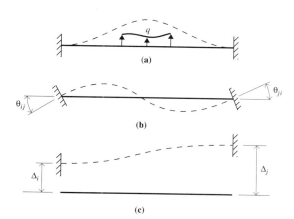

Figure 13.7

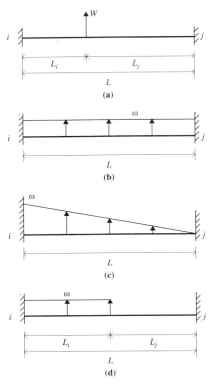

Figure 13.8

Figure 13.8a. The fixed end moments for a prismatic member, which is subjected to a positive concentrated force W at distances L_i from end i and L_j from end j, are

$$M_{ij}^F = -\frac{WL_iL_j^2}{L^2} \qquad (13.1a)$$

$$M_{ji}^F = \frac{WL_i^2L_j}{L^2} \qquad (13.1b)$$

Figure 13.8b. The fixed end moments for a prismatic member, which is subjected to a positive uniform distributed force over its entire length, are

$$M_{ij}^F = -\frac{\omega L^2}{12} \qquad (13.2a)$$

$$M_{ji}^F = \frac{\omega L^2}{12} \qquad (13.2b)$$

These moments are the opposite of the equivalent joint load moments which were given in Eq. (6.12) in Chapter 6 for use in the program SDPFRAME and verified in Example Problem 8.6 in Chapter 8.

Basic Principles of the Slope-Deflection Method

Figure 13.8c. The fixed end moments for a prismatic member, which is subjected to a positive triangular distributed force which varies linearly from a magnitude of ω at end *i* to a magnitude of zero at end *j*, are

$$M_{ij}^F = -\frac{\omega L^2}{20} \tag{13.3a}$$

$$M_{ji}^F = \frac{\omega L^2}{30} \tag{13.3b}$$

These moments are the opposite of the equivalent joint load moments which were given in Eqs. (6.17) and (6.18) in Chapter 6 for use in the program SDPFRAME.

Figure 13.8d. The fixed end moments for a prismatic member, which is subjected to a positive partial uniform force ω which extends over a distance L_i from end *i*, are

$$M_{ij}^F = -\frac{\omega L_i^2}{12}\left[6 - 8\left(\frac{L_i}{L}\right) + 3\left(\frac{L_i}{L}\right)^2\right] \tag{13.4a}$$

$$M_{ji}^F = \frac{\omega L_i^2}{12}\left[4\left(\frac{L_i}{L}\right) - 3\left(\frac{L_i}{L}\right)^2\right] \tag{13.4b}$$

Equations for other loadings or for nonprismatic members can be derived using the Method of Consistent Deformations which was discussed in Chapter 11 to analyze the statically indeterminate fixed-fixed member.

Contribution of Member End Rotations. Expressions for the contribution to the end moments by the member end rotations can be easily derived using the Moment Area Theorems. For convenience, each end rotation will be considered separately.

First, consider a prismatic member which is subjected to a positive rotation θ_{ij} at end *i*, with end *i* restrained against translation and end *j* restrained against translation and rotation, as shown in Figure 13.9a. The moments which are required to hold the member in this deformed shape will be designated as M'_{ij} and M'_{ji}. Figure 13.9b shows the bending moment diagram for this member using the designer sign convention for the moments. Although it is possible to use the bending moment diagram in this form with the Moment Area Theorems, it will be more convenient if it is redrawn to consider the effects of each of the end moments separately, as shown in Figure 13.9c. Using this diagram, we can now easily compute the relationship between the end moments and the end rotation θ_{ij}.

Applying the Second Moment Area Theorem, by taking moments of the M_z/EI_z diagram about end *i*, gives

$$t_{i/j} = -\frac{M'_{ij}L^2}{6EI_z} + \frac{M'_{ji}L^2}{3EI_z} \tag{13.5}$$

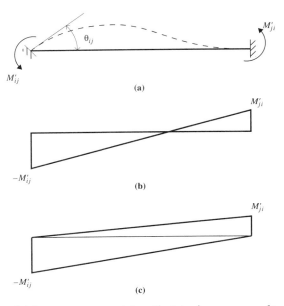

Figure 13.9

which, upon recognizing that $t_{i/j}$ is zero, results in

$$M'_{ji} = \frac{M'_{ij}}{2} \qquad (13.6)$$

Next, the rotation angle θ_j of the member at end j can be expressed in terms of the rotation angle θ_i at end i as

$$\theta_j = \theta_i + \Delta\theta_{(i/j)z} \qquad (13.7)$$

in which the change in the slope angle over the length of the member, $\Delta\theta_{(i/j)z}$, can be computed using the First Moment Area Theorem, by computing the area under the M_z/EI_z diagram, as

$$\Delta\theta_{(i/j)z} = -\frac{M'_{ij}L}{2EI_z} + \frac{M'_{ji}L}{2EI_z} \qquad (13.8)$$

Combining Eqs. (13.6), (13.7) and (13.8) and recognizing that θ_i is equal to θ_{ij} and θ_j is zero results in

$$M'_{ij} = \frac{4EI_z}{L}\theta_{ij} \qquad (13.9)$$

and

$$M'_{ji} = \frac{2EI_z}{L}\theta_{ij} \qquad (13.10)$$

Basic Principles of the Slope-Deflection Method

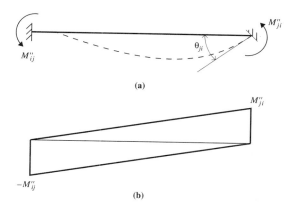

Figure 13.10

The positive signs for the end moments mean that they are counterclockwise.

Next, consider the same member subjected to a positive rotation θ_{ji} at end j, with end j restrained against translation and end i restrained against translation and rotation, as shown in Figure 13.10a. The end moments which are required to hold the member in this deformed shape will be designated M''_{ij} and M''_{ji}. The bending moment diagram for this case can be drawn in the form shown in Figure 13.10b. If we now apply the Moment Area Theorems in a similar manner to that used in the previous derivation, we will obtain

$$M''_{ij} = \frac{2EI_z}{L} \theta_{ji} \qquad (13.11)$$

and

$$M''_{ji} = \frac{4EI_z}{L} \theta_{ji} \qquad (13.12)$$

The verification of these equations is left as an exercise for the reader. As with the previous moments, the positive signs mean that the moments are counterclockwise.

Contribution of Member End Translations. To derive an expression relating the member end moments to the end translations, consider a prismatic member which is subjected to positive translations Δ_i and Δ_j at ends i and j while the ends are restrained against rotation, as shown in Figure 13.11a. The moments which are required to hold the member in this deformed shape will be designated as M'''_{ij} and M'''_{ji}. The bending moment diagram for this case can be drawn in the form shown in Figure 13.11b.

Applying the First Moment Area Theorem gives the change in slope over the length of the member as

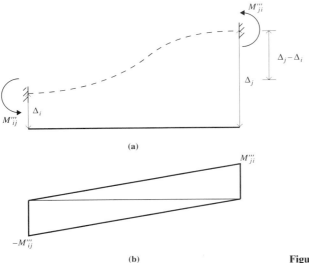

Figure 13.11

$$\Delta\theta_{(i/j)z} = -\frac{M'''_{ij}L}{2EI_z} + \frac{M'''_{ji}L}{2EI_z} \quad (13.13)$$

from which, upon recognizing that the slopes at the two ends are equal, gives

$$M'''_{ji} = M'''_{ij} \quad (13.14)$$

Next, applying the Second Moment Area Theorem by taking moments of the M_z/EI_z diagram about end j, gives

$$t_{j/i} = -\frac{M'''_{ij}L^2}{3EI_z} + \frac{M'''_{ji}L^2}{6EI_z} \quad (13.15)$$

Combining Eqs. (13.14) and (13.15) and recognizing that $t_{j/i}$ is equal to the relative translation between the ends i and j, since the slopes at the two ends are equal, gives

$$M'''_{ij} = M'''_{ji} = -\frac{6EI_z}{L^2}(\Delta_j - \Delta_i) \quad (13.16)$$

The negative signs for the moments means that they are clockwise.

Slope Deflection Equations. The final form of the Slope Deflection Equations for a prismatic plane frame member can be obtained by adding the contributions to the member end moments due to each of the previous effects:

$$M_{ij} = M^F_{ij} + M'_{ij} + M''_{ij} + M'''_{ij} \quad (13.17)$$

$$M_{ji} = M^F_{ji} + M'_{ji} + M''_{ji} + M'''_{ji} \quad (13.18)$$

Basic Principles of the Slope-Deflection Method

from which, on substituting the expressions in Eqs. (13.9), (13.10), (13.11), (13.12) and (13.16) gives

$$M_{ij} = M_{ij}^F + \frac{4EI_z}{L}\theta_{ij} + \frac{2EI_z}{L}\theta_{ji} - \frac{6EI_z}{L^2}(\Delta_j - \Delta_i) \qquad (13.19)$$

$$M_{ji} = M_{ji}^F + \frac{2EI_z}{L}\theta_{ij} + \frac{4EI_z}{L}\theta_{ji} - \frac{6EI_z}{L^2}(\Delta_j - \Delta_i) \qquad (13.20)$$

The Slope Deflection Equations for a nonprismatic member can be computed in a similar manner by modifying the shape of the M_z/EI_z diagram to consider the variation of EI_z over the length of the member. The equations will be in the same form, but the constants will vary, depending on the specific manner in which EI_z varies over the member.

Equilibrium Relationships

The third condition which must be satisfied during the analysis of a structure by the Slope Deflection Method is equilibrium. A set of equations must be generated, in terms of the moments at the ends of the members, which ensure that each joint and the overall structure are in equilibrium. The number of independent equilibrium equations which can be generated in terms of the member end moments will be equal to the number of independent displacement degrees of freedom for the structure when considering the axial deformations in all members to be zero.

If the structure only has rotation degrees of freedom at the joints, the complete set of equilibrium equations can be generated by summing moments at each joint which is unrestrained against rotation. However, if the structure also has translation degrees of freedom, additional equations will be required over and above those available by only considering the equilibrium of the joints. These additional equations must be generated by isolating specific portions of the structure and summing forces and/or moments to ensure that these portions are in equilibrium. Unfortunately, the procedure which can be used to generate these equations will vary from structure to structure depending on the structure geometry. No specific rules for generating these additional equations can be developed at this time. The process can be best demonstrated by several example problems.

Example Problem 13.1

Analyze the beam shown in Figure 13.12a by the Slope Deflection Method. Consider two different cases for the analysis:

Case I. The three supports remain at the original elevations as the beam is loaded.

Case II. The right support translates down 0.1 inch as the beam is loaded.

The mathematical model of the beam for either case is shown in Figure 13.12b. It consists of three joints and two members.

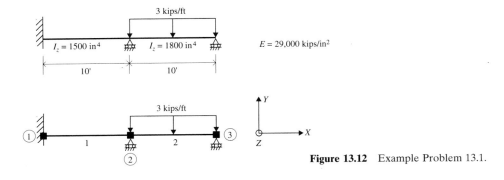

Figure 13.12 Example Problem 13.1.

Case I. The geometric compatibility conditions for Case I are

$$\theta_{12} = \theta_1 = 0$$
$$\theta_{21} = \theta_{23} = \theta_2$$
$$\theta_{32} = \theta_3$$
$$\Delta_{1X} = \Delta_{2X} = \Delta_{3X} = 0$$
$$\Delta_{1Y} = \Delta_{2Y} = \Delta_{3Y} = 0$$

Therefore, the beam has two degrees of freedom, which consist of the rotation θ_2 at joint 2 and the rotation θ_3 at joint 3.

The two independent equilibrium equations for this beam can be obtained by summing moments at joints 2 and 3. The moments acting on joint 2 will be M_{21} and M_{23}, which are applied to the joint by member 1 and member 2, respectively, while the only moment acting on joint 3 will be M_{32}, which is applied to the joint by member 2. Therefore, the equilibrium equations will be

$$\text{At joint 2, } \Sigma_2 M_Z \rightarrow M_{21} + M_{23} = 0$$
$$\text{At joint 3, } \Sigma_3 M_Z \rightarrow M_{32} = 0$$

An equation is not written for joint 1 since the moment M_{12} will be exactly balanced by the reactive moment for the fixed support.

The next step is to compute the fixed end moments for each member which is subjected to intermediate active loads. These moments can be computed for member 2 by Eqs. (13.2a) and (13.2b).

$$M_{23}^F = -\frac{\omega L^2}{12} = -\frac{(-3)(10)^2}{12} = 25 \text{ kip-feet} = 300 \text{ kip-inches}$$

$$M_{32}^F = \frac{\omega L^2}{12} = \frac{(-3)(10)^2}{12} = -25 \text{ kip-feet} = -300 \text{ kip-inches}$$

With these values known, the Slope-Deflection Equations now can be generated for each member by Eqs. (13.19) and (13.20), in kip and inch units, as follows.

Basic Principles of the Slope-Deflection Method

Member 1 (Joint 1 to Joint 2; $\theta_{ij} = \theta_1$; $\theta_{ji} = \theta_2$; $\Delta_i = \Delta_{1Y}$; $\Delta_j = \Delta_{2Y}$):

$$M_{12} = 25E\theta_2$$

$$M_{21} = 50E\theta_2$$

Member 2 (Joint 2 to Joint 3; $\theta_{ij} = \theta_2$; $\theta_{ji} = \theta_3$; $\Delta_i = \Delta_{2Y}$; $\Delta_j = \Delta_{3Y}$):

$$M_{23} = 300 + 60E\theta_2 + 30E\theta_3$$

$$M_{32} = -300 + 30E\theta_2 + 60E\theta_3$$

The numerical values for I_z have been substituted into the equations since they are different for each member, but, for convenience, the constant value for E is being carried through the analysis in symbolic form.

The final equations, which can be used to compute the joint rotations θ_2 and θ_3, can be obtained by substituting the Slope-Deflection Equations into the equilibrium equations. This results in the following set of linear simultaneous equations with θ_2 and θ_3 as the unknowns:

$$\Sigma_2 M_Z \rightarrow 110E\theta_2 + 30E\theta_3 = -300$$

$$\Sigma_3 M_Z \rightarrow 30E\theta_2 + 60E\theta_3 = 300$$

from which we can obtain

$$\theta_2 = -\frac{4.7368}{E} = -0.0001633 \text{ radians}$$

$$\theta_3 = \frac{7.3684}{E} = 0.0002541 \text{ radians}$$

The final values for the end moments for the members now can be obtained by substituting these values back into the Slope-Deflection Equations:

$$M_{12} = -118.420 \text{ kip-inches} = -9.868 \text{ kip-feet}$$

$$M_{21} = -236.840 \text{ kip-inches} = -19.737 \text{ kip-feet}$$

$$M_{23} = 236.844 \text{ kip-inches} = 19.737 \text{ kip-feet}$$

$$M_{32} = 0$$

The slight differences in the moments M_{21} and M_{23} in kip inch units is due to roundoff error in solving the simultaneous equations. Note that this difference disappears when the moments are converted to kip feet and rounded to three decimal places. Some amount of small roundoff error usually can be expected in the final values for the member end moments.

The shear force at the ends of the members now can be computed by summing moments about each end of each member. The final results of the analysis are shown in kip and feet units in the free body diagrams for each member in Figure 13.13a. Note that the axial forces in each member are zero. Therefore, the assumption that the effects of axial deformations can be ignored is a correct assumption for this structure. The member end loads in Figure 13.13a now can be used to determine the re-

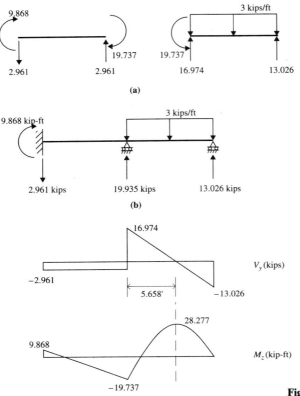

Figure 13.13 Example Problem 13.1, Case I.

active forces and moments for the beam, as shown in Figure 13.13b, from which the shear force and bending moment diagrams shown in Figure 13.13c can be developed. Any slight discrepancies which might occur in the closure of these diagrams are due to roundoff errors in computing the shear forces for the members.

Case II. The geometric compatibility conditions for Case II are

$$\theta_{12} = \theta_1 = 0$$

$$\theta_{21} = \theta_{23} = \theta_2$$

$$\theta_{32} = \theta_3$$

$$\Delta_{1X} = \Delta_{2X} = \Delta_{3X} = 0$$

$$\Delta_{1Y} = \Delta_{2Y} = 0$$

$$\Delta_{3Y} = -0.1 \text{ inch}$$

This beam also has two degrees of freedom which consist of the rotations θ_2 and θ_3 at joints 2 and 3, respectively. The vertical translation at joint 3 is not a degree of freedom since its magnitude is fixed.

Basic Principles of the Slope-Deflection Method

The independent equilibrium equations, the fixed end moments and the Slope-Deflection Equations for member 1 will be the same for this case as for the previous case. The difference in the two analyses will be in the Slope-Deflection Equations for member 2 due to the vertical translation of joint 3. The Slope-Deflection Equations for this member, in kip and inch units, are as follows:

Member 2 (Joint 2 to Joint 3; $\theta_{ij} = \theta_2$; $\theta_{ji} = \theta_3$; $\Delta_i = \Delta_{2Y}$; $\Delta_j = \Delta_{3Y}$):

$$M_{23} = 2475 + 60E\theta_2 + 30E\theta_3$$

$$M_{32} = 1875 + 30E\theta_2 + 60E\theta_3$$

The constant value for E has been used as a symbol in the coefficients for θ_2 and θ_3, but its numeric value has been used in computing the effect of the vertical translation of joint 3 for member 2.

As in Case I, the final equations which can be used to compute the joint rotations θ_2 and θ_3 can be obtained by substituting the Slope-Deflection Equations into the equilibrium equations. This results in the following set of linear simultaneous equations:

$$\Sigma_2 M_Z \rightarrow 110E\theta_2 + 30E\theta_3 = -2475$$

$$\Sigma_3 M_Z \rightarrow 30E\theta_2 + 60E\theta_3 = -1875$$

from which we can obtain

$$\theta_2 = -\frac{16.1842}{E} = -0.0005581 \text{ radians}$$

$$\theta_3 = -\frac{23.1579}{E} = -0.0007985 \text{ radians}$$

These displacements now can be substituted back into the Slope-Deflection Equations to obtain

$$M_{12} = -404.605 \text{ kip-inches} = -33.717 \text{ kip-feet}$$

$$M_{21} = -809.210 \text{ kip-inches} = -67.434 \text{ kip-feet}$$

$$M_{23} = 809.210 \text{ kip-inches} = 67.434 \text{ kip-feet}$$

$$M_{32} = 0$$

from which the reactions for the beam and the shear force and bending moment diagrams shown in Figure 13.14 can be determined.

A comparison of the results of the analyses for these two cases shows that the translation of the support has a significant effect on the final reactions and the shear forces and bending moments in the beam. A structural engineer must be very careful during the design of any structure to consider the possible effects of support displacements. Translations or rotations of the supports, which might at first seem to be insignificant due to their magnitudes, could have a significant effect upon the final stresses in the structure.

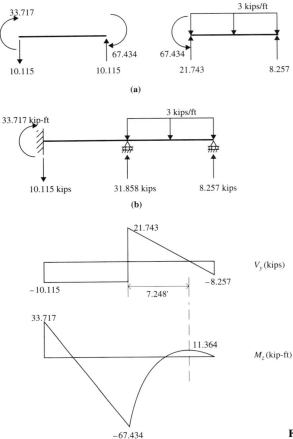

Figure 13.14 Example Problem 13.1, Case II.

These two solutions can be verified using the program F2DII in FATPAK II. Figures 13.15 and 13.16 show the outputs from the program for Case I and Case II, respectively, in kip and inch units. The results were printed in exponential form since the rotations of the joints in radians are very small numbers. The very small moments which are listed at the right end of member 2 for each case are due to roundoff error in the computer analyses. Figures 13.17 and 13.18 show the two deformed shapes for the beam with a displacement multiplication factor of 500 for each plot. Even though the vertical translation of 0.1 inch at joint 3 for Case II is small compared to the length of the beam, it can be seen from these plots that it is actually much larger than the vertical translations which occur at any interior point in the beam for Case I. This explains why it is not insignificant.

Basic Principles of the Slope-Deflection Method

```
FATPAK II -- Frame and Truss Package -- Student Edition
(C) Copyright 1995 by Structural Software Systems
Program F2DII -- Plane Frame Analysis

Example Problem 13.1 - Case I - Analysis by FATPAK II
Joint Displacements

Joint        X-Tran          Y-Tran              Z-Rot

  1       0.000000E+00    0.000000E+00       0.000000E+00
  2       0.000000E+00    0.000000E+00      -1.633394E-04
  3       0.000000E+00    0.000000E+00       2.540835E-04

Member End Loads

Member   Joint         Sx               Vy               Mz

  1        1       0.000000E+00    -2.960526E+00     -1.184211E+02
           2       0.000000E+00     2.960526E+00     -2.368421E+02
  2        2       0.000000E+00     1.697368E+01      2.368421E+02
           3       0.000000E+00     1.302632E+01     -3.649620E-06

Reactions

Joint               RX               RY                MZ

  1             0.000000E+00    -2.960526E+00     -1.184211E+02
  2             0.000000E+00     1.993421E+01      0.000000E+00
  3             0.000000E+00     1.302632E+01      0.000000E+00
```

Figure 13.15 Example Problem 13.1, Case I F2DII output.

Example Problem 13.2

Analyze the plane frame shown in Figure 13.19a by the Slope-Deflection Method.

The mathematical model for the frame is shown in Figure 13.19b. Note that a joint has not been placed at the location of the intermediate concentrated force on member 3 since this joint would introduce two additional degrees of freedom into the analysis. The concentrated force will be considered by including its fixed end moments in the Slope-Deflection Equations for member 3.

The geometric compatibility conditions for this mathematical model are

$$\theta_{12} = \theta_{13} = \theta_{14} = \theta_{15} = \theta_1$$

$$\theta_{21} = \theta_2 = 0$$

$$\theta_{31} = \theta_3 = 0$$

$$\theta_{41} = \theta_4$$

$$\theta_{51} = \theta_5 = 0$$

$$\Delta_{1X} = \Delta_{2X} = \Delta_{3X} = \Delta_{4X} = \Delta_{5X} = 0$$

$$\Delta_{1Y} = \Delta_{2Y} = \Delta_{3Y} = \Delta_{4Y} = \Delta_{5Y} = 0$$

Therefore, the frame has two degrees of freedom, which consist of the rotation θ_1 at joint 1 and the rotation θ_4 at joint 4.

```
FATPAK II -- Frame and Truss Package -- Student Edition
(C) Copyright 1995 by Structural Software Systems
Program F2DII -- Plane Frame Analysis

Example Problem 13.1 - Case II - Analysis by FATPAK II

Joint Displacements

Joint        X-Tran           Y-Tran           Z-Rot

  1       0.000000E+00     0.000000E+00     0.000000E+00
  2       0.000000E+00     0.000000E+00    -5.580763E-04
  3       0.000000E+00    -1.000000E-01    -7.985481E-04

Member End Loads

Member    Joint         Sx               Vy               Mz

  1         1       0.000000E+00    -1.011513E+01    -4.046053E+02
            2       0.000000E+00     1.011513E+01    -8.092106E+02
  2         2       0.000000E+00     2.174342E+01     8.092105E+02
            3       0.000000E+00     8.256579E+00    -1.124572E-05

Reactions

Joint              RX               RY               MZ

  1           0.000000E+00    -1.011513E+01    -4.046053E+02
  2           0.000000E+00     3.185855E+01     0.000000E+00
  3           0.000000E+00     8.256579E+00     0.000000E+00
```

Figure 13.16 Example Problem 13.1, Case II F2DII output.

Example Problem 13.1 - Case I - Analysis by FATPAK II
Displacement Multiplication Factor = 500

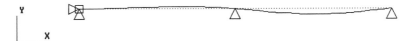

Figure 13.17 Example Problem 13.1, Case I, deformed geometry plot from F2DII.

Example Problem 13.1 - Case II - Analysis by FATPAK II
Displacement Multiplication Factor = 500

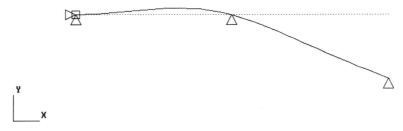

Figure 13.18 Example Problem 13.1, Case II, deformed geometry plot from F2DII.

Basic Principles of the Slope-Deflection Method

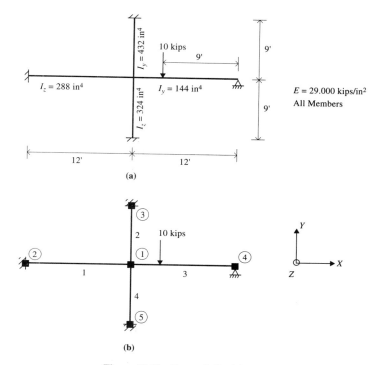

Figure 13.19 Example Problem 13.2.

The two independent equilibrium equations can be obtained by summing moments at joints 1 and 4. The resulting equations are

$$\text{At Joint 1,} \quad \Sigma_1 M_Z \rightarrow M_{12} + M_{13} + M_{14} + M_{15} = 0$$

$$\text{At Joint 4,} \quad \Sigma_4 M_Z \rightarrow M_{41} = 0$$

Equations are not written for joints 2, 3 and 5 since the moments applied to the joints by the members will be exactly balanced by the reactive moments for the fixed supports.

The fixed end moments for member 3 can be computed by Eqs. (13.1a) and (13.1b) by letting joint 1 correspond to end i and joint 4 correspond to end j:

$$M_{14}^F = -\frac{(-10)(3)(9)^2}{(12)^2} = 16.875 \text{ kip-feet} = 202.500 \text{ kip-inches}$$

$$M_{41}^F = \frac{(-10)(9)(3)^2}{(12)^2} = -5.625 \text{ kip-feet} = -67.500 \text{ kip-inches}$$

The Slope-Deflection Equations can be generated for each member by Eqs. (13.19) and (13.20), in kip and inch units, as follows.

Member 1 (Joint 2 to Joint 1; $\theta_{ij} = \theta_2$; $\theta_{ji} = \theta_1$; $\Delta_i = \Delta_{2Y}$; $\Delta_j = \Delta_{1Y}$):

$$M_{21} = 4E\theta_1$$

$$M_{12} = 8E\theta_1$$

Member 2 (Joint 1 to Joint 3; $\theta_{ij} = \theta_1$; $\theta_{ji} = \theta_3$; $\Delta_i; = -\Delta_{1X}$; $\Delta_j = -\Delta_{3X}$):

$$M_{13} = 16E\theta_1$$

$$M_{31} = 8E\theta_1$$

Member 3 (Joint 1 to Joint 4; $\theta_{ij} = \theta_1$; $\theta_{ji} = \theta_4$; $\Delta_i = \Delta_{1Y}$; $\Delta_j = \Delta_{4Y}$):

$$M_{14} = 202.500 + 4E\theta_1 + 2E\theta_4$$

$$M_{41} = -67.500 + 2E\theta_1 + 4E\theta_4$$

Member 4 (Joint 5 to Joint 1; $\theta_{ij} = \theta_5$; $\theta_{ji} = \theta_1$; $\Delta_i = -\Delta_{5X}$; $\Delta_j = -\Delta_{1X}$):

$$M_{51} = 6E\theta_1$$

$$M_{15} = 12E\theta_1$$

The numerical values for I_z have been used in the equations since they are different for each member. For convenience, the constant value of E has been represented in symbolic form.

The final equations, which can be used to compute the joint rotations θ_1 and θ_4, can be obtained by substituting the Slope-Deflection Equations into the equilibrium equations. The resulting set of linear simultaneous equations is

$$\Sigma_1 M_Z \to 40E\theta_1 + 2E\theta_4 = -202.500$$

$$\Sigma_4 M_Z \to 2E\theta_1 + 4E\theta_4 = 67.500$$

from which we can obtain

$$\theta_2 = -\frac{6.058}{E} = 0.0002089 \text{ radians}$$

$$\theta_4 = \frac{19.904}{E} = 0.0006863 \text{ radians}$$

The final values for the end moments for the members now can be obtained by substituting these values back into the Slope-Deflection Equations

$$M_{12} = -48.464 \text{ kip-inches} = -4.039 \text{ kip-feet}$$

$$M_{21} = -24.232 \text{ kip-inches} = -2.019 \text{ kip-feet}$$

$$M_{13} = -96.928 \text{ kip-inches} = -8.077 \text{ kip-feet}$$

$$M_{31} = -48.464 \text{ kip-inches} = -4.039 \text{ kip-feet}$$

$$M_{14} = 218.076 \text{ kip-inches} = 18.173 \text{ kip-feet}$$

$$M_{41} = 0$$

Basic Principles of the Slope-Deflection Method

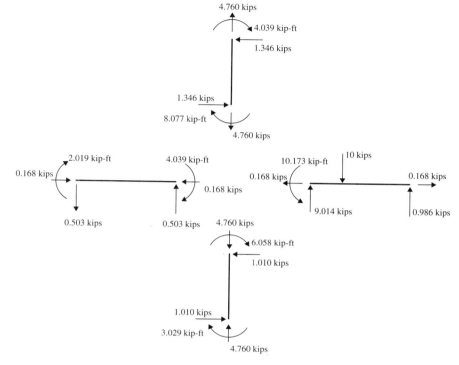

Figure 13.20 Example Problem 13.2.

$$M_{15} = -72.696 \text{ kip-inches} = -6.058 \text{ kip-feet}$$

$$M_{51} = -36.348 \text{ kip-inches} = -3.029 \text{ kip-feet}$$

The shear forces at the ends of the members can now be computed by summing moments about each end of each member. The results are shown in kip and feet units in the free body diagrams for the members shown in Figure 13.20. The axial forces in the members were computed by distributing the excess shear forces in each direction at joint 1 equally to each member in that direction since the members have the same length. The reactions for the frame now can be computed from the end loads of the members attached to each support joint, as shown in Figure 13.21. Finally, the bending moment diagram for each member can be drawn, as shown in the exploded view of the frame in Figure 13.22. The bending moments are plotted on the compression side of each member.

Figure 13.23 shows a plot of the geometry of the mathematical model for this frame which was used in an analysis by the program F2DII in FATPAK II. The cross section area which was assigned to all of the members was 10 square inches since this

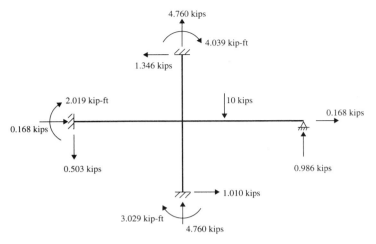

Figure 13.21 Example Problem 13.2.

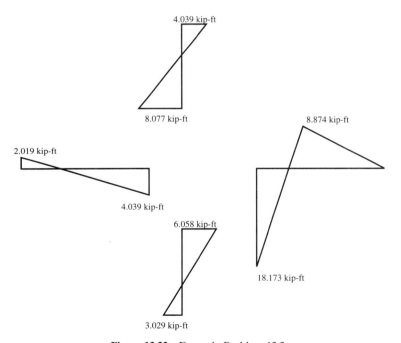

Figure 13.22 Example Problem 13.2.

is the order of magnitude of the areas for steel wide flange sections with values for I_z corresponding to the members in this frame. The results of the analysis by F2DII in kip and inch units are shown in Figure 13.24 on page 480 and a plot of the deformed frame with a displacement multiplication factor of 1500 is shown in Figure 13.25 on page 481. Comparing these member end moments to those obtained by the Slope-

Basic Principles of the Slope-Deflection Method

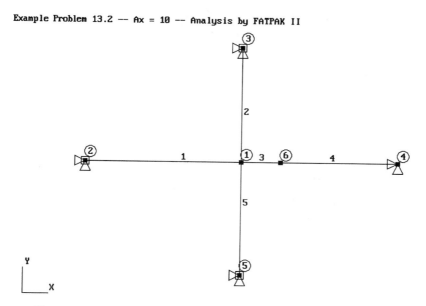

Figure 13.23 Example Problem 13.2, original geometry plot from F2DII.

Deflection Method shows that there are significant differences, particularly for the moments M_{12} and M_{21}, where the differences are 8% and 19% respectively. The reason for these differences becomes obvious when we look at the deformed shape of the frame, since the vertical translation of -0.00176 inches for joint 1 is of the same order of magnitude as the maximum interior vertical translation of member 1. Figure 13.26 shows the results of another analysis by F2DII in which an area of 10,000 square inches was assigned to the cross section areas for all of the members. The member end moments for this analysis are much closer to those obtained by the Slope-Deflection Method since the axial deformations in the members and the resulting horizontal and vertical translations of joint 1 are much smaller.

Note that the analysis of this frame was much easier by the Slope-Deflection Method than if it were analyzed by the Method of Consistent Deformations. For the Slope-Deflection Method analysis, it was only necessary to generate and solve two linear simultaneous equations, whereas a total of eight equations would have to be solved by the Method of Consistent Deformations since the frame is eight degrees statically indeterminate. There is no specific relationship between the number of degrees of freedom for a structure and the number of degrees of static indeterminacy. A structural engineer should carefully study each type of structure to determine which type of analysis should be performed.

Example Problem 13.3

Analyze the plane frame whose mathematical model is shown in Figure 13.27 by the Slope-Deflection Method. All members in the frame are prismatic and have the same cross section. This frame has three degrees of freedom, consisting of the rotation θ_2

```
FATPAK II -- Frame and Truss Package -- Student Edition
(C) Copyright 1995 by Structural Software Systems
Program F2DII -- Plane Frame Analysis

Example Problem 13.2 -- Ax = 10 -- Analysis by FATPAK II

Joint Displacements

Joint         X-Tran          Y-Tran          Z-Rot

  1  -8.055241E-05  -1.761155E-03  -2.118230E-04
  2   0.000000E+00   0.000000E+00   0.000000E+00
  3   0.000000E+00   0.000000E+00   0.000000E+00
  4   0.000000E+00   0.000000E+00   7.061532E-04
  5   0.000000E+00   0.000000E+00   0.000000E+00
  6  -6.041431E-05  -2.625313E-02  -6.830529E-04

Member End Loads

Member  Joint         Sx              Vy              Mz

  1       1      1.622236E-01    -4.527927E-01    -4.488681E+01
          2     -1.622236E-01     4.527927E-01    -2.031534E+01
  2       1     -4.729026E+00    -1.355468E+00    -9.776675E+01
          3      4.729026E+00     1.355468E+00    -4.862381E+01
  3       1     -1.622236E-01     9.005259E+00     2.167573E+02
          6      1.622236E-01    -9.005259E+00     1.074319E+02
  4       6     -1.622236E-01    -9.947402E-01    -1.074319E+02
          4      1.622236E-01     9.947402E-01     7.121616E-07
  5       1      4.729026E+00    -1.031021E+00    -7.410373E+01
          5     -4.729026E+00     1.031021E+00    -3.724653E+01

Reactions

Joint            RX              RY              MZ

  2          1.622236E-01    -4.527927E-01    -2.031534E+01
  3         -1.355468E+00     4.729026E+00    -4.862381E+01
  4          1.622236E-01     9.947402E-01     0.000000E+00
  5          1.031021E+00     4.729026E+00    -3.724653E+01
```

Figure 13.24 Example Problem 13.2, F2DII output for $A_x = 10$ in^2.

at joint 2, the rotation θ_3 at joint 3 and the equal horizontal translations Δ of joints 2 and 3. The horizontal translations of these joints must be equal since the axial deformation in member 2 is assumed to be zero. Therefore, the geometric compatibility conditions for this mathematical model are

$$\theta_{12} = \theta_1 = 0$$
$$\theta_{21} = \theta_{23} = \theta_2$$
$$\theta_{32} = \theta_{34} = \theta_3$$
$$\theta_{43} = \theta_4 = 0$$
$$\Delta_{1X} = \Delta_{4X} = 0$$

Basic Principles of the Slope-Deflection Method

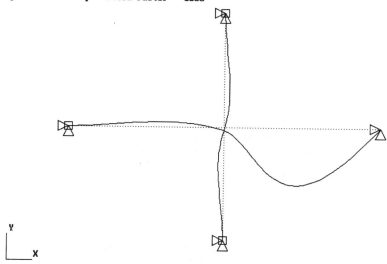

Figure 13.25 Example Problem 13.2, deformed geometry plot from F2DII for $A_x = 10$ in².

$$\Delta_{2X} = \Delta_{3X} = \Delta$$

$$\Delta_{1Y} = \Delta_{2Y} = \Delta_{3Y} = \Delta_{4Y} = 0$$

Two of the independent equilibrium equations can be obtained by summing moments at joints 2 and 3. The resulting equations are

$$\text{At Joint 2,} \quad \Sigma_2 M_Z \rightarrow M_{21} + M_{23} = 0$$

$$\text{At Joint 3,} \quad \Sigma_3 M_Z \rightarrow M_{32} + M_{34} = 0$$

The third independent equilibrium equation can be obtained by using the free body diagrams of members 1 and 3 shown in Figure 13.28. The shear forces V_1 and V_3 can be expressed in terms of the member end moments by summing moments for each member

$$V_1 = \frac{M_{12} + M_{21}}{5}$$

$$V_3 = \frac{M_{34} + M_{43}}{10}$$

For horizontal equilibrium of the frame,

$$\Sigma F_X \rightarrow V_1 + V_3 = 20$$

from which, after substituting the expressions for V_1 and V_3 and simplifying, gives

$$2M_{12} + 2M_{21} + M_{34} + M_{43} = 200$$

The Slope-Deflection Equations for each member can be generated by Eqs. (13.19) and (13.20), in kilonewton and meter units, as

```
FATPAK II -- Frame and Truss Package -- Student Edition
(C) Copyright 1995 by Structural Software Systems
Program F2DII -- Plane Frame Analysis

Example Problem 13.2 -- Ax = 10000 -- Analysis by FATPAK II
Joint Displacements

Joint        X-Tran           Y-Tran            Z-Rot

  1    -8.355121E-08   -1.772535E-06    -2.088888E-04
  2     0.000000E+00    0.000000E+00     0.000000E+00
  3     0.000000E+00    0.000000E+00     0.000000E+00
  4     0.000000E+00    0.000000E+00     6.863592E-04
  5     0.000000E+00    0.000000E+00     0.000000E+00
  6    -6.266340E-08   -2.457562E-02    -6.900623E-04

Member End Loads

Member   Joint        Sx               Vy                Mz

  1        1      1.682628E-01    -5.047551E-01    -4.845792E+01
           2     -1.682628E-01     5.047551E-01    -2.422682E+01
  2        1     -4.759584E+00    -1.346162E+00    -9.692388E+01
           3      4.759584E+00     1.346162E+00    -4.846167E+01
  3        1     -1.682629E-01     9.014413E+00     2.180755E+02
           6      1.682629E-01    -9.014413E+00     1.064432E+02
  4        6     -1.682628E-01    -9.855857E-01    -1.064433E+02
           4      1.682628E-01     9.855857E-01     4.667894E-06
  5        1      4.759584E+00    -1.009637E+00    -7.269372E+01
           5     -4.759584E+00     1.009637E+00    -3.634706E+01

Reactions

Joint            RX               RY                MZ

  2         1.682628E-01    -5.047551E-01    -2.422682E+01
  3        -1.346162E+00     4.759584E+00    -4.846167E+01
  4         1.682628E-01     9.855857E-01     0.000000E+00
  5         1.009637E+00     4.759584E+00    -3.634706E+01
```

Figure 13.26 Example Problem 13.2, F2DII output for $A_x = 10{,}000$ in^2.

Member 1 (Joint 1 to Joint 2; $\theta_{ij} = \theta_1$; $\theta_{ji} = \theta_2$; $\Delta_i = -\Delta_{1X}$; $\Delta_j = -\Delta_{2X}$):

$$M_{12} = 0.4EI_z\theta_2 + 0.24EI_z\Delta$$

$$M_{21} = 0.8EI_z\theta_2 + 0.24EI_z\Delta$$

Member 2 (Joint 2 to Joint 3; $\theta_{ij} = \theta_2$; $\theta_{ji} = \theta_3$; $\Delta_i = \Delta_{2Y}$; $\Delta_j = \Delta_{3Y}$):

$$M_{23} = 0.4EI_z\theta_2 + 0.2EI_z\theta_3$$

$$M_{32} = 0.2EI_z\theta_2 + 0.4EI_z\theta_3$$

Basic Principles of the Slope-Deflection Method

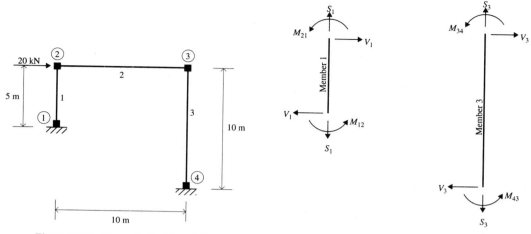

Figure 13.27 Example Problem 13.3.

Figure 13.28 Example Problem 13.3.

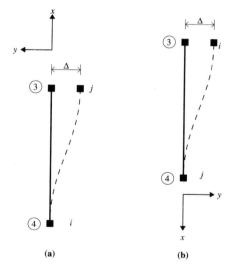

(a)   (b)   **Figure 13.29** Example Problem 13.3.

Member 3 (Joint 4 to Joint 3; $\theta_{ij} = \theta_4$; $\theta_{ji} = \theta_3$; $\Delta_i = -\Delta_{4X}$; $\Delta_j = -\Delta_{3X}$):

$$M_{43} = 0.2EI_z\theta_3 + 0.06EI_z\Delta$$

$$M_{34} = 0.4EI_z\theta_3 + 0.06EI_z\Delta$$

The constant value of EI_z for each member has been carried through symbolically in the equations. As we will see later, the final member end moments will be independent of the specific value for this quantity.

Either end of a member may be designated as end i when generating the Slope-Deflection Equations. For example, Figure 13.29a shows the orientation of the local member axes for member 3 for the previous designations for i and j. The detailed calculations for the Slope-Deflection Equations for the member this case are

Member 3 (Joint 4 to Joint 3; $\theta_{ij} = \theta_4$; $\theta_{ji} = \theta_3$; $\Delta_i = -\Delta_{4X}$; $\Delta_j = -\Delta_{3X}$):

$$M_{43} = \frac{4EI_z}{10}(0) + \frac{2EI_z}{10}(\theta_3) - \frac{6EI_z}{(10)^2}(-\Delta - 0) = 0.2EI_z\theta_3 + 0.06EI_z\Delta$$

$$M_{34} = \frac{2EI_z}{10}(0) + \frac{4EI_z}{10}(\theta_3) - \frac{6EI_z}{(10)^2}(-\Delta - 0) = 0.4EI_z\theta_3 + 0.06EI_z\Delta$$

Figure 13.29b shows the orientation of the local member axes for member 3 if the designations for i and j are reversed. Since the positive direction of local y is reversed, the positive direction for Δ_i and Δ_j will change. The detailed calculations for the Slope-Deflection Equations for the member for this case are

Member 3 (Joint 3 to Joint 4; $\theta_{ij} = \theta_3$; $\theta_{ji} = \theta_4$; $\Delta_i = \Delta_{3X}$; $\Delta_j = \Delta_{4X}$):

$$M_{34} = \frac{4EI_z}{10}(\theta_3) + \frac{2EI_z}{10}(0) - \frac{6EI_z}{(10)^2}(0 - \Delta) = 0.4EI_z\theta_3 + 0.06EI_z\Delta$$

$$M_{43} = \frac{4EI_z}{10}(0) + \frac{2EI_z}{10}(\theta_3) - \frac{6EI_z}{(10)^2}(0 - \Delta) = 0.2EI_z\theta_3 + 0.06EI_z\Delta$$

Therefore, the same results are obtained for either designation for i and j.

The final equations which can be used to compute θ_2, θ_3, and Δ can be obtained by substituting the Slope-Deflection Equations into the equilibrium equations. The resulting set of linear simultaneous equations is

$$\Sigma_2 M_Z \rightarrow 1.2EI_z\theta_2 + 0.2EI_z\theta_3 + 0.24EI_z\Delta = 0$$

$$\Sigma_3 M_Z \rightarrow 0.2EI_z\theta_2 + 0.8EI_z\theta_3 + 0.06EI_z\Delta = 0$$

$$\Sigma F_X \rightarrow 2.4EI_z\theta_2 + 0.6EI_z\theta_3 + 1.08EI_z\Delta = 200$$

which can be solved by the program SEQSOLVE, which is described in the ASCII disk file SEQSOLVE.DOC, to obtain

$$\theta_2 = -\frac{65.789}{EI_z}$$

$$\theta_3 = -\frac{8.772}{EI_z}$$

$$\Delta = \frac{336.257}{EI_z}$$

The final values for the member end moments now can be obtained by substituting these values back into the Slope-Deflection Equations. Note that the quantity EI_z will cancel out during these calculations.

$$M_{12} = 54.386 \text{ kN-m}$$

$$M_{21} = 28.070 \text{ kN-m}$$

Basic Principles of the Slope-Deflection Method

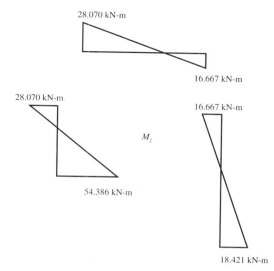

Figure 13.30 Example Problem 13.3.

$$M_{23} = -28.070 \text{ kN-m}$$
$$M_{32} = -16.667 \text{ kN-m}$$
$$M_{34} = 16.667 \text{ kN-m}$$
$$M_{43} = 18.421 \text{ kN-m}$$

These values now can be used to plot the bending moment diagrams for each member as shown in the exploded view of the frame in Figure 13.30. The moments are plotted on the compression side of the members.

The reader can verify this solution with the program F2DII in FATPAK II. It would be interesting to try several different combinations of A_x and I_z for the member cross sections to determine the effect of the axial deformations in the members on the displacements and member end moments for this frame. This is left as an exercise for the reader.

Example Problem 13.4

Analyze the plane frame whose mathematical model is shown in Figure 13.31a by the Slope-Deflection Method. All members in the frame are prismatic and have the same cross section. This frame has three degrees of freedom, consisting of the rotation θ_2 at joint 2, the rotation θ_3 at joint 3 and the equal horizontal translations Δ of joints 2 and 3. The vertical translations of joints 2 and 3 are not independent degrees of freedom since they can be expressed in terms of the horizontal translations. The final location of joint 2 must be on an arc centered at joint 1, with a radius equal to the length of member 1, and the final location of joint 3 must be on an arc centered at joint 4, with a radius equal to the length of member 3. If the translations are very small compared to the overall dimensions of the frame, it can be assumed that joint 2 translates perpendicular to member 1 and joint 3 translates perpendicular to member 3. By using this assumption, the translations of joints 2 and 3 can be expressed as shown in Fig-

486 The Slope-Deflection Method   Chap. 13

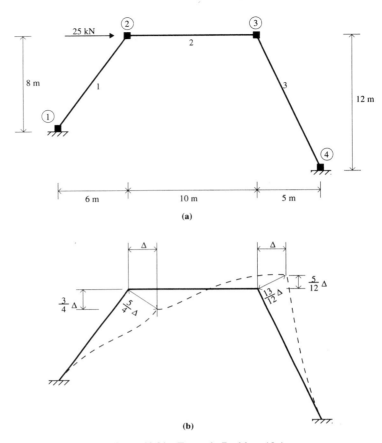

Figure 13.31 Example Problem 13.4.

ure 13.31b. By using these relationships, the geometric compatibility conditions for this mathematical model are

$$\theta_{12} = \theta_1 = 0$$
$$\theta_{21} = \theta_{23} = \theta_2$$
$$\theta_{32} = \theta_{34} = \theta_3$$
$$\theta_{43} = \theta_4 = 0$$
$$\Delta_{1X} = \Delta_{4X} = 0$$
$$\Delta_{2X} = \Delta_{3X} = \Delta$$
$$\Delta_{1Y} = \Delta_{4Y} = 0$$
$$\Delta_{2Y} = -3/4\Delta_{2X}$$
$$\Delta_{3Y} = 5/12\Delta_{3X}$$

Basic Principles of the Slope-Deflection Method

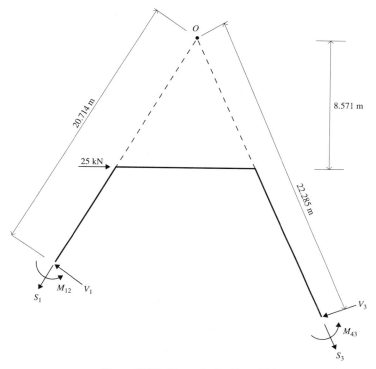

Figure 13.32 Example Problem 13.4.

Two of the independent equilibrium equations can be obtained by summing moments at joints 2 and 3. The resulting equations are

$$\text{At Joint 2,} \quad \Sigma_2 M_Z \rightarrow M_{21} + M_{23} = 0$$

$$\text{At Joint 3,} \quad \Sigma_3 M_Z \rightarrow M_{32} + M_{34} = 0$$

The third independent equilibrium equation can be obtained by using the free body diagram of the frame shown in Figure 13.32. The shear forces V_1 and V_3 for members 1 and 3 can be expressed in terms of the member end moments, in the same manner as in the previous example problem, by summing moments for each member

$$V_1 = \frac{M_{12} + M_{21}}{10}$$

$$V_3 = \frac{M_{34} + M_{43}}{13}$$

The overall moment equilibrium of the frame now can be ensured by summing moments about point O, at the intersection of the extensions of members 1 and 3, to obtain

$$\Sigma_O M_Z \rightarrow M_{12} + M_{43} - 20.714 V_1 - 22.285 V_3 + (8.571)(25) = 0$$

which, on substituting the expressions for V_1 and V_3, gives

$$-1.0714M_{12} - 2.0714M_{21} - 1.7142M_{34} - 0.7142M_{43} + 214.275 = 0$$

The Slope-Deflection Equations for each member can be generated by Eqs. (13.17) and (13.18), in kN and meter units, as

Member 1 (Joint 1 to Joint 2; $\theta_{ij} = \theta_1$; $\theta_{ji} = \theta_2$; $\Delta_i = -(5/4)\Delta_{1X}$; $\Delta_j = -(5/4)\Delta_{2X}$):

$$M_{12} = 0.2EI_z\theta_2 + 0.075EI_z\Delta$$

$$M_{21} = 0.4EI_z\theta_2 + 0.075EI_z\Delta$$

Member 2 (Joint 2 to Joint 3; $\theta_{ij} = \theta_2$; $\theta_{ji} = \theta_3$; $\Delta_i = -(3/4)\Delta_{2X}$; $\Delta_j = (5/12)\Delta_{3X}$):

$$M_{23} = 0.4EI_z\theta_2 + 0.2EI_z\theta_3 - 0.07EI_z\Delta$$

$$M_{32} = 0.2EI_z\theta_2 + 0.4EI_z\theta_3 - 0.07EI_z\Delta$$

Member 3 (Joint 3 to Joint 4; $\theta_{ij} = \theta_3$; $\theta_{ji} = \theta_4$; $\Delta_i = (13/12)\Delta_{3X}$; $\Delta_j = (13/12)\Delta_{4X}$):

$$M_{34} = 0.3077EI_z\theta_3 + 0.0385EI_z\Delta$$

$$M_{43} = 0.1538EI_z\theta_3 + 0.0385EI_z\Delta$$

The final equations which can be used to compute θ_2, θ_3, and Δ can be obtained by substituting the Slope-Deflection Equations into the equilibrium equations. The resulting set of linear simultaneous equations is

$$\Sigma_2 M_Z \rightarrow 0.8EI_z\theta_2 + 0.2EI_z\theta_3 + 0.005EI_z\Delta = 0$$

$$\Sigma_3 M_Z \rightarrow 0.2EI_z\theta_2 + 0.7077EI_z\theta_3\ 0.0315EI_z\Delta = 0$$

$$\Sigma_O M_Z \rightarrow 1.0429EI_z\theta_2 + 0.6373EI_z\theta_3 + 0.3293EI_z\Delta = 214.275$$

which can be solved by the program SEQSOLVE to obtain

$$\theta_2 = -\frac{11.731}{EI_z}$$

$$\theta_3 = \frac{31.241}{EI_z}$$

$$\Delta = \frac{627.391}{EI_z}$$

The final values for the end moments for the members can be obtained by substituting these values back into the Slope-Deflection Equations:

$$M_{12} = 44.708 \text{ kN-m}$$

$M_{21} = 42.362$ kN-m

$M_{23} = -42.362$ kN-m

$M_{32} = -33.767$ kN-m

$M_{34} = 33.767$ kN-m

$M_{43} = 28.959$ kN-m

The reader can verify this solution and plot the shear force and bending moment diagrams for each member with the program F2DII in FATPAK II

A similar procedure to that shown in these example problems can be used to analyze any beam or plane frame by the Slope Deflection Method. The primary difficulties which are encountered are in determining the number and the nature of the independent degrees of freedom of the structure and the development of the set of independent equilibrium equations for structures with translation degrees of freedom.

SUGGESTED PROBLEMS

SP13.1 to SP13.9 Compute the end moments for each member in the mathematical models for the structures shown in Figures SP13.1 to SP13.9 by the Slope-Deflection Method. Draw the bending moment diagrams for each structure. Treat the quantity EI_z as a symbol during the analysis.

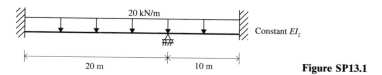

Figure SP13.1

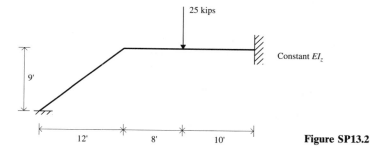

Figure SP13.2

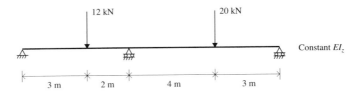

Figure SP13.3

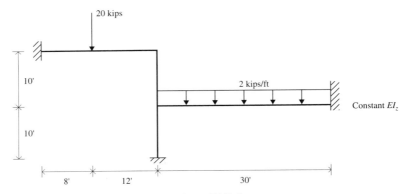

Figure SP13.4

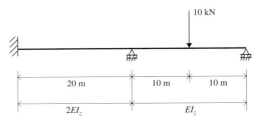

Figure SP13.5

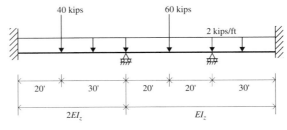

Figure SP13.6

Suggested Problems

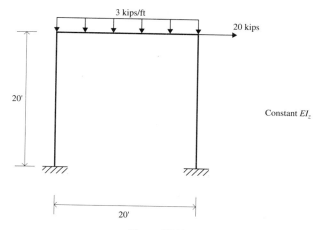

Figure SP13.7

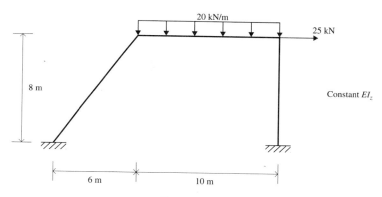

Figure SP13.8

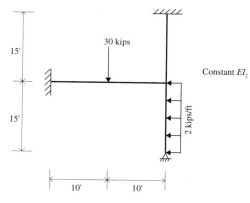

Figure SP13.9

SP13.10 Reanalyze the plane frame in Figure SP13.9 if the left support translates down 0.2 inches as the frame is loaded. Use the following values for E and I_z.

$$E = 29{,}000 \text{ kips/in}^2$$
$$I_z = 1000 \text{ in}^4$$

SP13.11 The Slope-Deflection Equations for an elastic plane frame member are in the form

$$M_{ij} = M_{ij}^F = C_1\theta_{ij} + C_2\theta_{ji} + C_3(\Delta_j - \Delta_i)$$
$$M_{ji} = M_{ji}^F = C_4\theta_{ij} + C_5\theta_{ji} + C_6(\Delta_j - \Delta_i)$$

Derive expressions for the coefficients C_1 through C_6 for the nonprismatic member shown in Figure SP13.11 using the Moment Area Theorems.

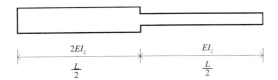

Figure SP13.11

SP13.12 Derive expressions for the fixed end moments for the member in Figure SP13.11 for a positive uniform force over the entire length of the member. Use the Method of Consistent Deformations to perform the analysis of the fixed-fixed member.

14

The Moment Distribution Method

BASIC ANALYSIS PROCEDURE

The *Moment Distribution Method* is a procedure for computing the member end moments in statically indeterminate beams and plane frames. The basic process is to assume an initial set of values for the member end moments which do not satisfy equilibrium at the joints, and then to incrementally correct these moments by a series of iterations, using the geometric compatibility and load-deformation characteristics of the structure, until equilibrium is satisfied within the desired accuracy. This analysis procedure was first introduced by Professor Hardy Cross of the University of Illinois in the 1920s. It was probably the most popular analysis procedure available to structural engineers until the introduction of computers into the field of structural engineering. Its primary advantage over the Method of Consistent Deformations and the Slope-Deflection Method is that it eliminates the requirement of solving a set of simultaneous equations for determining the member end moments for any statically indeterminate beam or plane frame in which the joints are restrained against translation. It began to lose popularity, along with the other classical manual analysis procedures, when computer programs became available which could analyze essentially any type of statically indeterminate structure. At the present time, it might be used primarily for the preliminary analysis of a simple beam or frame during the initial stages of design. One other use that a structural engineer might find for this procedure is to solve problems in a professional engineering examination, since at the present time computers are not permitted for use in these examinations by any examining body in the United States. It is for this reason that this procedure is being discussed here. However, the discussion will be limited to the application of the Moment Distribution Method to the analysis of beams and simple frames in which only rotation degrees of freedom exist at the joints.

As in the Slope-Deflection Method, only the effects of bending deformations in the members are included during the analysis of the mathematical model of any structure by the Moment Distribution Method. The axial deformations and the shear deformations are not considered. Therefore, the translation and rotation degrees of freedom for the structure can be determined by the same process described previously in Chapter 13. Although this analysis procedure can also be used for the analysis of structures with joint translations, the analysis procedure is much more complicated and will not be considered here. It is improbable that any engineer today would attempt to analyze manually any type of large structure with translation degrees of freedom by this method. Gere (1963) contains a very complete discussion of the application of this analysis procedure to structures of this type for those readers who might be interested.

The analysis of the mathematical model for any beam or frame by the Moment Distribution Method, in which the joints are restrained against translation, will consist of the following steps:

1. Compute the moments at the ends of each member in the structure, due to the active loads on the members, assuming that all joints which are free to rotate are restrained by a fictitious rotation restraint.

2. Sum the member end moments at each joint to determine the unbalanced moment which is being resisted by the fictitious rotation restraint. It is not necessary to compute unbalanced moments at any support joint, which has a rotation restraint, since the member end moments will be automatically balanced by the reactive moment.

3. Remove the fictitious rotation restraint at each joint in turn, assuming that the other joints are still restrained against rotation, and distribute the unbalanced moment at the joint to the ends of the members which are rigidly attached to the joint. In addition, distribute a percentage of the moment along each member to its far joint to keep that joint from rotating. The geometric compatibility conditions for the structure and the load-deformation characteristics of the members must be satisfied during this distribution process.

4. Repeat the distribution process in Step 3 for each joint as many times as necessary until the member end moments are balanced at each joint within an acceptable value. The final member end moments will correspond to the correct end moments for this structure and loading condition.

5. Compute the axial forces and the shear forces in the members by using the member equilibrium equations, and compute the reactive forces and reactive moments at the support joints by using the joint equilibrium equations.

The number of times that the distribution process must be performed for each joint will depend on the specific properties of the structure and the type of loading. For

Basic Analysis Procedure

some structures, an acceptable equilibrium condition for each joint will be reached very quickly, whereas for other structures, particularly those with many rotation degrees of freedom, the distribution procedure might have to be repeated many times. The engineer performing the analysis must finally decide on what is considered to be an acceptable equilibrium condition.

Before we can proceed with the application of the Moment Distribution Method to any specific structures, it is necessary to define the sign convention and the various properties of the structure which will be used in the analysis.

Sign Convention

All loads acting on the members in the mathematical model of a beam or a plane frame, which is to be analyzed by the Moment Distribution Method, will be expressed as components in the individual local member coordinate systems. For consistency, the definition for these local coordinate systems will be the same as used previously for the analysis of plane frames by the Slope Deflection Method and in the computer programs SDPFRAME and F2DII. Therefore, the member end moments, which are applied to each member by its end joints, will be positive in a counterclockwise direction, and all active transverse intermediate member forces will be positive in the direction of the local member y axis.

Fixed End Moments

The initial member end moments, which are computed assuming that the member ends are restrained against translation and rotation, correspond to the fixed end moments described in Chapter 13 for use in the Slope-Deflection Method. Equations (13.1) through (13.4) contain expressions which can be used to compute these moments for prismatic members for several different types of member loads.

Member Stiffness

The definition for the *member stiffness*, K_n, at the end of any member n in the structure, as used in the Moment Distribution Method, corresponds to the magnitude of the moment which must be applied to that end of the member to produce a unit rotation at the end with both ends restrained against translation. Figure 14.1

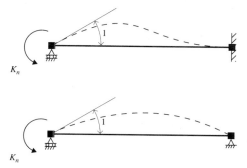

Figure 14.1 Member stiffness.

shows two different cases for this member stiffness. The stiffness of a prismatic member, with the far end restrained against rotation, as shown in Figure 14.1a can be determined from Eq. (13.9) in Chapter 13, after setting θ_{ij} to a unit value, to be

$$K_n = \frac{4EI_z}{L} \qquad (14.1)$$

for either end of the member. By using a similar process to that used to derive Eq. (13.9), the stiffness of a prismatic member with the far end unrestrained against rotation can be determined to be

$$K_n = \frac{3EI_z}{L} \qquad (14.2)$$

The actual derivation of this expression is left as an exercise for the reader. Similar expressions can be derived for nonprismatic members by modifying the shape of the M_z/EI_z diagram during the derivation to correspond to the variation of EI_z over the length of the member. For these members, the stiffness will be different at each end.

The moment required at the end of a member to produce a rotation angle θ can be expressed in terms of its stiffness as

$$M_n = K_n \theta \qquad (14.3)$$

while, for a joint with N members rigidly connected to it, the total moment required to rotate the joint through an angle θ will be equal to the sum of the moments which are required to rotate the end of each member:

$$M = K_1\theta + K_2\theta + \cdots + K_i\theta + \cdots + K_N\theta \qquad (14.4)$$

or

$$M = \sum_{i=1}^{N} K_i \theta = \theta \sum_{i=1}^{N} K_i \qquad (14.5)$$

Distribution Factor

As the fictitious rotation restraint is removed at each joint, the angle through which the joint will rotate can be determined from Eq. (14.5), in terms of the unbalanced moment M_u acting on the joint from the member ends, as

$$\theta = \frac{M_u}{\sum_{i=1}^{N} K_i} \qquad (14.6)$$

Since each member which is rigidly attached to the joint must rotate through this same angle, the moment which will be transmitted to the end of any member n can be determined from Eq. (14.3). Solving this equation for θ gives

$$\theta = \frac{M_n}{K_n} \qquad (14.7)$$

which, on equating to Eq. (14.6) and solving for M_n, gives

$$M_n = \frac{K_n}{\sum_{i=1}^{N} K_i} M_u = DF_n M_u \qquad (14.8)$$

in which the quantity

$$DF_n = \frac{K_n}{\sum_{i=1}^{N} K_i} \qquad (14.9)$$

is known as the *distribution factor* for member n at that joint. This factor corresponds to the fractional portion of the total applied moment at a joint which will be resisted by any member n as the joint rotates through an angle θ.

Carry Over Factor

When a moment is applied to the near end of any member during the distribution process, whose far end is restrained against rotation, a moment must be applied to the far end by the joint at that end to resist rotation of the member end. This moment is called the *carry o er moment* and the ratio of the far end moment to the near end moment is known as the *carry o er factor*. This factor, which will be represented by the symbol CO_n for any member n, can be determined for a prismatic member from Eq. (13.6) in Chapter 13 to be

$$CO_n = \tfrac{1}{2} \qquad (14.10)$$

for either end. The positive sign for the carry over factor means that the moment at the far end of the member will be in the same direction as the moment at the near end. The carry over factors for each end of a nonprismatic member, whose far end is restrained against rotation, can be determined by using the Second Moment Area Theorem in a similar manner to that used in deriving Eq. (13.6). The carry over factor for either a prismatic or a nonprismatic member whose far end is unrestrained against rotation will be zero.

APPLICATION OF THE MOMENT DISTRIBUTION METHOD TO THE ANALYSIS OF BEAMS

The Moment Distribution Method is a very convenient procedure for computing the member end moments in the mathematical model of a statically indeterminate beam. The convergence of the iterative process to an acceptable equilibrium condition for the joints is usually fairly rapid for a beam with only a few rotation degrees of freedom. After the member end moments have been determined, the member shear forces and the reactive forces and the reactive moments can be determined by an equilibrium analysis of the individual members and the support joints, as shown previously in the example problems for the Slope-Deflection Method in Chapter 13.

Example Problem 14.1

Compute the member end moments for the prismatic beam shown in Figure 14.2. The mathematical model of the beam is shown in Figure 14.3a. A joint has not been placed at the location of the concentrated force in the left span since the effect of this force will be considered during the computation of the fixed end moments for the two members in the mathematical model. Only the three joints shown in the mathematical model are required to perform the analysis by the Moment Distribution Method.

The fixed end moments can be computed by using Eqs. (13.1a) and (13.1b) for member 1 and Eqs. (13.2a) and (13.2b) for member 2. The resulting moments are

$$M^F_{12} = 64 \text{ kN-m}$$
$$M^F_{21} = -256 \text{ kN-m}$$
$$M^F_{23} = 187.5 \text{ kN-m}$$
$$M^F_{32} = -187.5 \text{ kN-m}$$

where the subscripts have the same meaning as in the Slope-Deflection Method in Chapter 13. These moments are shown acting on the member ends and on the joints in Figure 14.3b with the fictitious rotation restraint at joint 2. The rotations of joints 1 and 3 are restrained by the fixed supports. If we now sum the moments which are applied to joint 2 by the member ends, we will see that there is an unbalanced counterclockwise moment of 68.5 kN m at this joint which is being resisted by the fictitious rotation restraint. There is no need to sum moments at joint 1 and joint 3 since the reactive moments due to the actual support restraints will balance any moments which are applied by the member ends to these joints. If we now remove the fictitious rotation restraint at joint 2, the counterclockwise moment of 68.5 kN will be distributed to the ends of members 1 and 2 as the joint rotates. The moments which will be distributed to each member end can be determined from by Eq. (14.8), in which the member stiffnesses K_1 and K_2 of members 1 and 2, respectively, can be computed by Eq. (14.1) as

$$K_1 = \frac{4EI_z}{50} = 0.08EI_z$$

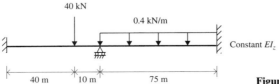

Figure 14.2 Example Problem 14.1.

Application of the Moment Distribution Method to the Analysis of Beams

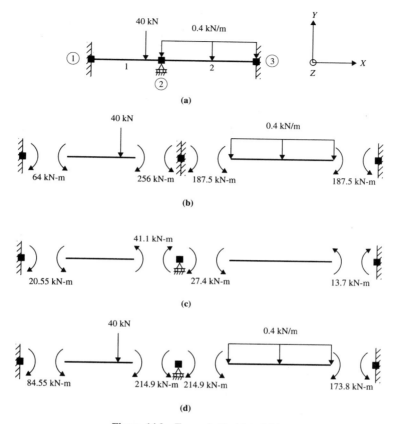

Figure 14.3 Example Problem 14.1.

and

$$K_2 = \frac{4EI_z}{75} = 0.0533EI_z$$

and the distribution factors DF_1 and DF_2 for each member can be determined from Eq. (14.9) as

$$DF_1 = \frac{K_1}{K_1 + K_2} = 0.6$$

and

$$DF_2 = \frac{K_2}{K_1 + K_2} = 0.4$$

Note that the quantity EI_z cancels out since the distribution factors are dimensionless. The distributed moments and the carry over moments are shown for each member

(1)	Joint	1	2		3
(2)	Moment	M_{12}	M_{21}	M_{23}	M_{32}
(3)	K		$0.08EI_z$	$0.0533EI_z$	
(4)	DF	0	0.6	0.4	0
(5)	FEM	64	−256	187.5	−187.5
(6)	Dist.		41.1	27.4	
(7)	CO	20.55			13.7
(8)	Final Moments	84.55	−214.9	214.9	−173.8

Figure 14.4 Example Problem 14.1.

and each joint in Figure 14.3c. Joint 2 is now in equilibrium since the clockwise moments of 41.1 and 27.4 kN-m, which are applied to it by the ends of members 1 and 2, exactly balance the 68.5 kN-m counterclockwise moment. By now superimposing the moments which are applied to the ends of the members by the joints and to the joints by the ends of the members in Figures 14.3b and 14.3c, the final moments for each member end and each joint can be determined, as shown in Figure 14.3d. Note that the final moments applied to either side of joint 2 are equal in magnitude and in opposite directions, as required for equilibrium of the joint. The unbalanced moment at joint 2 has been removed by the distribution process.

The previous calculations can be set up in a simple tabular form, as shown in Figure 14.4. The rows in the table have been numbered for identification.

Row 1. The columns in the table are assigned for each joint. Two columns are assigned for joint 2 since there are two member end moments at that joint.

Row 2. The columns in the table are assigned for each member end moment. The notation which is used to identify each end moment is the same as used previously in the Slope Deflection Method. Moment M_{ij} is the moment which is applied to end i of the member which is attached to joints i and j and M_{ji} is the moment which is applied to end j of the same member.

Row 3. The member stiffness corresponding to each member end moment at joint 2 is computed. The member stiffnesses corresponding to M_{21} and M_{23} are computed for members 1 and 2 by Eq. (14.1) . Member stiffnesses are not required for M_{12} or M_{32} since no distributions will be made for the moments at joint 1 and joint 3. The member end moments at these joints will be resisted by the support rotation restraints.

Row 4. The distribution factors corresponding to each member end moment at joint 2 are computed by Eq. (14.9). The distribution factors for M_{12} and M_{32} are assigned zero values since no distributions will be made for the moments at joint 1 and joint 3.

Row 5. The fixed end moments for each member are computed. During this process, it is assumed that joint 2 is restrained against rotation. These member end moments correspond to those shown in Figure 14.3b.

Application of the Moment Distribution Method to the Analysis of Beams

Row 6. A moment with the opposite sign of the sum of the member end moments at joint 2 is distributed to M_{21} and M_{23}. The distributed moments have the opposite sign of the sum of the member end moments since the moments applied to joint 2 by the members are in the opposite directions of the member end moments. The purpose of this operation is to place joint 2 into equilibrium by removing the unbalanced moment at the joint and distributing it to the members. These distributed moments correspond to the member end moments adjacent to joint 2 in Figure 14.3c.

Row 7. The carry over moments at the opposite ends of members 1 and 2 are computed using a carry over factor of 1/2 for the prismatic members. These carry over moments correspond to the member end moments adjacent to joint 1 and joint 3 in Figure 14.3c. The slanted arrows are included in the table to indicate the direction of the carry over.

Row 8. The moments are summed in each column. These moments correspond to the member end moments shown in Figure 14.3d. Equilibrium at joint 2 is satisfied since the member end moments on either side of the joint have equal magnitudes and opposite signs.

For most beams, these calculations can be performed with a hand calculator and recorded directly on each row of the table without writing out any equations or recording any intermediate numbers. Only one distribution and one carry over cycle were required in this analysis since the beam had only one rotation degree of freedom.

Example Problem 14.2

Compute the member end moments for the beam shown in Figure 14.5a. The beam is composed of only one material, but has a different cross section moment of inertia in each span. The mathematical model for the beam is shown in Figure 14.5b.

The analysis of this beam can be performed in several different ways using the Moment Distribution Method. We will consider three different cases here. The calculations will be performed to three decimal place accuracy for comparison to the analysis of this same beam in Chapter 13 by the Slope-Deflection Method.

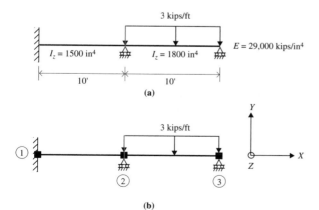

Figure 14.5 Example Problem 14.2.

Case I. Figure 14.6 shows a table for the analysis of the beam in which joint 3 is considered to have a fictitious rotation restraint during both the computation of the fixed end moments and also during the distribution and carry over operations.

Row 1 and Row 2. The columns are assigned for each joint and each member end moment.

Row 3. The member stiffness corresponding to each member end moment at joint 2 is computed in kip and inch units. The member stiffnesses for both M_{21} for member 1 and M_{23} for member 2 are computed using Eq. (14.1) since joint 3 will be considered to be restrained against rotation during the distribution of the unbalanced moments at joint 2. The modulus of elasticity E is carried through as a symbol for convenience during the evaluation of Eq. (14.1) since it will cancel out during the computations for the distribution factors. A member stiffness is not required for M_{32} since only one member is attached to joint 3.

Row 4. The distribution factors corresponding to each member end moment at joint 2 are computed for members 1 and 2 by Eq. (14.9). The distribution factors have been expressed as fractions to reduce roundoff error since they do not divide out evenly. It is a simple task to either perform continuous divide and multiply operations on a hand calculator during each distribution calculation or to compute the factors at the start of the calculations and store them in the calculator memories for retrieval as needed. The factor E cancels out since the distribution factors are dimensionless. The distribution factor for M_{12} is assigned a zero value since no distributions will be made at joint 1 and the distribution factor for M_{32} is assigned a unit value since only one member is attached to joint 3.

Row 5. The fixed end moments for each member are computed. Both joints 2 and 3 are considered to be fixed for the fixed end moments for member 2. The units which are being used for the moments are kips and feet. The fact that the stiffnesses in line 3 were computed in kip and inch units is not important. The units canceled out in computing the distribution factors since they are dimensionless.

Row 6. The unbalanced moments at both joints 2 and 3 are distributed to the member ends. Although the fictitious rotation restraints must be removed and then replaced separately at each joint, the results of both distributions are shown in the same row for simplicity.

Row 7. The carry over moments at the opposite ends of members 1 and 2 are computed using a carry over factor of 1/2 for each member. Joint 3 is assumed to be fixed during this carry over. Although the carry over for each distribution will occur as the distribution is performed, both carry over operations are shown in the same row for simplicity. Note that these carry over moments result in new unbalanced moments at joints 2 and 3.

Rows 8 through 26. The distribution and carry over calculations are repeated for each new set of unbalanced moments at joints 2 and 3 until the process converges to a point that any additional carry over moments are beyond the accuracy to which the calculations are being performed. The carry over process is not performed for the final distribution.

Application of the Moment Distribution Method to the Analysis of Beams

(1)	Joint	1	2		3
(2)	Moment	M_{12}	M_{21}	M_{23}	M_{32}
(3)	K		50E	60E	
(4)	DF	0	50/110	60/110	1
(5)	FEM	0	0	25	−25
(6)	Dist.		−11.364	−13.636	25
(7)	CO	−5.682		12.5	−6.818
(8)	Dist.		−5.682	−6.818	6.818
(9)	CO	−2.841		3.409	−3.409
(10)	Dist.		−1.550	−1.859	3.409
(11)	CO	−0.775		1.705	−0.930
(12)	Dist.		−0.775	−0.930	0.930
(13)	CO	−0.388		0.465	−0.465
(14)	Dist.		−0.211	−0.254	0.465
(15)	CO	−0.106		0.233	−0.127
(16)	Dist.		−0.106	−0.127	0.127
(17)	CO	−0.053		0.064	−0.064
(18)	Dist.		−0.029	−0.035	0.064
(19)	CO	−0.015		0.032	−0.017
(20)	Dist.		−0.015	−0.017	0.017
(21)	CO	−0.008		0.009	−0.009
(22)	Dist.		−0.004	−0.005	0.009
(23)	CO	−0.002		0.005	−0.003
(24)	Dist.		−0.002	−0.003	0.003
(25)	CO	−0.001		0.002	−0.002
(26)	Dist.		−0.001	−0.001	0.002
(27)	Final Moments	−9.871	−19.739	19.739	0

Figure 14.6 Example Problem 14.2, Case I.

Row 27. The moments are summed in each column to obtain the final member end moments.

An analysis of this type can be performed to any desired accuracy by terminating the distribution and carry over operations when it is felt that the results are acceptable. Although roundoff errors can occur during each distribution and carry over, these errors usually balance out, so that the total roundoff error is within acceptable limits.

(1)	Joint	1	2		3
(2)	Moment	M_{12}	M_{21}	M_{23}	M_{32}
(3)	K		50E	45E	
(4)	DF	0	50/95	45/95	1
(5)	FEM	0	0	25	−25
(6)	Dist.				25
(7)	CO			12.5	
(8)	Interim Moments	0	0	37.5	0
(9)	Dist.		−19.737	−17.763	
(10)	CO	−9.869			0
(11)	Final Moments	−9.869	−19.737	19.737	0

Figure 14.7 Example Problem 14.2, Case II.

Case II. Figure 14.7 shows a table for the analysis of the beam in which joint 3 is considered to have a fictitious rotation restraint during the computation of the fixed end moments and then to be unrestrained against rotation during the distribution and carry over operations.

Rows 1 and 2. The columns are assigned for the each joint and each member end moment.

Row 3. The member stiffness corresponding to each member end moment at joint 2 is computed. The member stiffness for M_{21} for member 1 is computed using Eq. (14.1), while the member stiffness for M_{23} for member 2 is computed using Eq. (14.2) since joint 3 will be considered to be unrestrained against rotation when the unbalanced moment is distributed at joint 2. A member stiffness is not required for M_{32} at joint 3 for this analysis.

Row 4. The distribution factors corresponding to each member end moment are computed or assigned.

Row 5. The fixed end moments for each member are computed. Both joints 2 and 3 are considered to be fixed for the fixed end moments for member 2.

Row 6. The unbalanced moment at joint 3 is distributed.

Row 7. The distributed moment at the end of member 2 at joint 3 is carried over to the end of member 2 at joint 2 assuming that joint 2 is restrained against rotation.

Row 8. The intermediate member end moments at this stage of the analysis are computed by summing the moments in each column. From this point on in the calculations, joint 3 will be assumed to be unrestrained against rotation.

Row 9. The unbalanced moment at joint 2 is distributed.

Row 10. The carry over moments at the opposite ends of members 1 and 2 are computed using a carry over factor of 1/2 for member 1 and a carry over factor of 0 for member 2.

Application of the Moment Distribution Method to the Analysis of Beams

(1)	Joint	1		2		3
(2)	Moment	M_{12}	M_{21}		M_{23}	M_{32}
(3)	K		50E		45E	
(4)	DF	0	50/95		45/95	0
(5)	FEM	0	0		37.5	0
(6)	Dist.		−19.737		−17.763	
(7)	CO	−9.869				0
(8)	Final Moments	−9.869	−19.737		19.737	0

Figure 14.8 Example Problem 14.2, Case III.

Row 11. The moments are summed in each column to obtain the final member end moments.

Case III. Figure 14.8 shows a table for the analysis of the beam in which joint 3 is considered to be unrestrained against rotation during both the computation of the fixed end moments and the distribution and carry over operations at joint 2.

Rows 1 and 2. The columns are assigned for each joint and each member end moment.

Rows 3 and 4. The member stiffnesses and distribution factors are computed for the member end moments at joint 2 assuming joint 3 is unrestrained against rotation.

Row 5. The fixed end moments for each member are computed. Joint 2 is considered to be fixed and joint 3 is considered to be pinned for the fixed end moments for member 2. The moment M_{21}^F can be computed using the expression for the bending moment at any point in the beam which was developed in Example Problem 8.5 in Chapter 8.

$$M_{23}^F = -\frac{\omega L^2}{8} = -\frac{(-3)(10)^2}{8} = 37.5 \text{ kip feet}$$

Note that this moment corresponds to the intermediate moment M_{23} in Line 8 of the analysis in Case II. The moment M_{32}^F is zero at the pinned joint.

Rows 6 through 8. These rows are identical to Rows 9 through 11 in Case II since the same unbalanced moment exists at joint 2.

The final member end moments for Case II and Case III are identical while the member end moments for Case I differ slightly due to roundoff errors in the numerous distribution and carry over operations. It is obvious that Case III requires the fewer number of calculations, but it has the disadvantage of requiring an expression for the fixed end moments for a fixed-pinned beam. Most engineers who are experienced in using the Moment Distribution Method would probably use the analysis procedure in Case II since expressions for the fixed end moments for various loading conditions are more readily available in design manuals and handbooks for fixed-fixed beams rather than for fixed-pinned beams..

The Moment Distribution Method   Chap. 14

The final end moments for this beam agree with those obtained in Example Problem 13.1 in Chapter 13 where the same beam was analyzed by the Slope-Deflection Method.

Example Problem 14.3

Compute the member end moments for the beam in the previous example problem if the right support translates down 0.1 inch as the beam is loaded.

Figure 14.9 shows a table for the analysis of this beam by the Moment Distribution Method. This analysis is similar to Case II in Example Problem 14.2.

Rows 1 through 5. These rows are identical to Rows 1 through 5 in Case II in Example Problem 14.2. The fixed end moments in Row 5 correspond to those produced by the distributed force on member 2.

Row 6. The fixed end moments are computed due to the downward translation of joint 3 by Eq. (13.16) in Chapter 13:

$$M_{ij}^F = M_{ji}^F = -\frac{6EI_z}{L^2}(\Delta_j - \Delta_i)$$

which, on substituting the properties of member 2, in kip and inch units, gives

$$M_{23}^F = M_{32}^F = -\frac{6(29000)(1800)}{(120)^2}(-0.1 - 0) = 2175 \text{ kip inches} = 181.25 \text{ kip feet}$$

Row 7. The total fixed end moments due to the combined member load and support translation are determined by summing the moments in each column.

	Joint	1	2		3
(1)					
(2)	Moment	M_{12}	M_{21}	M_{23}	M_{32}
(3)	K		50E	45E	
(4)	DF		50/95	45/95	1
(5)	FEM$_W$	0	0	25	−25
(6)	FEM$_\Delta$	0	0	181.25	181.25
(7)	FEM	0	0	206.25	156.25
(8)	Dist.				−156.25
(9)	CO			−78.125	
(10)	Interim Moments	0	0	128.125	0
(11)	Dist.		−67.434	−60.691	
(12)	CO	−33.717			0
(13)	Final Moments	−33.717	−67.434	63.434	0

Figure 14.9 Example Problem 14.3.

Rows 8 through 13. The distribution and carry over operations for joint 3 and then joint 2 are performed in the same manner as in Case II in Example Problem 14.2. The moments in Row 13 correspond to the final member end moments.

The final member end moments for this analysis agree with those obtained for this same beam in Example Problem 13.1 in Chapter 13.

APPLICATION OF THE MOMENT DISTRIBUTION METHOD TO THE ANALYSIS OF FRAMES

The Moment Distribution Method also can be used to analyze plane frames. The primary difference between this type of analysis and the analysis of beams is that there can be more than two members attached to a joint. Otherwise, the analysis procedure is essentially the same as shown in the previous beam examples.

Example Problem 14.4

Compute the member end moments for the plane frame shown in Figure 14.10a. The frame is composed of only one material but has a different cross section moment of inertia for each member.

The mathematical model for the frame is shown in Figure 14.10b. A joint has not been placed at the location of the concentrated force since its effect will be con-

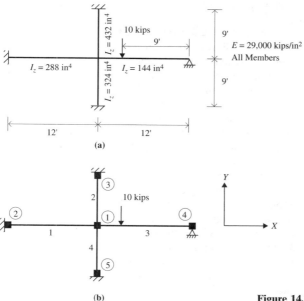

Figure 14.10 Example Problem 14.4.

(1)	Joint	1				2	3	4	5
(2)	Moment	M_{12}	M_{13}	M_{14}	M_{15}	M_{21}	M_{31}	M_{41}	M_{51}
(3)	Carry-Over	$\rightarrow M_{21}$	$\rightarrow M_{31}$	$\rightarrow M_{41}$	$\rightarrow M_{51}$			$\leftarrow M_{14}$	
(4)	K	8E	16E	3E	12E				
(5)	DF	8/39	16/39	3/39	12/39	0	0	1	0
(6)	FEM	0	0	16.875	0	0	0	−5.625	0
(7)	Dist.							5.625	
(8)	CO			2.813					
(9)	Interim Moments	0	0	19.688	0	0	0	0	0
(10)	Dist.	−4.039	−8.077	−1.514	−6.058				
(11)	CO					−2.019	−4.039	0	−3.029
(12)	Final Moments	−4.038	−8.077	18.174	−6.058	−2.019	−4.039	0	−3.029

Figure 14.11 Example Problem 14.4.

sidered in the initial fixed end moments for member 3. The calculations will be performed to an accuracy of three decimal places for comparison to the results of the analysis of this same frame by the Slope-Deflection Method in Example Problem 13.2 in Chapter 13.

Figure 14.11 shows a table for the analysis of this frame in which joint 4 is considered to be restrained against rotation during the computation of the fixed end moments for member 3 and then to be unrestrained against rotation during the distribution and carry over operations. The analysis process which is used here is the same as that used in Case II of Example Problem 14.2. The primary difference in this table is that Row 3 has been added to show how the moments are carried over from each joint since the columns in the table corresponding to the member end moments do not occur in the nice sequential order in a frame as in a beam. This row shows that M_{12} carries over to M_{21}, M_{13} carries over to M_{31}, M_{14} carries over to M_{41}, M_{15} carries over to M_{51} and M_{41} carries over to M_{14}. There will be no carry over from M_{21}, M_{31} or M_{51} since they occur at support joints which are restrained against rotation. Row 6 shows the fixed end moments M_{14}^F and M_{41}^F in kip feet for member 3. Row 9 shows the unbalanced moment at joint 1 after the unbalanced moment at joint 4 is distributed and carried over to joint 1. The unbalanced moment at joint 1 is then distributed in Row 10 and carried over to the other joints in Row 11. The carry over factor is 1/2 for members 1, 2 and 4, while it is zero for member 3 since joint 4 is now considered to be unrestrained against rotation. The final member end moments in Row 12 agree with those obtained in the analysis of this same frame by the Slope-Deflection Method in Example Problem 13.2 in Chapter 13.

Application of the Moment Distribution Method to the Analysis of Frames

Example Problem 14.5

Compute the member end moments for the frame in Figure 14.12a. The members are all prismatic with the same cross section and material.

The mathematical model of the frame is shown in Figure 14.12b. The calculations will be carried out to an accuracy of two decimal places for comparison to an analysis by FATPAK II.

Figure 14.13 shows a table for the analysis of this frame. Each row in the table should be self explanatory. Six distributions were required to obtain the desired accuracy. We can see that the frame is in equilibrium since the sum of the final member end moments at joint 2 and joint 4 are zero. Figure 14.14 shows a listing of the input data for the program F2DII in FATPAK II for this frame. Arbitrary values of 1.0 have been used for E and I_z for each member since the actual magnitude of these quantities will not affect the final member end moments or the support reactive forces or moments. A value of 10,000 was used for the cross section area in order to eliminate the effects of axial deformations in the members. Shear deformations in the members were ignored by assigning a zero value to the effective shear area, as descrtibed in the FATPAK II instructions. Figure 14.15 shows the results of the analysis by F2DII. The member end moments agree with those obtained by the Moment Distribution Method.

The calculations to analyze any frame with only rotation degrees of freedom at the joints by the Moment Distribution Procedure can be set up in a similar format to that shown in the previous example problems. The primary difficulty which is encountered is very slow convergence of the procedure if the frame has more than a few degrees of freedom.

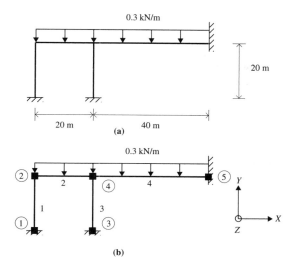

Figure 14.12 Example Problem 14.5.

(1)	Joint	1	2		3	4			5
(2)	Moment	M_{12}	M_{21}	M_{24}	M_{34}	M_{42}	M_{43}	M_{45}	M_{54}
(3)	Carry-Over		← M_{12}	→ M_{42}		← M_{24}	← M_{34}	→ M_{54}	
(4)	K		$0.2EI_z$	$0.2EI_z$		$0.2EI_z$	$0.2EI_z$	$0.1EI_z$	
(5)	DF		0.5	0.5		0.4	0.4	0.2	
(6)	FEM	0	0	10	0	−10	0	40	−40
(7)	Dist.		−5	−5		−12	−12	−6	
(8)	CO	−2.5		−6	−6	−2.5			−3
(9)	Dist.		3	3		1	1	0.5	
(10)	CO	1.5		0.5	0.5	1.5			0.25
(11)	Dist.		−0.25	−0.25		−0.6	−0.6	−0.3	
(12)	CO	−0.13		−0.3	−0.3	−0.13			−0.15
(13)	Dist.		0.15	0.15		0.05	0.05	0.03	
(14)	CO	0.08		0.03	0.03	0.08			0.02
(15)	Dist.		−0.02	−0.01		−0.03	−0.03	−0.02	
(16)	CO	−0.01		−0.02	−0.02	−0.01			−0.01
(17)	Dist.		0.01	0.01		0.01	0	0	
(18)	Final Moments	−1.06	−2.11	2.11	−5.79	−22.63	−11.58	34.21	−42.89

Figure 14.13 Example Problem 14.5.

```
Example Problem 14.5 - Analysis by FATPAK II
5,4,1,1,0,3,0,0,2,0,0
Joint Coordinates
1,0.0,0.0
2,0.0,20.0
3,20.0,0.0
4,20.0,20.0
5,60.0,20.0
Material Data
1,1.0,1.0
Cross Section Data
1,10000.0,0.0,1.0
Member Data
1,1,2
2,2,4
3,3,4
4,4,5
Support Restraints
1,1,1,1
3,1,1,1
5,1,1,1
Member Distributed Loads
2,0.0,0.0,-0.3,-0.3,0
4,0.0,0.0,-0.3,-0.3,0
```

Figure 14.14 Example Problem 14.5, F2DII input.

```
FATPAK II -- Frame and Truss Package -- Student Edition
(C) Copyright 1995 by Structural Software Systems
Program F2DII -- Plane Frame Analysis

Example Problem 14.5 - Analysis by FATPAK II

Joint Displacements

Joint         X-Tran          Y-Tran          Z-Rot

  1           0.000           0.000           0.00000
  2           0.004          -0.004         -10.52699
  3           0.000           0.000           0.00000
  4           0.004          -0.020         -57.89505
  5           0.000           0.000           0.00000

Member End Loads

Member   Joint         Sx              Vy              Mz

  1        1          1.974          -0.158           -1.05
           2         -1.974           0.158           -2.11
  2        2          0.158           1.974            2.11
           4         -0.158           4.026          -22.63
  3        3          9.809          -0.868           -5.79
           4         -9.809           0.868          -11.58
  4        4          1.026           5.783           34.21
           5         -1.026           6.217          -42.89

Reactions

Joint             RX              RY              MZ

  1              0.158           1.974           -1.05
  3              0.868           9.809           -5.79
  5             -1.026           6.217          -42.89
```

Figure 14.15 Example Problem 14.5, F2DII output.

REFERENCE

GERE, JAMES M. (1963). *Moment Distribution*. New York: D. Van Nostrand.

SUGGESTED PROBLEMS

Compute the end moments for each member in the mathematical models for the structures described in the following Suggested problems at the end of Chapter 13 by the Moment Distribution Method.

SP14.1   Problem SP13.1
SP14.2   Problem SP13.2
SP14.3   Problem SP13.3
SP14.4   Problem SP13.4
SP14.5   Problem SP13.5
SP14.6   Problem SP13.6
SP14.7   Problem SP13.9
SP14.8   Problem SP13.10
SP14.9   Compute the member stiffness and the carry over factor for each end of the non-prismatic plane frame member shown Figure SP14.9.   Hint: See the solution for Problem SP13.11 at the end of Chapter 13.

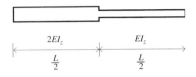

Figure SP14.9

15

The Stiffness Method

BASIC PRINCIPLES OF THE STIFFNESS METHOD

The *Stiffness Method* is an analysis procedure which can be used to compute the joint displacements and the member end loads for a linear elastic structure. The basic analysis process is similar to that used in the Slope-Deflection Method since a set of equilibrium equations is generated with the displacements corresponding to the degrees of freedom at the joints of the mathematical model of the structure as the unknowns. However, any direct correspondence between the two procedures ends here, since essentially any type of two dimensional or three dimensional elastic structural system can be analyzed by the Stiffness Method, rather than only beams and plane frames as in the Slope-Deflection Method. In addition, the degrees of freedom at each joint may correspond to all of the global translation and rotation components of the joints which are consistent with the type of structure which is being analyzed, and all of the possible contributions to the deformations in the members can be considered during the analysis. Therefore, the procedure is equally applicable to the analysis of plane trusses, space trusses, plane frames and space frames. However, in this chapter, we will restrict our discussion to the application of the Stiffness Method to the analysis of plane trusses and plane frames. A description of the application of this procedure to the analysis of space trusses and space frames can be found in Fleming (1986, 1989).

The basic operations for the analysis of any structure by the Stiffness Method can be expressed as a simple set of matrix operations which are the same for any type of structure. The differences for various structure types are in the specific definitions of the individual elements in the matrices. Due to this matrix formulation, this analysis procedure is ideally suited for implementation in a computer program. Essentially all of the commercial structural analysis programs which are available use the Stiffness Method. This analysis procedure is used in each of the individual analysis programs in FATPAK II - Student Edition.

Before the basic equations can be developed for the analysis procedure which is used in the Stiffness Method, it is necessary to define several matrices which will be used to describe the properties of the structure and the applied loads. Several of these matrices are very similar to those defined previously in Chapter 12 in the discussion of the Flexibility Method. The same notation will be used here to represent a total matrix and any individual element in the matrix, as described in Chapter 12.

Global Joint Load Matrix and Global Joint Displacement Matrix

The active and reactive loads for any structure, which is to be analyzed by the Stiffness Method, must consist only of concentrated loads acting on the joints of the mathematical model of the structure. These loads are expressed as components in the right hand orthogonal global coordinate system for the structure. Any distributed forces acting on the members must be converted into equivalent concentrated joint loads before the analysis is performed. The process for computing these equivalent joint loads is the same as described in the discussion of the Flexibility Method in Chapter 12.

As in the Flexibility Method, the individual concentrated loads acting on the joints will be represented by the *global joint load matrix* $\{W\}$ and the corresponding joint displacements at the location of each of the joint loads will be represented by the *global joint displacement matrix* $\{D\}$. However, one major difference between the Stiffness Method and the Flexibility Method is that $\{D\}$ must now contain an element representing each independent displacement degree of freedom for the structure and $\{W\}$ must contain a corresponding joint load. Therefore, $\{W\}$ and $\{D\}$ for any structure, which is to be analyzed by the Stiffness Method, will have the form

$$\{W\} = \begin{Bmatrix} w(1) \\ w(2) \\ \vdots \\ w(\text{NDOF}) \end{Bmatrix} \tag{15.1}$$

and

$$\{D\} = \begin{Bmatrix} d(1) \\ d(2) \\ \vdots \\ d(\text{NDOF}) \end{Bmatrix} \tag{15.2}$$

Basic Principles of the Stiffness Method

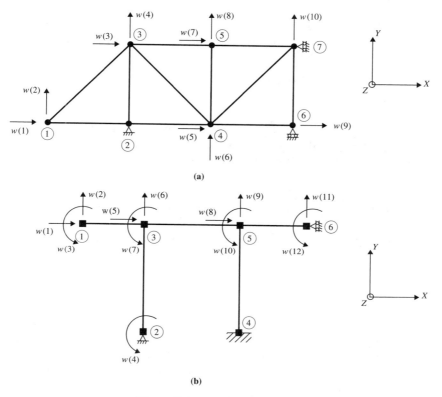

Figure 15.1 Degrees of freedom.

where NDOF is the number of degrees of freedom, as defined previously in Eqs. (12.3) through (12.5) in Chapter 12 for various types of structures. Although it is possible to number the elements in $\{W\}$ and $\{D\}$ in any desired order, a very specific numbering scheme will be used in this chapter for ease in implementing the Stiffness Method in a computer program. This scheme consists of numbering the degrees of freedom consecutively for each joint in turn in the mathematical model of the structure, as shown by the numbering for the elements in $\{W\}$ in Figure 15.1a for a plane truss with 10 degrees of freedom and in Figure 15.1b for a plane frame with 12 degrees of freedom. Elements have not been included in $\{W\}$ at the support joints in the directions of the support restraints since the displacements in these restrained directions do not correspond to degrees of freedom for the structures. The rotations at the joints in a frame are always expressed in radians.

Global Stiffness Matrix

The joint loads for any structure can be related to the joint displacements by the *global stiffness coefficients* in a set of equations of the form

$$\begin{aligned}
k(1,1)d(1) &+ k(1,2)d(2) + \cdots + k(1,\text{NDOF})d(\text{NDOF}) = w(1) \\
k(2,1)d(1) &+ k(2,2)d(2) + \cdots + k(2,\text{NDOF})d(\text{NDOF}) = w(2) \\
&\vdots \\
k(\text{NDOF},1)d(1) &+ k(\text{NDOF},2)d(2) + \cdots + k(\text{NDOF},\text{NDOF})d(\text{NDOF}) = w(\text{NDOF})
\end{aligned}$$

(15.3)

where any stiffness coefficient $k(i,j)$ is equal to the magnitude of the load $w(i)$ which is required to produce a unit displacement $d(j)$ with all other elements in $\{D\}$ being zero. This set of equations can be expressed in matrix form as

$$\begin{bmatrix} k(1,1) & k(1,2) & \cdots & k(1,\text{NDOF}) \\ k(2,1) & k(2,2) & \cdots & k(2,\text{NDOF}) \\ \vdots & \vdots & \ddots & \vdots \\ k(\text{NDOF},1) & k(\text{NDOF},2) & \cdots & k(\text{NDOF},\text{NDOF}) \end{bmatrix} \begin{Bmatrix} d(1) \\ d(2) \\ \vdots \\ d(\text{NDOF}) \end{Bmatrix} = \begin{Bmatrix} w(1) \\ w(2) \\ \vdots \\ w(\text{NDOF}) \end{Bmatrix} \quad (15.4)$$

or

$$[K]\{D\} = \{W\} \tag{15.5}$$

where $[K]$ is the *global stiffness matrix* for the structure. This matrix has several important properties: first, it is always square since the number of elements in both $\{W\}$ and $\{D\}$ is always equal; second, it is always symmetric as a result of Maxwell's Theorem; and third, it will never have a zero or negative element on the main diagonal for a stable structure.

The global stiffness matrix $[K]$ can be related to the global flexibility matrix $[F]$ for any structure by substituting the expression for $\{D\}$ in Eq. (12.8) in Chapter 12 into Eq. (15.5) to obtain

$$[K][F]\{W\} = \{W\} \tag{15.6}$$

which can only be satisfied, for any arbitrary set of values for the elements in $\{W\}$, if

$$[K] = [F]^{-1} \tag{15.7}$$

and

$$[F] = [K]^{-1} \tag{15.8}$$

Therefore, the matrices $[K]$ and $[F]$ are not independent. If one of the matrices is known for a stable structure, the other can be easily determined by the simple operation of matrix inversion. As we will see later in this chapter, the Stiffness Method will provide a simple and convenient procedure for generating the global stiffness matrix $[K]$ for any type of statically determinate or statically indeterminate structure.

Basic Principles of the Stiffness Method

Local Member End Load Matrix and Local Member Deformation Matrix

The loads which are applied to any member n in a structure by the joints at its two ends will be represented by a matrix $\{S\}_n$ and the corresponding member deformations will be represented by a matrix $\{U\}_n$. The elements in these matrices will consist of components of the member end loads and member deformations in the local member coordinate system. The local member axes for a plane truss or a plane frame member will be oriented with the x and y axes in the plane of the structure and the z axis extending outward from the plane, as described previously for the Flexibility Method. The elements in the *local member end load matrix* $\{S\}_n$ may be defined to consist of either a statically independent set of local end load components, as used in the Flexibility Method, or all of the local end load components for any member. The corresponding elements in the *local member deformation matrix* $\{U\}_n$ will consist of either the relative local displacement components or the absolute local displacement components of the member ends. Either definition will lead to the same results for the analysis of the structure. For convenience, we will consider the elements in $\{S\}_n$ to consist of all of the local end load components throughout all of the equations in this chapter. Examples of the equations when a statically independent set of end loads is used can be found in Fleming (1989).

Plane Truss Member. The end loads for a plane truss member consist of a force at each end of the member with the positive directions corresponding to the positive direction of the local member x axis, as shown in Figure 15.2, in which the angles θ_X and θ_Y represent the orientation of the member with respect to the global X and Y axes. Therefore, the matrix $\{S\}_n$ for this member will contain two elements:

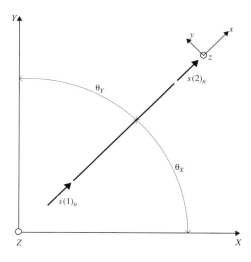

Figure 15.2 Plane truss local member end loads.

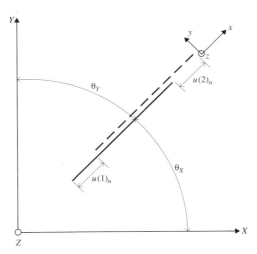

Figure 15.3 Plane truss local member deformations.

$$\{S\}_n = \begin{Bmatrix} s(1) \\ s(2) \end{Bmatrix}_n \quad (15.9)$$

Since the forces at each end of the member must be equal and act in opposite directions, the two elements in $\{S\}_n$ will always have equal magnitudes and opposite signs. Therefore, only one of the end loads for a plane truss member is statically independent, as described in Chapter 12 in the development of the Flexibility Method. A positive value for $s(2)_n$ will correspond to tension in the member while a negative value will correspond to compression.

The elements in $\{U\}_n$ will correspond to the components of the translations of the two ends of the member along the undeformed orientation of the local member x axis as shown in Figure 15.3, in which the solid line represents the original position of the member and the dashed line represents its deformed position. Therefore, the matrix $\{U\}_n$ will also contain two elements

$$\{U\}_n = \begin{Bmatrix} u(1) \\ u(2) \end{Bmatrix}_n \quad (15.10)$$

If we assume that the rigid body displacements of the member are small, the change in length of the member can be expressed as

$$\Delta L_n = u(2)_n - u(1)_n \quad (15.11)$$

Plane Frame Member. The end loads for a plane frame member will consist of an axial force along x, a transverse force along y and a moment about z at each end of the member, as shown in Figure 15.4. Therefore, the matrix $\{S\}_n$ will have six elements:

Basic Principles of the Stiffness Method

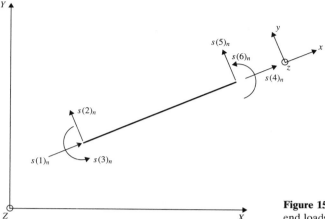

Figure 15.4 Plane frame local member end loads.

$$\{S\}_n = \begin{Bmatrix} s(1) \\ s(2) \\ s(3) \\ s(4) \\ s(5) \\ s(6) \end{Bmatrix}_n \tag{15.12}$$

It can be easily determined that the pair of elements $s(1)_n$ and $s(4)_n$, which represent the axial forces at each end, and the pair of elements $s(2)_n$ and $s(5)_n$, which represent the shear forces at each end, must have equal magnitudes with opposite signs by summing forces in the local x and y directions for the member. An additional equation also can be developed relating the shear forces to the end moments by summing moments about any point:

$$s(2)_n = -s(5)_n = \frac{s(3)_n + s(6)_n}{L} \tag{15.13}$$

where L is the length of the member. Therefore, only three of the six end loads for a plane frame member are statically independent.

The corresponding elements in $\{U\}_n$ will consist of the translation component along x, the translation component along y and the rotation about z at each end of the member as shown in Figure 15.5. Therefore, the matrix $\{U\}_n$ will also have six elements:

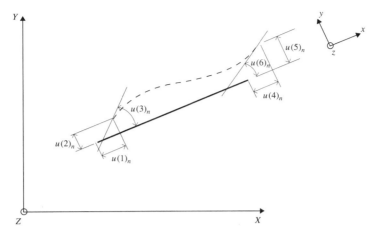

Figure 15.5 Plane frame local member deformations.

$$\{U\}_n = \begin{Bmatrix} u(1) \\ u(2) \\ u(3) \\ u(4) \\ u(5) \\ u(6) \end{Bmatrix}_n \quad (15.14)$$

The end rotations will always be expressed in radians.

Local Member Stiffness Matrix

The local member end loads can be expressed in terms of the local member deformations, for any member n in the structure, by the *local member stiffness matrix* $[K_m]_n$, as

$$\{S\}_n = [K_m]_n \{U\}_n \quad (15.15)$$

where any element $k_m(i, j)_n$ in $[K_m]_n$ is equal to the magnitude of $s(i)_n$ which is required to produce a unit $u(j)_n$ with all other elements in $\{U\}_n$ being zero. The matrix $[K_m]_n$ will always be square since the number of elements in $\{S\}_n$ and $\{U\}_n$ are always equal. It will also always be a symmetric matrix as a result of Maxwell's Theorem.

Plane Truss Member. The elements in $[K_m]_n$ for a prismatic plane truss member can be computed by using Eq. (12.15) in Chapter 12 to express the change in length of the member in terms of the two equal and opposite end loads.

Basic Principles of the Stiffness Method

$$\Delta L_n = u(2)_n - u(1)_n = -\frac{s(1)_n L}{A_x E} \qquad (15.16a)$$

$$\Delta L_n = u(2)_n - u(1)_n = \frac{s(2)_n L}{A_x E} \qquad (15.16b)$$

which can be rewritten as

$$s(1)_n = -[u(2)_n - u(1)_n]\left(\frac{A_x E}{L}\right)_n \qquad (15.17a)$$

$$s(2)_n = [u(2)_n - u(1)_n]\left(\frac{A_x E}{L}\right)_n \qquad (15.17b)$$

This pair of equations now can be written in matrix form as

$$\begin{Bmatrix} s(1) \\ s(2) \end{Bmatrix}_n = \begin{bmatrix} \dfrac{A_x E}{L} & -\dfrac{A_x E}{L} \\ -\dfrac{A_x E}{L} & \dfrac{A_x E}{L} \end{bmatrix}_n \begin{Bmatrix} u(1) \\ u(2) \end{Bmatrix}_n \qquad (15.18)$$

from which we can conclude from Eq. (15.15) that

$$[K_m]_n = \begin{bmatrix} \dfrac{A_x E}{L} & -\dfrac{A_x E}{L} \\ -\dfrac{A_x E}{L} & \dfrac{A_x E}{L} \end{bmatrix}_n \qquad (15.19)$$

Note that the determinant of this matrix is zero since the rows are linearly dependent (i.e.; the second row is equal to the first row multiplied by -1). Therefore, the inverse of $[K_m]_n$, which would correspond to the local member flexibility matrix $[F_m]_n$ which was defined in Chapter 12 in the Flexibility Method, does not exist for the definitions for $\{S\}_n$ and $\{U\}_n$ which have been used here. This further verifies why $\{S\}_n$ could only contain the statically independent end loads for any analysis which was performed by the Flexibility Method in Chapter 12.

Plane Frame Member. The local member stiffness matrix for a prismatic plane frame member, in which the effects of the shear deformations are ignored, is

$$[K_m]_n = \begin{bmatrix} \dfrac{A_x E}{L} & 0 & 0 & -\dfrac{A_x E}{L} & 0 & 0 \\ 0 & \dfrac{12EI_z}{L^3} & \dfrac{6EI_z}{L^2} & 0 & -\dfrac{12EI_z}{L^3} & \dfrac{6EI_z}{L^2} \\ 0 & \dfrac{6EI_z}{L^2} & \dfrac{4EI_z}{L} & 0 & -\dfrac{6EI_z}{L^2} & \dfrac{2EI_z}{L} \\ -\dfrac{A_x E}{L} & 0 & 0 & \dfrac{A_x E}{L} & 0 & 0 \\ 0 & -\dfrac{12EI_z}{L^3} & -\dfrac{6EI_z}{L^2} & 0 & \dfrac{12EI_z}{L^3} & -\dfrac{6EI_z}{L^2} \\ 0 & \dfrac{6EI_z}{L^2} & \dfrac{2EI_z}{L} & 0 & -\dfrac{6EI_z}{L^2} & \dfrac{4EI_z}{L} \end{bmatrix}_n \quad (15.20)$$

where I_z is the moment of inertia of the member cross section about the local z axis. The elements at the intersections of the first and fourth rows and columns are the same as the elements in the local member stiffness matrix for a plane truss member since they correspond to the effects of axial deformation in the member. The remaining elements in these rows and columns are zero since it is being assumed that there is no interaction between the axial and bending deformations in the member. Although this interaction can be important in some very special structures, it is beyond the scope of the present discussion since it requires a nonlinear analysis of the structure. Further information on this topic can be found in Fleming (1989).

The elements in the second column in $[K_m]_n$ correspond to the member end loads which are required to produce a unit value of $u(2)_n$ with all other elements in $\{U\}_n$ being zero, as shown in Figure 15.6. The elements $k_m(3,2)_n$ and $k_m(6,2)_n$

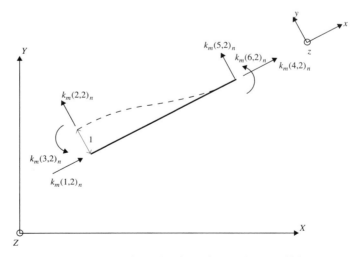

Figure 15.6 Plane frame local member stiffness coefficients.

The Assembly Process

can be determined from Eq. (13.16) in Chapter 13, in which Δ_i corresponds to the value of $u(2)_n$ and Δ_j corresponds to the value of $u(5)_n$, after which the elements $k_m(2,2)_n$ and $k_m(5,2)_n$ can be determined by summing moments about each end of the member. The elements in the fifth column can be determined by the same process by introducing a unit value for $u(5)_n$ into the member. The elements in the third and sixth columns can be determined in a similar manner by using Eqs. (13.9) and (13.10) in Chapter 13 and the equilibrium requirements for the member.

It is also possible to modify the expression for $[K_m]_n$ in Eq. (15.20) to include the effects of shear deformation in a plane frame member. There are several different forms in which the matrix can be written. One of the most popular forms is

$$[K_m]_n = \begin{bmatrix} \frac{A_x E}{L} & 0 & 0 & -\frac{A_x E}{L} & 0 & 0 \\ 0 & \frac{12EI_z}{L^3(1+\beta_y)} & \frac{6EI_z}{L^2(1+\beta_y)} & 0 & -\frac{12EI_z}{L^3(1+\beta_y)} & \frac{6EI_z}{L^2(1+\beta_y)} \\ 0 & \frac{6EI_z}{L^2(1+\beta_y)} & \frac{(4+\beta_y)EI_z}{L(1+\beta_y)} & 0 & -\frac{6EI_z}{L^2(1+\beta_y)} & \frac{(2-\beta_y)EI_z}{L(1+\beta_y)} \\ -\frac{A_x E}{L} & 0 & 0 & \frac{A_x E}{L} & 0 & 0 \\ 0 & -\frac{12EI_z}{L^3(1+\beta_y)} & -\frac{6EI_z}{L^2(1+\beta_y)} & 0 & \frac{12EI_z}{L^3(1+\beta_y)} & -\frac{6EI_z}{L^2(1+\beta_y)} \\ 0 & \frac{6EI_z}{L^2(1+\beta_y)} & \frac{(2-\beta_y)EI_z}{L(1+\beta_y)} & 0 & -\frac{6EI_z}{L^2(1+\beta_y)} & \frac{(4+\beta_y)EI_z}{L(1+\beta_y)} \end{bmatrix}_n$$

(15.21)

in which the quantity β_y is known as the *shear deformation constant* for the member. This quantity can be expressed in terms of the member cross section properties and the material properties as

$$\beta_y = \frac{12EI_z}{GA_y L^2} \qquad (15.22)$$

where G is the shear modulus of the material and A_y is the effective shear area of the member cross section in the local y direction. Note that the matrix in Eq. (15.21) reduces to the form in Eq. (15.20) if the effective shear area of the cross section is infinite, which corresponds to the quantity β_y being zero.

THE ASSEMBLY PROCESS

If the global stiffness matrix $[K]$ for a structure can be determined, it is a simple task to invert $[K]$ to obtain the global flexibility matrix $[F]$, after which the joint displacements can be easily computed by

$$\{D\} = [F]\{W\} \tag{15.23}$$

Since the individual elements in the matrix $[K]$ represent the stiffness of the structure at the joints, and since the stiffness at each joint must be equal to the sum of the stiffnesses of the members connected to that joint, we should be able to compute the elements in $[K]$ by summing the stiffnesses of the members. However, this process cannot be performed directly at this time since the elements in the matrix $[K]$ are expressed in terms of components in the global coordinate system while the elements in the stiffness matrix $[K_m]_n$ for each member is expressed in terms of components in the local coordinate system for that member. Therefore, we must first develop a procedure for transforming the matrix $[K_m]_n$ for each member from the local coordinate system to the global coordinate system. An expression to accomplish this transformation can be derived by considering the work which is required to deform a member. However, before we can derive this expression, we must define several additional matrices for a member.

Global Member End Load Matrix and Global Member Deformation Matrix

The member end loads and the member deformations for any member n in a structure can also be expressed as components in the global coordinate system. The quantities will be represented by the *global member end load matrix* $\{W\}_n$ and the *global member deformation matrix* $\{D\}_n$.

Plane Truss Member. The global member end load matrix for a plane truss member will contain four elements:

$$\{W\}_n = \begin{Bmatrix} w(1) \\ w(2) \\ w(3) \\ w(4) \end{Bmatrix}_n \tag{15.24}$$

which will consist of a force component along X and Y at each end of the member, as defined in Figure 15.7. The corresponding global member deformation matrix will also contain four elements:

$$\{D\}_n = \begin{Bmatrix} d(1) \\ d(2) \\ d(3) \\ d(4) \end{Bmatrix}_n \tag{15.25}$$

which will consist of a translation component along X and Y at each end, as defined in Figure 15.8.

The Assembly Process

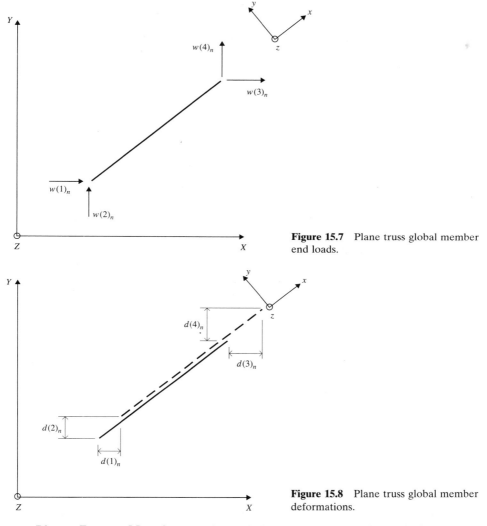

Figure 15.7 Plane truss global member end loads.

Figure 15.8 Plane truss global member deformations.

Plane Frame Member. The global member end load matrix for a plane frame member will contain six elements:

$$\{W\}_n = \begin{Bmatrix} w(1) \\ w(2) \\ w(3) \\ w(4) \\ w(5) \\ w(6) \end{Bmatrix}_n \tag{15.26}$$

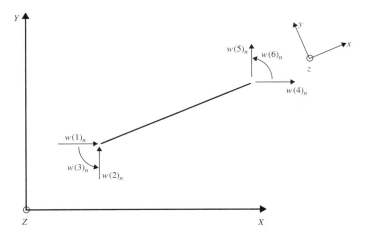

Figure 15.9 Plane frame global member end loads.

which will consist of a force component along X and Y and a moment about Z at each end of the member, as defined in Figure 15.9. The corresponding global member deformation matrix will also contain six elements:

$$\{D\}_n = \begin{Bmatrix} d(1) \\ d(2) \\ d(3) \\ d(4) \\ d(5) \\ d(6) \end{Bmatrix}_n \tag{15.27}$$

which will consist of a translation component along X and Y and a rotation about Z at each end, as defined in Figure 15.10.

Global Member Stiffness Matrix

The global member end loads can be related to the global member deformations, for any member n in a structure, by the relationship

$$\{W\}_n = [K]_n \{D\}_n \tag{15.28}$$

where $[K]_n$ is the *global member stiffness matrix* for the member. Any element $k(i, j)_n$ in $[K]_n$ is equal to the magnitude of $w(i)_n$, which is required to produce a unit $d(j)_n$ with all other elements in $\{D\}_n$ being zero. A simple expression for computing the elements in $[K]_n$ for any type of member can be developed by first defining the *coordinate transformation matrix* $[A]_n$, which transforms the global member deformations into the local member deformations by the relationship

$$\{U\}_n = [A]_n \{D\}_n \tag{15.29}$$

The Assembly Process

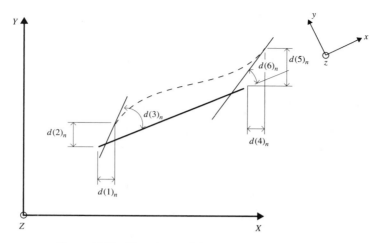

Figure 15.10 Plane frame global member deformations.

Any element $a(i,j)_n$ in $[A]_n$ is equal to the magnitude of $u(i)_n$, which will be produced if a unit displacement corresponding to $d(j)_n$ is introduced at the end of the member with all other elements in $\{D\}_n$ being zero. The elements in $[A]_n$ can be computed using the orientation angles θ_X and θ_Y of the member, as defined previously in Figure 15.2.

The first step in developing the expression for computing the elements in $[K]_n$ is to compute the work which must be performed to deform the member in terms of quantities in the local coordinate system:

$$\text{Work}_{n,\text{local}} = \frac{\{S\}_n^T \{U\}_n}{2} \tag{15.30}$$

which, on substituting the expression for $\{S\}_n$ in Eq. (15.15) and recognizing that since $[K_m]_n$ is symmetric, it is equal to its transpose, becomes

$$\text{Work}_{n,\text{local}} = \frac{\{U\}_n^T [K_m]_n \{U\}_n}{2} \tag{15.31}$$

If we now substitute the expression for $\{U\}_n$ in Eq. (15.29), the final expression for the local work will be

$$\text{Work}_{n,\text{local}} = \frac{\{D\}_n^T [A]_n^T [K_m]_n [A]_n \{D\}_n}{2} \tag{15.32}$$

The second step is to compute the work which must be performed to deform the member in terms of quantities in the global coordinate system:

$$\text{Work}_{n,\text{global}} = \frac{\{W\}_n^T \{D\}_n}{2} \tag{15.33}$$

which, on substituting the expression for $\{W\}_n$ in Eq. (15.28) and recognizing that $[K]_n$ is also symmetric, becomes

$$\text{Work}_{n,\,\text{global}} = \frac{\{D\}_n^T [K]_n \{D\}_n}{2} \tag{15.34}$$

Finally, since work is a scaler quantity, it must be the same no matter what coordinate system it is expressed in. Therefore, equating the expression for the local work in Eq. (15.32) and the expression for the global work in Eq. (15.34) results in the following expression relating $[K]_n$ and $[K_m]_n$:

$$[K]_n = [A]_n^T [K_m]_n [A]_n \tag{15.35}$$

The global member stiffness matrix $[K]_n$ is often called the *transformed member stiffness matrix* and Eq. (15.35) is often called the *coordinate transformation equation*.

Plane Truss Member. The local member deformations can be expressed in terms of the global member deformations for a plane truss member by the geometric relationships

$$u(1)_n = d(1)_n \cos\theta_X + d(2)_n \cos\theta_Y \tag{15.36a}$$

$$u(2)_n = d(3)_n \cos\theta_X + d(4)_n \cos\theta_Y \tag{15.36b}$$

which can be written in matrix form as

$$\begin{Bmatrix} u(1) \\ u(2) \end{Bmatrix}_n = \begin{bmatrix} \cos\theta_X & \cos\theta_Y & 0 & 0 \\ 0 & 0 & \cos\theta_X & \cos\theta_Y \end{bmatrix}_n \begin{Bmatrix} d(1) \\ d(2) \\ d(3) \\ d(4) \end{Bmatrix}_n \tag{15.37}$$

If we now compare this expression to the expression in Eq. (15.29), we can conclude that the matrix $[A]_n$ for a plane truss member is

$$[A]_n = \begin{bmatrix} \cos\theta_X & \cos\theta_Y & 0 & 0 \\ 0 & 0 & \cos\theta_X & \cos\theta_Y \end{bmatrix}_n \tag{15.38}$$

which can be substituted in Eq. (15.35) to obtain the global member stiffness matrix for a plane truss member as

$$[K]_n = \begin{bmatrix} C_X^2 & C_X C_Y & -C_X^2 & -C_X C_Y \\ C_X C_Y & C_Y^2 & -C_X C_Y & -C_Y^2 \\ -C_X^2 & -C_X C_Y & C_X^2 & C_X C_Y \\ -C_X C_Y & -C_Y^2 & C_X C_Y & C_Y^2 \end{bmatrix} \left(\frac{A_x E}{L}\right)_n \tag{15.39}$$

where the symbols C_X and C_Y correspond to $\cos\theta_X$ and $\cos\theta_Y$, respectively.

Plane Frame Member. The local member deformations can be expressed in terms of the global member deformations for a plane frame member by the geometric relationships

$$u(1)_n = d(1)_n \cos \theta_X + d(2)_n \cos \theta_Y \quad (15.40a)$$

$$u(2)_n = -d(1)_n \cos \theta_Y + d(2)_n \cos \theta_X \quad (15.40b)$$

$$u(3)_n = d(3)_n \quad (15.40c)$$

$$u(4)_n = d(4)_n \cos \theta_X + d(5)_n \cos \theta_Y \quad (15.40d)$$

$$u(5)_n = -d(4)_n \cos \theta_Y + d(5)_n \cos \theta_X \quad (15.40e)$$

$$u(6)_n = d(6)_n \quad (15.40f)$$

from which we can see that the matrix $[A]_n$ is

$$[A]_n = \begin{bmatrix} \cos \theta_X & \cos \theta_Y & 0 & 0 & 0 & 0 \\ -\cos \theta_Y & \cos \theta_X & 0 & 0 & 0 & 0 \\ 0 & 0 & 1 & 0 & 0 & 0 \\ 0 & 0 & 0 & \cos \theta_X & \cos \theta_Y & 0 \\ 0 & 0 & 0 & -\cos \theta_Y & \cos \theta_X & 0 \\ 0 & 0 & 0 & 0 & 0 & 1 \end{bmatrix}_n \quad (15.41)$$

This matrix can now be used to generate $[K]_n$ for a plane frame member. The resulting matrix, which includes the effects of shear deformation in the member, is shown in Eq. (15.42) in Figure 15.11. The effects of the shear deformation in any member in the structure can be ignored by setting the value for β_y for that member to zero, as described previously.

Generation of the Global Stiffness Matrix.

The elements in the global stiffness matrix corresponding to any joint in the structure can be computed by summing the elements in the global member stiffness matrices for all of the members connected to that joint. This overall process for generating $[K]$ can be represented symbolically by the expression

$$[K] = \sum_{n=1}^{NM} [K]_n \quad (15.43)$$

where NM is the total number of members in the structure. Of course, this equation is not quite correct from a pure mathematical point of view since the matrix $[K]$ for the structure and the matrix $[K]_n$ for any member are not the same size. Therefore, a scheme must be developed for inserting the elements in each $[K]_n$ into the correct locations in $[K]$. This process will be demonstrated later by an example problem. This process of assembling the global stiffness matrix for the structure from the individual global member stiffness matrices for the members is

$$[K]_n = \begin{bmatrix} C_X^2 \dfrac{A_x E}{L} + C_Y^2 \dfrac{12EI_z}{L^3(1+\beta_y)} & C_X C_Y \dfrac{A_x E}{L} - C_X C_Y \dfrac{12EI_z}{L^3(1+\beta_y)} & -C_Y \dfrac{6EI_z}{L^2(1+\beta_y)} & -C_X^2 \dfrac{A_x E}{L} - C_Y^2 \dfrac{12EI_z}{L^3(1+\beta_y)} & -C_X C_Y \dfrac{A_x E}{L} + C_X C_Y \dfrac{12EI_z}{L^3(1+\beta_y)} & -C_Y \dfrac{6EI_z}{L^2(1+\beta_y)} \\ C_X C_Y \dfrac{A_x E}{L} - C_X C_Y \dfrac{12EI_z}{L^3(1+\beta_y)} & C_Y^2 \dfrac{A_x E}{L} + C_X^2 \dfrac{12EI_z}{L^3(1+\beta_y)} & C_X \dfrac{6EI_z}{L^2(1+\beta_y)} & -C_X C_Y \dfrac{A_x E}{L} + C_X C_Y \dfrac{12EI_z}{L^3(1+\beta_y)} & -C_Y^2 \dfrac{A_x E}{L} - C_X^2 \dfrac{12EI_z}{L^3(1+\beta_y)} & C_X \dfrac{6EI_z}{L^2(1+\beta_y)} \\ -C_Y \dfrac{6EI_z}{L^2(1+\beta_y)} & C_X \dfrac{6EI_z}{L^2(1+\beta_y)} & \dfrac{(4+\beta_y)EI_z}{L(1+\beta_y)} & C_Y \dfrac{6EI_z}{L^2(1+\beta_y)} & -C_X \dfrac{6EI_z}{L^2(1+\beta_y)} & \dfrac{(2-\beta_y)EI_z}{L(1+\beta_y)} \\ -C_X^2 \dfrac{A_x E}{L} - C_Y^2 \dfrac{12EI_z}{L^3(1+\beta_y)} & -C_X C_Y \dfrac{A_x E}{L} + C_X C_Y \dfrac{12EI_z}{L^3(1+\beta_y)} & C_Y \dfrac{6EI_z}{L^2(1+\beta_y)} & C_X^2 \dfrac{A_x E}{L} + C_Y^2 \dfrac{12EI_z}{L^3(1+\beta_y)} & C_X C_Y \dfrac{A_x E}{L} - C_X C_Y \dfrac{12EI_z}{L^3(1+\beta_y)} & C_Y \dfrac{6EI_z}{L^2(1+\beta_y)} \\ -C_X C_Y \dfrac{A_x E}{L} + C_X C_Y \dfrac{12EI_z}{L^3(1+\beta_y)} & -C_Y^2 \dfrac{A_x E}{L} - C_X^2 \dfrac{12EI_z}{L^3(1+\beta_y)} & -C_X \dfrac{6EI_z}{L^2(1+\beta_y)} & C_X C_Y \dfrac{A_x E}{L} - C_X C_Y \dfrac{12EI_z}{L^3(1+\beta_y)} & C_Y^2 \dfrac{A_x E}{L} + C_X^2 \dfrac{12EI_z}{L^3(1+\beta_y)} & -C_X \dfrac{6EI_z}{L^2(1+\beta_y)} \\ -C_Y \dfrac{6EI_z}{L^2(1+\beta_y)} & C_X \dfrac{6EI_z}{L^2(1+\beta_y)} & \dfrac{(2-\beta_y)EI_z}{L(1+\beta_y)} & C_Y \dfrac{6EI_z}{L^2(1+\beta_y)} & -C_X \dfrac{6EI_z}{L^2(1+\beta_y)} & \dfrac{(4+\beta_y)EI_z}{L(1+\beta_y)} \end{bmatrix}_n$$

Figure 15.11 Plane frame global member stiffness matrix.

Computation of Member End Loads

usually called the *Assembly Process*. This analysis procedure is equally applicable to both statically determinate and statically indeterminate structures. The degree of static indeterminacy has no effect upon the complexity of the analysis. The only requirement for the structure is that it must be stable. Otherwise, the global stiffness matrix $[K]$ will be singular and it will not be possible to determine a finite set of joint displacements for any set of joint loads $\{W\}$.

COMPUTATION OF MEMBER END LOADS

The next step in the analysis of any structure, after the global stiffness matrix has been assembled and the joint displacements have been computed, is to compute the member end loads. An expression for computing these end loads can be developed by substituting Eq. (15.29) into Eq. (15.15) to obtain

$$\{S\}_n = [K_m]_n [A]_n \{D\}_n \tag{15.44}$$

Plane Truss Member. Substituting the expression for $[K_m]_n$ in Eq. (15.19) and the expression for $[A]_n$ in Eq. (15.38) into Eq. (15.44) gives the following expressions for the member end loads for a prismatic plane truss member:

$$s(1)_n = -\frac{A_x E}{L}([d(3)_n - d(1)_n]C_X + [d(4)_n - d(2)_n]C_Y) \tag{15.45a}$$

$$s(2)_n = \frac{A_x E}{L}([d(3)_n - d(1)_n]C_X + [d(4)_n - d(2)_n]C_Y) \tag{15.45b}$$

As expected, the two member end loads have the same magnitudes and opposite signs. Since it is only necessary to evaluate one of these equations, it is usually more convenient to use Eq. (15.45b) since a positive sign for $s(2)_n$ corresponds to tension in the member. The elements in $\{D\}_n$ correspond to the global displacements of the joints in the matrix $\{D\}$ at the ends of the member. A zero value should be used for any of the elements in $\{D\}_n$ which correspond to displacements in restrained directions at support joints since these displacements are not included in $\{D\}$.

Plane Frame Member. Expressions for the member end loads for a prismatic plane frame member can be obtained by substituting the expression for $[K_m]_n$ in Eq. (15.21) and the expression for $[A]_n$ in Eq. (15.41) into Eq. (15.44):

$$s(1)_n = -\frac{A_x E}{L}([d(4)_n - d(1)_n]C_X + [d(5)_n - d(2)_n]C_Y) \tag{15.46a}$$

$$s(2)_n = -\frac{12EI_z}{L^3(1+\beta_y)}([d(5)_n - d(2)_n]C_X - [d(4)_n - d(1)_n]C_Y)$$

$$+ \frac{6EI_z}{L^2(1+\beta_y)}[d(3)_n + d(6)_n] \qquad (15.46b)$$

$$s(3)_n = -\frac{6EI_z}{L^2(1+\beta_y)}([d(5)_n - d(2)_n]C_X - [d(4)_n - d(1)_n]C_Y)$$

$$+ \frac{(4+\beta_y)EI_z}{L(1+\beta_y)}d(3)_n + \frac{(2-\beta_y)EI_z}{L(1+\beta_y)}d(6)_n \qquad (15.46c)$$

$$s(4)_n = \frac{A_x E}{L}([d(4)_n - d(1)_n]C_X + [d(5)_n - d(2)_n]C_Y) \qquad (15.46d)$$

$$s(5)_n = \frac{12EI_z}{L^3(1+\beta_y)}([d(5)_n - d(2)_n]C_X - [d(4)_n - d(1)_n]C_Y)$$

$$- \frac{6EI_z}{L^2(1+\beta_y)}[d(3)_n + d(6)_n] \qquad (15.46e)$$

$$s(6)_n = -\frac{6EI_z}{L^2(1+\beta_y)}([d(5)_n - d(2)_n]C_X - [d(4)_n - d(1)_n]C_Y)$$

$$+ \frac{(2-\beta_y)EI_z}{L(1+\beta_y)}d(3)_n + \frac{(4+\beta_y)EI_z}{L(1+\beta_y)}d(6)_n \qquad (15.46f)$$

Note that the pair of axial forces $s(1)_n$ and $s(4)_n$ and the pair of shear forces $s(2)_n$ and $s(5)_n$ have equal magnitudes and opposite signs, as required for equilibrium of the member. If the effects of shear distortion in the member was originally ignored in computing $[K_m]_n$ for the member during the generation of $[K]$ to determine the joint displacements, then the quantity β_y should be set to zero in these equations when computing the member end loads.

EXAMPLE PROBLEM

The basic calculation procedure for the analysis of a structure by the Stiffness Method can be demonstrated by analyzing the plane truss whose mathematical model is shown in Figure 15.12a. This truss has six degrees of freedom which correspond to the global X and Y translation components at joints 2, 3 and 5. There are no degrees of freedom at joints 1 and 4 due to the support restraints.

Example Problem

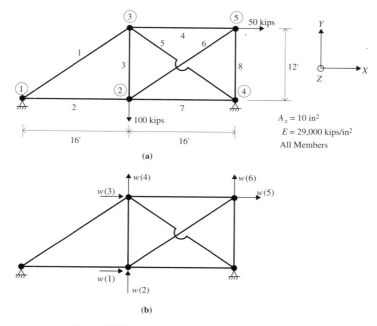

Figure 15.12 Example Problem, assembly process.

Figure 15.12b shows the definition of the elements in the global joint load matrix $\{W\}$ which correspond to the elements in the global joint displacement matrix $\{D\}$. The degrees of freedom have been numbered consecutively by joint, as described previously.

The first step in the generation of the global stiffness matrix $[K]$ is to compute the elements in the individual global member stiffness matrices $[K]_n$ for each plane truss member by Eq. (15.39). The results of these calculations in kip and inch units are shown in Figure 15.13. Only the elements corresponding to the degrees of freedom have been computed since the elements in the matrices which correspond to the support restraints will not be used in the Assembly Process. The identifying numbers for each row and column for each matrix correspond to the numbers for the corresponding degrees of freedom as an aid for assembling the individual elements into the correct locations in $[K]$. For example, the degrees of freedom at the first joint for member 6 are $d(1)$ and $d(2)$ while the degrees of freedom at the second joint for the member are $d(5)$ and $d(6)$. Either end joint for a member may be designated as the first joint as long as the signs of C_X and C_Y are computed properly by using θ_X and θ_Y at the first joint.

The next step is to add the elements in the global stiffness matrices to generate the elements in $[K]$. This operation can be easily performed by inserting each element in each global member stiffness matrix in turn into a table of the form

534 The Stiffness Method Chap. 15

Member 1

$$[K]_1 = \begin{array}{c} \\ \\ 3 \\ 4 \end{array} \begin{bmatrix} & & 3 & 4 \\ - & - & - & - \\ - & - & - & - \\ - & - & 773.333 & 580 \\ - & - & 580 & 435 \end{bmatrix}$$

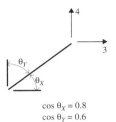
$\cos\theta_X = 0.8$
$\cos\theta_Y = 0.6$

Member 2

$$[K]_2 = \begin{array}{c} \\ \\ 1 \\ 2 \end{array} \begin{bmatrix} & & 1 & 2 \\ - & - & - & - \\ - & - & - & - \\ - & - & 1510.417 & 0 \\ - & - & 0 & 0 \end{bmatrix}$$

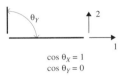
$\cos\theta_X = 1$
$\cos\theta_Y = 0$

Member 3

$$[K]_3 = \begin{array}{c} 1 \\ 2 \\ 3 \\ 4 \end{array} \begin{bmatrix} 1 & 2 & 3 & 4 \\ 0 & 0 & 0 & 0 \\ 0 & 2013.889 & 0 & -2013.889 \\ 0 & 0 & 0 & 0 \\ 0 & -2013.889 & 0 & 2013.889 \end{bmatrix}$$

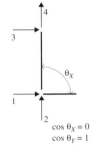
$\cos\theta_X = 0$
$\cos\theta_Y = 1$

Member 4

$$[K]_4 = \begin{array}{c} 3 \\ 4 \\ 5 \\ 6 \end{array} \begin{bmatrix} 3 & 4 & 5 & 6 \\ 1510.417 & 0 & -1510.417 & 0 \\ 0 & 0 & 0 & 0 \\ -1510.417 & 0 & 1510.417 & 0 \\ 0 & 0 & 0 & 0 \end{bmatrix}$$

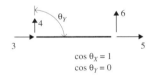
$\cos\theta_X = 1$
$\cos\theta_Y = 0$

Member 5

$$[K]_5 = \begin{array}{c} 3 \\ 4 \\ \\ \end{array} \begin{bmatrix} 3 & 4 & & \\ 773.333 & -580 & - & - \\ -580 & 435 & - & - \\ - & - & - & - \\ - & - & - & - \end{bmatrix}$$

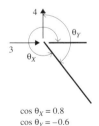
$\cos\theta_X = 0.8$
$\cos\theta_Y = -0.6$

Figure 15.13 Example Problem, assembly process-global member stiffness matrices.

Example Problem

Member 6

$$[K]_6 = \begin{array}{c} 1 \\ 2 \\ 5 \\ 6 \end{array} \begin{array}{cccc} 1 & 2 & 5 & 6 \\ \left[\begin{array}{cccc} 773.333 & 580 & -773.333 & -580 \\ 580 & 435 & -580 & -435 \\ -773.333 & -580 & 773.333 & 580 \\ -580 & -435 & 580 & 435 \end{array}\right] \end{array}$$

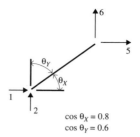

$\cos \theta_X = 0.8$
$\cos \theta_Y = 0.6$

Member 7

$$[K]_7 = \begin{array}{c} 1 \\ 2 \end{array} \begin{array}{cc} 1 & 2 \\ \left[\begin{array}{cccc} 1510.417 & 0 & - & - \\ 0 & 0 & - & - \\ - & - & - & - \\ - & - & - & - \end{array}\right] \end{array}$$

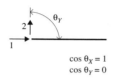

$\cos \theta_X = 1$
$\cos \theta_Y = 0$

Member 8

$$[K]_8 = \begin{array}{c} \\ \\ 5 \\ 6 \end{array} \begin{array}{cccc} & & 5 & 6 \\ \left[\begin{array}{cccc} - & - & - & - \\ - & - & - & - \\ - & - & 0 & 0 \\ - & - & 0 & 2013.889 \end{array}\right] \end{array}$$

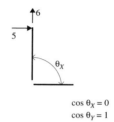

$\cos \theta_X = 0$
$\cos \theta_Y = 1$

Figure 15.13 *(cont.)*

shown in Figure 15.14, where the row and column numbers for each location in the table correspond to the row and column numbers which have been assigned to the elements in $[K]_n$. The elements in $[K]$ then can be computed by summing the numbers in each location in the table after all of the matrix elements have been inserted into the correct rows and columns. The final result of this operation is

$$[K] = \left[\begin{array}{cccccc} 3794.167 & 580 & 0 & 0 & -773.333 & -580 \\ 580 & 2448.889 & 0 & -2013.889 & -580 & -435 \\ 0 & 0 & 3057.083 & 0 & -1510.417 & 0 \\ 0 & -2013.889 & 0 & 2883.889 & 0 & 0 \\ -773.333 & -580 & -1510.417 & 0 & 2283.75 & 580 \\ -580 & -435 & 0 & 0 & 580 & 2448.889 \end{array}\right]$$

Note that the matrix $[K]$ is symmetric as required by Maxwell's Theorem.

The global flexibility matrix for the structure can be determined by inverting the global stiffness matrix. Performing this operation gives

	1	2	3	4	5	6
1	1510.417 773.333 1510.417	580.000			−773.333	−580.000
2	580.000	2013.889 435.000		−2013.889	−580.000	−435.000
3			773.333 1510.417 773.333	580.000 −580.000	−1510.417	
4		−2013.889	580.000 −580.000	435.000 2013.889 435.000		
5	−773.333	−580.000	−1510.417		1510.417 773.333	580.000
6	−580.000	−435.000			580.000	435.000 2013.889

Figure 15.14 Example Problem, assembly process for $[K]$.

$$[F] = [K]^{-1} = \begin{bmatrix} 0.0003046 & -0.0000988 & 0.0000517 & -0.0000690 & 0.0001047 & 0.0000298 \\ -0.0000988 & 0.0012777 & 0.0001929 & 0.0008922 & 0.0003904 & 0.0001111 \\ 0.0000517 & 0.0001929 & 0.0005455 & 0.0001347 & 0.0004421 & -0.0000582 \\ -0.0000690 & 0.0008922 & 0.0001347 & 0.0009698 & 0.0002726 & 0.0000776 \\ 0.0001047 & 0.0003904 & 0.0004421 & 0.0002726 & 0.0008948 & -0.0001178 \\ 0.0000298 & 0.0001111 & -0.0000582 & 0.0000776 & -0.0001178 & 0.0004630 \end{bmatrix}$$

Note that this matrix is also symmetric, as expected. With this matrix now known, the joint displacements can be found by Eq. (15.23) using the joint load matrix $\{W\}$ corresponding to the active loads on the structure

$$\{W\} = \begin{Bmatrix} 0 \\ -100 \\ 0 \\ 0 \\ 50 \\ 0 \end{Bmatrix}$$

This results in the displacements

Example Problem

$$\{D\} = [F]\{W\} = \begin{Bmatrix} 0.0151 \\ -0.1082 \\ 0.0028 \\ -0.0756 \\ 0.0057 \\ -0.0170 \end{Bmatrix}$$

Since all calculations have been performed in kip and inch units, the elements in $\{D\}$ represent the joint translations in inches.

Even for very small structures such as this truss, it will be found that the calculations are very tedious to perform manually. The process can be simplified by using a spreadsheet program to invert the matrix $[K]$ to obtain $[F]$ and to perform the matrix multiplication to compute the joint displacement matrix $\{D\}$. A printout of a spreadsheet which was used to perform these operations for this structure is shown in Figure 15.15. Any of the numerous commercial spreadsheet programs can be used since they all have the capability to perform the operations of matrix inversion and matrix multiplication.

The final step in the analysis is to compute the member end loads. Since the two end loads for a truss member have equal magnitudes and act in opposite directions, we will only compute that end load corresponding to Eq. (15.45b) since the sign of this end load agrees with our previously defined definition for tension and compression in a truss member. As an example of these calculations, we will compute the end load for member 6 for which

$$C_X = 0.8$$
$$C_Y = 0.6$$
$$d(1)_6 = d(1) = 0.0151$$
$$d(2)_6 = d(2) = -0.1082$$
$$d(3)_6 = d(5) = 0.0057$$
$$d(4)_6 = d(6) = -0.0170$$

Substituting these values into Eq. (15.45b) along with the values for A_x, E and L, in kip and inch units, gives

$$s(2)_6 = \frac{(10)(29{,}000)}{(240)} \{[(0.0057) - (0.0151)](0.8) + [(-0.0170) - (-0.1082)](0.6)\}$$

$$= 57.033 \text{ kips}$$

The forces in the other members can be determined in a similar manner. The final values for all of the member forces are

Example Problem for Assembly Process - Spreadsheet Calculations

Global Stiffness Matrix [K]

3794.167	580	0	0	-773.333	-580
580	2448.889	0	-2013.889	-580	-435
0	0	3057.083	0	-1510.417	0
0	-2013.889	0	2883.889	0	0
-773.333	-580	-1510.417	0	2283.75	580
-580	-435	0	0	580	2448.889

Global Flexibility Matrix [F]

0.00030455	-9.876E-05	5.1724E-05	-6.897E-05	0.00010469	2.9793E-05
-9.876E-05	0.00127769	0.00019289	0.00089224	0.00039041	0.0001111
5.1724E-05	0.00019289	0.00054553	0.0001347	0.00044208	-5.819E-05
-6.897E-05	0.00089224	0.0001347	0.00096983	0.00027263	7.7586E-05
0.00010469	0.00039041	0.00044208	0.00027263	0.00089477	-0.0001178
2.9793E-05	0.0001111	-5.819E-05	7.7586E-05	-0.0001178	0.00046303

Global Load Matrix {W}

0
-100
0
0
50
0

Global Displacement Matrix {D}

0.01511034
-0.1082487
0.00281519
-0.0755927
0.00569795
-0.0169991

Figure 15.15 Example Problem, assembly process-spreadsheet calcualtions.

$$s(2)_1 = -52.103 \text{ kips}$$
$$s(2)_2 = 22.807 \text{ kips}$$
$$s(2)_3 = 65.653 \text{ kips}$$
$$s(2)_4 = 4.380 \text{ kips}$$
$$s(2)_5 = -57.517 \text{ kips}$$
$$s(2)_6 = 57.033 \text{ kips}$$
$$s(2)_7 = -22.807 \text{ kips}$$
$$s(2)_8 = -34.236 \text{ kips}$$

The reactive forces for the truss now can be easily computed by summing forces in the global X and Y directions at each support joint.

Computer Programs PTRUSS and PFRAME

The accuracy of the computed values for the member forces is highly dependent on how many significant figures are carried through in the numbers during the assembly of $[K]$, the inversion of $[K]$ to obtain $[F]$ and the multiplication of $[F]$ and $\{W\}$ to obtain $\{D\}$. A considerable amount of roundoff error can accumulate during the matrix inversion if a sufficient number of significant figures is not retained throughout the calculations. The values shown here for the member forces agree within expected roundoff error with the values which were computed in Example Problem 11.3 in Chapter 11 by the Method of Consistent Deformations.

This same procedure can be used for any type of structure as long as expressions are available for computing the elements in the global member stiffness matrices for the members. Expressions are given in Fleming (1986, 1989) for the elements in the 6-by-6 global member stiffness matrix for a space truss member and the 12-by-12 global member stiffness matrix for a space frame member.

COMPUTER PROGRAMS PTRUSS AND PFRAME

The QBasic source codes for a computer program named PTRUSS and a computer program named PFRAME, which will analyze a plane truss and a plane frame, respectively, by the Stiffness Method, is contained in the ASCII disk files PTRUSS.BAS and PFRAME.BAS. These programs have been supplied for this book to demonstrate how the analysis steps for the Stiffness Method can be easily implemented in a computer program. The advantage of these two programs over the program FLEX, which was presented in Chapter 12 for the Flexibility Method, is that essentially no preprocessing is required to generate the input data. However, it is possible to modify the program FLEX to eliminate the need for preprocessing as described at the end of Chapter 12. Instructions for using the programs and a description of the format required for the input data files are contained in the ASCII disk files PTRUSS.DOC and PFRAME.DOC. The input data format is very similar to that for the programs T2DII and F2DII in FATPAK II. The executable files for the programs are in the disk files PTRUSS.EXE and PFRAME.EXE.

The program PTRUSS will analyze the mathematical model of a statically determinate or statically indeterminate plane truss of arbitrary geometry to determine the joint translations, the member forces and the support reactive forces. All members in the truss must be prismatic. The truss may be subjected to any combination of concentrated forces on the joints.

The program PFRAME will analyze the mathematical model of a statically determinate or statically indeterminate plane frame of arbitrary geometry to determine the joint translations and rotations, the axial forces, shear forces and moments at each end of each member and the reactive forces and moments. All members in the frame must be prismatic. Only axial deformations and bending deformations are considered in the members during the analysis. Therefore, it is not necessary to specify a value for the effective shear area A_y and the shear modulus of the material G for each member, as required in the programs F2DII and F3DII in FATPAK II. In addition, in order to simplify the source code, the frame may only be subjected to concentrated forces and moments on the joints. Distributed forces on the members are not permitted in this program unless they are first converted into equivalent joint loads, as described previously in Example Problem 12.4 in Chapter 12 for the Flexibility Method.

The analysis procedure which is used in each program is essentially the same as that used in the previous example problem, except that rather than using an indirect approach to compute the joint displacements, by inverting the matrix $[K]$ to obtain the matrix $[F]$, the stiffness equations are solved directly for the displacements by the Gauss Elimination Procedure. The SUB Procedure which is used in each program is identical to the one used in the programs SDPTRUSS, SDSTRUSS, SDPFRAME and SDSFRAME in previous chapters.

The size of the structure which can be analyzed by either program is limited by the size of the global stiffness matrix which can be stored in the computer memory. This results in a maximum number of degrees of freedom for the mathematical model for any structure of approximately 127 since the QBasic language has a limit of 64 kilobytes for any numeric array. There are some programming tricks which can be used to analyze much larger structures, but they are beyond the scope of this book. The source listing for a FORTRAN program which will analyze essentially any size of plane truss is shown in Fleming (1989).

Sample Plane Truss for Program PTRUSS

Figure 15.16 shows a listing of the input data file in kip and inch units for the analysis of the plane truss shown in Figure 15.12a by PTRUSS. This data file is supplied in the disk file PTRUSS.PTR. The output from the program is shown in Figure 15.17. The member forces in the output agree with those obtained in the previous solution of this truss within expected roundoff error. The member forces which were computed by PTRUSS are probably more accurate than those in the manual solution since more significant figures were carried through for the joint displacements in the calculations.

Sample Plane Frame for Program PFRAME

Figure 15.18 shows a sample plane frame which will be analyzed by the program PFRAME. The listing of the input data in kip and inch units is shown in Fig-

Computer Programs PTRUSS and PFRAME

```
Sample Plane Truss for Program PTRUSS
5,8,1,1,2,2
Joint Coordinates
1,0.0,0.0
2,192.0,0.0
3,192.0,144.0
4,384.0,0.0
5,384.0,144.0
Material Data
1,29000.0
Cross Section Data
1,10.0
Member Data
1,1,3
2,1,2
3,2,3
4,3,5
5,3,4
6,2,5
7,2,4
8,4,5
Support Restraints
1,1,1
4,1,1
Joint Loads
2,0.0,-100.0
5,50.0,0.0
```

Figure 15.16 Sample plane truss, PTRUSS input.

```
Sample Plane Truss for Program PTRUSS

Joint Displacements
```

Joint	X-Tran	Y-Tran
1	0.000	0.000
2	0.015	-0.108
3	0.003	-0.076
4	0.000	0.000
5	0.006	-0.017

Member Forces (Tension Positive)

Member	Force
1	-52.083
2	22.823
3	65.766
4	4.354
5	-57.526
6	57.057
7	-22.823
8	-34.234

Reactions

Joint	RX	RY
1	18.844	31.250
4	-68.844	68.750

Figure 15.17 Sample plane truss, PTRUSS output.

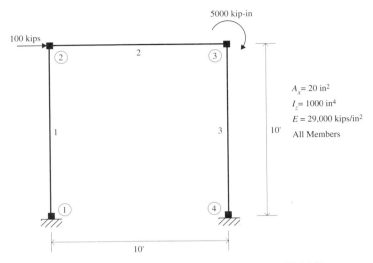

Figure 15.18 Sample plane frame for program PFRAME.

```
Sample Plane Frame for Program PFRAME
4,3,1,1,2,2
Joint Coordinates
1,0.0,0.0
2,0.0,120.0
3,120.0,120.0
4,120.0,0.0
Material Data
1,29000.0
Member Cross Section Data
1,20.0,1000.0
Member Data
1,1,2
2,2,3
3,3,4
Support Restraints
1,1,1,1
4,1,1,1
Joint Loads
2,100.0,0.0,0.0
3,0.0,0.0,-5000.0
```

Figure 15.19 Sample plane frame, PFRAME input.

ure 15.19 and the results of the analysis are shown in Figure 15.20. This input data file is in the disk file PFRAME.PFR.

MOMENT END RELEASES FOR PLANE FRAME MEMBERS

The program PFRAME is limited to the analysis of plane frames in which the members are rigidly attached to the joints at each end. However, there are many situations where it is necessary to analyze a frame in which at least one of the members is pinned to the joints at one or both ends. In order to consider this situation

Moment End Releases for Plane Frame Members

```
Sample Plane Frame for Program PFRAME
Joint Displacements
  Joint       X-Tran          Y-Tran          Z-Rot
    1         0.000           0.000           0.00000
    2         0.460           0.016          -0.00178
    3         0.454          -0.016          -0.00518
    4         0.000           0.000           0.00000
Member End Loads
  Member  Joint        Sx              Vy              Mz
    1       1        -77.647          71.134         4698.57
            2         77.647         -71.134         3837.52
    2       2         28.866         -77.647        -3837.52
            3        -28.866          77.647        -5480.13
    3       3         77.647          28.866          480.13
            4        -77.647         -28.866         2983.80
Reactions
  Joint           RX              RY              MZ
    1           -71.134         -77.647         4698.57
    4           -28.866          77.647         2983.80
```

Figure 15.20 Sample plane frame, PFRAME output.

in the analysis, it is necessary to modify the member stiffness matrix by considering the moment at each pinned end to be zero. It is a fairly simple task to compute the new elements in the local member stiffness matrix $[K_m]_n$ for a prismatic plane frame member for a pinned connection at one or both ends using the Moment Area Theorems. The resulting local member stiffness matrix for a prismatic member, in which shear deformations are ignored, with a pinned connection at the first joint and a rigid connection at the second joint, is

$$[K_m]_n = \begin{bmatrix} \dfrac{A_x E}{L} & 0 & 0 & -\dfrac{A_x E}{L} & 0 & 0 \\ 0 & \dfrac{3EI_z}{L^3} & 0 & 0 & -\dfrac{3EI_z}{L^3} & \dfrac{3EI_z}{L^2} \\ 0 & 0 & 0 & 0 & 0 & 0 \\ -\dfrac{A_x E}{L} & 0 & 0 & \dfrac{A_x E}{L} & 0 & 0 \\ 0 & -\dfrac{3EI_z}{L^3} & 0 & 0 & \dfrac{3EI_z}{L^3} & -\dfrac{3EI_z}{L^2} \\ 0 & \dfrac{3EI_z}{L^2} & 0 & 0 & -\dfrac{3EI_z}{L^2} & \dfrac{3EI_z}{L} \end{bmatrix}_n \quad (15.47)$$

Note that all of the elements in the third row and third column are zero since no moment is transmitted between the member and the joint at the pinned end. Therefore, the member end load $s(3)_n$ will be zero. A similar matrix can be developed for a member with a rigid connection at the first joint and a pinned connection at the second joint. For this matrix, the elements in the sixth row and sixth column will be zero since the member end load $s(6)_n$ will be zero.

The local member stiffness matrix for a prismatic member with a pinned connection at each end is

$$[K_m]_n = \begin{bmatrix} \frac{A_x E}{L} & 0 & 0 & -\frac{A_x E}{L} & 0 & 0 \\ 0 & 0 & 0 & 0 & 0 & 0 \\ 0 & 0 & 0 & 0 & 0 & 0 \\ -\frac{A_x E}{L} & 0 & 0 & \frac{A_x E}{L} & 0 & 0 \\ 0 & 0 & 0 & 0 & 0 & 0 \\ 0 & 0 & 0 & 0 & 0 & 0 \end{bmatrix}_n \quad (15.48)$$

which is very similar to the local member stiffness matrix for a prismatic plane truss member. The elements in the second, third, fifth and sixth rows and columns are all zero since the moments and the shear forces at each end of the member must be zero. The moments are zero due to the pinned ends while the shear forces must be zero to satisfy moment equilibrium for the member.

These local member stiffness matrices now can be transformed into the global coordinate system for use in the Assembly Process by using the coordinate transformation equation shown previously in Eq. (15.35). The coordinate transformation matrix $[A]_n$ will still be the same as shown in Eq. (15.41). The resulting global member stiffness matrices then could be used in a program such as PFRAME to perform the analysis of plane frames with pinned members. Expressions for computing the member end loads for these members, after the stiffness equations have been solved for the joint displacements, can be developed by substituting the local member stiffness matrices into Eq. (15.44).

The expressions which were used in the program F2DII in FATPAK II to generate the global member stiffness matrices and to compute the member end loads for members with various combinations of pinned ends were developed by the process described in the previous paragraph. The actual matrix manipulations to obtain the new expressions is left as an exercise for the reader. The resulting expressions can be checked against those shown in Fleming (1989). These expressions can be used to modify the program PFRAME to include the effect of pinned ends at one or both ends of the members in the plane frame.

REFERENCES

FLEMING, JOHN F. (1986), *Structural Engineering Analysis on Personal Computers*, New York: McGraw-Hill.

——— (1989), *Computer Analysis of Structural Systems*, New York: McGraw-Hill

SUGGESTED PROBLEMS

SP15.1 and **SP15.2** Analyze the plane truss and the plane frame whose mathematical models are shown in Figures SP15.1 and SP15.2 by the Stiffness Method.
(a) Generate the global stiffness matrix by the Assembly Process.
(b) Compute the joint displacements by solving the stiffness equations with the program SEQSOLVE. As an alternate approach, compute the joint displacements by a combination of matrix inversion and matrix multiplication using a spreadsheet program.

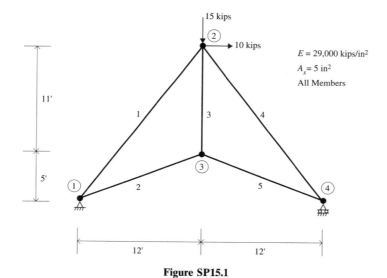

Figure SP15.1

Figure SP15.2

(c) Compute the member end loads corresponding to the computed joint displacements.

(d) Verify your analysis using the program PTRUSS for Figure SP15.1 and PFRAME for Figure SP15.2.

SP15.3 Derive expressions for the elements in the local member stiffness matrix for the plane frame member shown in Figure SP15.3 Neglect the effect of shear deformation.

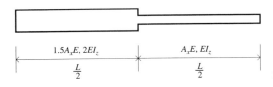

Figure SP15.3

INDEX

Assembly process, 523
AASHTO truck loading, 12
Beam:
   analysis by BEAMIL, 61, 164
   analysis by BEAMVM, 156, 281
   Conjugate Beam Method, 269
   differential equation of bending, 222
   influence lines for reactions, 50
   influence lines for shear force and bending moment, 158
   mathematical model, 5, 130
   maximum bending moment, 167
   Moment Area Method, 244
   nonprismatic, 241, 267
   shear force and bending moment by equilibrium analysis, 131
   shear force and bending moment by integration, 140
   shear force and bending moment diagrams, 146
   shear force and bending moment envelpes, 171
BEAMIL:
   input format, BEAMIL.DOC
   influence lines for beam reactions, 50
   influence lines for shear force and bending moment, 164
   program description, BEAMIL.DOC
BEAMVM:
   input format, BEAMVM.DOC
   program description, BEAMVM.DOC
   shear force and bending moment diagrams for beams, 156
   slope and deflection of beams, 281
Bending moment diagrams:
   beams, 146
   by cantilever parts, 255
   plane frame members, 180
   space frame members, 206
Bending moment envelope, 171
Betti's Law, 340

Castigliano's Theorems:
   First Theorem, 310
   Second Theorem, 309
Carry over factor, 497
Conjugate Beam Method, 269
   support restraints, 273
   statically indeterminate beams, 279
Conservation of energy, 291
Complex truss, 93
Compound truss, 88
Computer Programs:
   BEAMIL, BEAMIL.DOC
   BEAMVM, BEAMVM.DOC
   FATPAK II - Student Edition, FATPAKII.DOC
   FLEX, FLEX.DOC
   PFRAME, PFRAME.DOC
   PTRUSS, PTRUSS.DOC
   PTRUSSIL, PTRUSSIL.DOC
   SDPFRAME, SDPFRAME.DOC
   SDPTRUSS, SDPTRUSS.DOC

   SDSFRAME, SDSFRAME.DOC
   SDSTRUSS, SDSTRUSS.DOC
   SEQSOLVE, SEQSOLVE.DOC
Coordinate transformation matrix:
   plane frame member, 529
   plane truss member, 528

Dead loads, 10
Designer sign convention, 132
Differential equation of bending, 222
Direction cosines:
   plane member, 72
   space member, 117
Distribution factor, 496

Effective shear area, 305
Equations of condition, 34
Equations of equilibrium:
   plane structures, 15
   space structures, 17
External work:
   force, 292
   moment, 293

FATPAK II - Student Edition:
   influence lines for beams, 362
   influence lines for trusses, 370
   input format, FATPAKII.DOC
   program description, FATPAKII.DOC
Fixed end moments, 461, 495
FLEX:
   input format, FLEX.DOC
   program description, FLEX.DOC
   source code, FLEX.BAS
Flexibility coefficients, 344
Flexibility Method, 405
   computation of joint displacements, 425
   computation of member end loads, 425
   computation of redundant loads, 421
   geometric compatibility check, 426
   global equilibrium matrix, 417
   global flexibility matrix, 408
   global joint displacement matrix, 406
   global joint load matrix, 406
   local member deformation matrix:
      plane frame member, 411
      plane truss member, 409
   local member end load matrix:
      plane frame member, 411
      plane truss member, 409
   local member flexibility matrix:
      plane frame member, 413
      plane truss member, 413
   program FLEX, 429
   total structure member deformation matrix, 415
   total structure member end load matrix, 415
   total structure member flexibility matrix, 417

Global coordinate system, 7
Global equilibrium matrix, 417

Global flexibility matrix, 408
Global joint displacement matrix, 406, 514
Global joint load matrix, 406, 514
Global member deformation matrix:
   plane frame member, 525
   plane truss member, 524
Global member end load matrix:
   plane frame member, 525
   plane truss member, 524
Global member stiffness matrix:
   plane frame member, 529
   plane truss member, 528
Global stiffness matrix, 515

Influence lines:
   computation by BEAMIL, 61, 164
   computation by FATPAK II:
      beam reactions, 362
      plane truss member forces, 370
   computation by PTRUSSIL, 103
   definition, 51
   properties:
      concentrated force, 54
      concentrated moment, 56
      distributed force, 54
   reactions, 50, 58
   shear force and bending moment in beams, 158
   truss member forces, 99
Internal work:
   axial force, 294
   bending moment, 297
   shear force, 299
   twisting moment, 307
Impact factor, 14

Live loads:
   earthquake loads on buildings, 11
   horizontal wind loads on buildings, 11
   vertical loads on buildings, 11
   vertical loads on highway bridges, 12
Local member coordinate system:
   plane frame member, 42
   plane truss member, 70
   space frame member, 206
   space truss member, 117
Local member deformation matrix:
   plane frame member, 411, 518
   plane truss member, 409, 517
Local member end load matrix:
   plane frame member, 411, 518
   plane truss member, 409, 517
Local member flexibility matrix:
   plane frame member, 413
   plane truss member, 413
Local member stiffness matrix:
   plane frame member, 521
   plane truss member, 520

Mathematical model, 3
   beam, 5, 130
   plane frame, 5, 180
   plane truss, 5, 69
   space frame, 5, 206